Global Governance on Renewable Energy

Sybille Roehrkasten

Global Governance on Renewable Energy

Contrasting the Ideas of the German and the Brazilian Governments

With a foreword by Prof. Dr. Miranda Schreurs

Sybille Roehrkasten
Potsdam, Germany

Dissertation Freie Universität Berlin, Germany, 2014

ISBN 978-3-658-10479-5   ISBN 978-3-658-10480-1 (eBook)
DOI 10.1007/978-3-658-10480-1

Library of Congress Control Number: 2015943744

Springer VS
© Springer Fachmedien Wiesbaden 2015
This work is subject to copyright. All rights are reserved by the Publisher, whether the whole or part of the material is concerned, specifically the rights of translation, reprinting, reuse of illustrations, recitation, broadcasting, reproduction on microfilms or in any other physical way, and transmission or information storage and retrieval, electronic adaptation, computer software, or by similar or dissimilar methodology now known or hereafter developed.
The use of general descriptive names, registered names, trademarks, service marks, etc. in this publication does not imply, even in the absence of a specific statement, that such names are exempt from the relevant protective laws and regulations and therefore free for general use.
The publisher, the authors and the editors are safe to assume that the advice and information in this book are believed to be true and accurate at the date of publication. Neither the publisher nor the authors or the editors give a warranty, express or implied, with respect to the material contained herein or for any errors or omissions that may have been made.

Printed on acid-free paper

Springer VS is a brand of Springer Fachmedien Wiesbaden
Springer Fachmedien Wiesbaden is part of Springer Science+Business Media
(www.springer.com)

Foreword

This book makes an important contribution to our understanding of the links between global renewable energy governance and national energy politics. It compares the positions of two leaders in the field—Germany and Brazil. Germany has embarked on an *Energiewende*, a deep structural transformation of the country's energy system away from nuclear energy and fossil fuels and towards greater energy efficiency and renewable energy. Renewable energy accounts for an increasingly large share of its electricity consumption. Brazil is a leading producer and consumer of ethanol and has one of the worldwide highest shares of renewables in its energy mix. Brazil has also played a critical role in sustainability and renewable intiatives at the global level, hosting the 1992 United Nations Conference on Environment and Development and the 2012 United Nations Conference on Sustainable Development.

There are many similarities between the two countries. Both are keen to promote renewable energy internationally. Both are leaders in the domestic deployment of renewable energy. Both are taking an active stance in global renewable energy governance.

And yet, despite their common interests, Brazil and Germany have very different ideas about what kind of global governance is needed. Germany's focus has been on institution building as can be seen with its initiative in creating the International Renewable Energy Agency (IRENA). In setting up IRENA, the German government prioritized the development and diffusion of new renewable energies (primarily, wind and solar energy). It has, however, been hesitant about strongly promoting biofuels internationally, siting concerns about their environmental impact and consequences for food security. Brazil has been critical of Germany's position on biofuels. It has chosen not to join IRENA. Facilitating international trade on biofuels has been the major goal of its ethanol diplomacy.

North-South tensions that frequently surface in global negotiations on sustainable development are clearly evident in the diverging German and Brazilian positions on renewable energy. Whereas Brazil in supporting biofuels emphasizes the right to development and the needs of poorer communities, Germany in critiquing biofuels prioritizes environmental protection and raises concerns about food security, especially for less well off populations. The two countries concentrate on different economic sectors with their renewable energy strategies: agriculture and transport in the case of Brazil and the electricity sector in the case of Germany.

While Germany criticises discriminatory energy policies, Brazil points at discriminatory trade policies. Brazil is more focused on removing international trade barriers to biofuels, Germany on improving regulatory conditions for renewables. Whereas the German government supports global governance bodies emphasizing new renewable energies, the Brazilian government supports those bodies that are most likely to be favorable to biofuels.

This book is a fascinating read and significantly advances our understanding of the ways in which domestic political interests and concerns can shape positions towards global governance institutions.

Prof. Dr. Miranda Schreurs

Preface and Acknowledgement

This publication presents the results of my PhD research which started in 2010 and concluded with the submission and defense of my thesis in the course of 2014. The support of many people was crucial to the completion of this long journey. First of all, I would like to thank my supervisors at the Freie Universität Berlin, Professor Guenther Maihold and Professor Miranda Schreurs, for their critical comments, intellectual rigor and the freedom to develop this PhD project according to my own ideas.

The German Institute for International and Security Affairs (SWP) was the intellectual home of my PhD years. It provided an inspiring and supportive environment as well as close interactions with researchers and decision-makers alike. I would like to extend very special thanks to my SWP mentor Kirsten Westphal. With her expertise on global energy governance, she was a major source of inspiration for my PhD research. Joint projects – be it for academic purposes or policy advice – also gave me the opportunity to grow professionally. I would also like to express my gratitude to Claudia Zilla, who involved me in policy advice early on, supported my research and was always an inspiring discussion partner, not only on Brazilian matters. Thanks to Susanne Dröge, I had the freedom to pursue my PhD research and involve myself in many interesting SWP projects. Sabine Herrmann was a great source of moral support.

My research benefited considerably from the constructive PhD colloquia at SWP and the Environmental Policy Research Center at the Freie Universität Berlin. The International Futures Program of the Federal Foreign Office allowed for a very fruitful exchange of ideas with young professionals from different emerging countries on a range of global issues. I am particularly grateful to a number of colleagues, whose detailed and constructive feedback was of tremendous value to the success of this PhD project: Leonie Beining, Marianne Beisheim, Oliver Geden, Roberto Guimarães, Peer Krumrey, Johannes Plagemann, Henriette Rytz, Ursula Stiegler, Sonja Thielges, Kathrin Ulmer, Florian Wassenberg, and Clara Weinhardt. In addition, I would like to thank Isabel Faulhaber for sharing her quantitative expertise, Sandra Boyd for reliable and quick proofreading, and Inder Kumar Popat and the SWP library staff for all their support.

I am very thankful to the host institutions of my research stay in Brazil: the Brazilian Center for International Relations (CEBRI) in Rio de Janeiro and the Division for International Studies, Political and Economic Relations at the Institute for Applied Economic Research (IPEA-Dinte) in Brasília. They not only provided a working space, but also served as a door-opener and source of intellectual inspiration. Special thanks go to Selena Herrera and Guilherme Schmitz, who considerably enriched my research stay in Brazil. For the analysis of the German and the Brazilian ideas on global renewable energy governance, the insights of 75 interviews were of utmost importance. I would like to thank all my interview partners in Brazil and Germany for sharing their time and thoughts, making this PhD project a truly inspiring journey.

Last but not least, I would like to express my deepest gratitude to my friends and family, whose backing of all kinds was not only central for the completion of this thesis, but is my backbone in the ups and downs of life. Of this group of people, I would like to single out my uncle Michael Bunte, my parents Sabine and Burkhard Röhrkasten and my sisters Regine Sprenger and Miriam Röhrkasten, who – as always – deserve my deepest thanks. A special 'thank you' also goes to Alex Farid for his moral support and patience during the last stages of this journey – and for helping me getting my priorities right in life.

Sybille Röhrkasten

Contents

List of Tables 13
List of Figures 13
List of Abbreviations 15

1 Introduction 19
    1.1 Research Purpose and Questions 21
    1.2 State of the Art 24
    1.3 Theoretical-Analytical Framework 27
    1.4 Methodological Approach 29
    1.5 Outline 31

2 Theoretical-Analytical Framework: Contested Ideas in Global Governance 33
    2.1 Global Governance As a Perspective on Transboundary Policy-Making 33
        2.1.1 Contextualizing Global Governance Research 35
        2.1.2 Blurring the Boundaries between Domestic and Global Affairs 37
        2.1.3 Governments: Central, but not the Only Actors 38
        2.1.4 Value Added of Transboundary Cooperation 40
        2.1.5 Transboundary Policy-Making in the Absence of a Supreme Authority 42
    2.2 Integrating Contested Ideas into Global Governance Research 42
        2.2.1 The Role of Ideas in Global Governance 43
        2.2.2 Contestation and Power in Global Governance 46
        2.2.3 Global Governors 49
        2.2.4 Global Governance as a Concept of the OECD World? 51
    2.3 Actor-Centered Framework for the Analysis of Global Governance Ideas 55
        2.3.1 Policy Actors as Drafters and Carriers of Ideas 55
        2.3.2 Weak Cognitivism and its Basic Behavioral Assumptions 56
        2.3.3 Ideas as Causes and Results of Political Action 59
        2.3.4 Ideas on Global Renewable Energy Governance: Contents and Reasons Behind 61
    2.4 Methodology 63
        2.4.1 Interpretative Research Design 63
        2.4.2 Comparative Case Study Analysis 66
        2.4.3 Sources 68

3 Global Governance on Renewable Energy 73
    3.1 Tracing the Origins and Evolution 74
        3.1.1 The Origins of Global Energy Governance 74
        3.1.2 Initial Attempts to Promote Renewable Energy 79
        3.1.3 Transboundary Policy-Making on Renewables Taking Shape 82
    3.2 Global Challenges Addressed by Renewable Energy Promotion 89
        3.2.1 Energy Security 90

    3.2.2   Access to Energy 93
    3.2.3   Environmental Sustainability 94
    3.2.4   Trade-Offs Involved 96
3.3   Structural Characteristics 98
    3.3.1   Dominance of National Policy-Making 98
    3.3.2   Fragmentation.. 100
3.4   Global Governors and their Governance Activities 104
    3.4.1   IRENA 104
    3.4.2   IEA... 106
    3.4.3   REEEP 108
    3.4.4   REN21 109
    3.4.5   GBEP.. 111
    3.4.6   CEM.. 112
    3.4.7   UN Bodies and Agencies... 113
    3.4.8   G8 and G20... 114
3.5   A Snapshot at Contestation and Social Construction 115

4   German Ideas on Global Renewable Energy Governance 117
4.1   Tracing the Government's Action on Global Renewable Energy Governance 118
    4.1.1   The Renewables 2004 Conference and the Establishment of REN21 118
    4.1.2   Founding IRENA... 126
    4.1.3   Launch of the Renewables Club 137
4.2   Ideas on Global Renewable Energy Governance 139
    4.2.1   Global Challenges: Predominance of Climate Protection 139
    4.2.2   Renewable Energy Options: Sustainability and Electricity Markets 144
    4.2.3   Barriers: Markets and Policies Favoring Conventional Energy 149
    4.2.4   Tasks: Improving Domestic Regulatory Frameworks 152
    4.2.5   Global Governors: Providing Information and Advice on Renewable Energy 156
    4.2.6   Salient Features: the Responsibility of Industrialized Countries to Lead........ 165
4.3   The German Government's Action and Ideas at a Glance 168

5   Brazilian Ideas on Global Renewable Energy Governance 171
5.1   Tracing the Government's Action on Global Renewable Energy Governance 172
    5.1.1   Initial Steps: WSSD 2002 and the Renewables 2004 Conference................... 172
    5.1.2   Ethanol Diplomacy Begins: Time of Expansion 174
    5.1.3   Intensification of Ethanol Diplomacy in the Light of Growing International Criticism 182
    5.1.4   Continuation of Ethanol Diplomacy at a Lower Political Profile.................... 188
5.2   Ideas on Global Renewable Energy Governance 192
    5.2.1   Global Challenges: Predominance of Socio-Economic Development 192
    5.2.2   Renewable Energy Options: Biofuels and Competitiveness 199
    5.2.3   Barriers: Prejudices and Trade Restrictions 201
    5.2.4   Tasks: Creating an International Biofuels Market... 206
    5.2.5   Global Governors: Focusing on Biofuels and Trade Issues............................ 210
    5.2.6   Salient Features: North-South Conflicts, Power Asymmetries and Re-Balancing.......... 218
5.3   The Brazilian Government's Action and Ideas at a Glance 222

6	**Ideational Differences in Global Renewable Energy Governance**	**225**
	6.1 Identifying Ideational Differences	226
	6.1.1 Global Challenges: Environment Protection vs. Socio-Economic Development	226
	6.1.2 Renewable Energy Options: Electricity vs. Transport Sector	228
	6.1.3 Barriers: Discriminatory Energy Policies vs. Discriminatory Trade Policies	232
	6.1.4 Tasks: Improving Domestic Policies vs. Strengthening International Markets	234
	6.1.5 Global Governors: Diverging Relevance and Assessments	236
	6.1.6 Salient Features: Mutual Benefits vs. Conflicts and Power Asymmetries	237
	6.2 Reasons Behind Ideational Differences	240
	6.2.1 Coherency Between Ideas	240
	6.2.2 Domestic Policy Context of Renewable Energy Promotion	244
	6.2.3 Embeddedness in Transboundary Policy-Making	258
	6.2.4 Institutional Context: Division of Competencies	261
	6.2.5 Self-Interests	264
	6.3 Overview of Ideational Differences and Reasons Behind	274
7	**Conclusion**	**277**
	7.1 Ideational Differences in Global Renewable Energy Governance	277
	7.2 The Reasons Behind the Ideational Differences	284
	7.3 General Reflections and Suggestions for Further Research	292
Bibliography		**297**

List of Tables

Table 1: Overview of Different Energy Sources 24
Table 2: Typology of Global Public Goods 41
Table 3: Types of Power in Global Governance 48
Table 4: Weak Cognitivism as a Middle Way between Rational Choice and Strong Cognitivism 56
Table 5: Guiding Questions for the Case Study Analyses 68
Table 6: Arenas in Global Energy Governance 102
Table 7: Ideational Differences and the Reasons Behind 276

List of Figures

Figure 1: Crude Oil Prices, 1945–2012 76
Figure 2: Worldwide Supply of Renewable Energy Sources, 2000-2011 88
Figure 3: Development of Worldwide Energy Consumption, 1990-2011 91
Figure 4: Electrification Access Deficit on a Global Scale, 2010 93
Figure 5: Global Energy-Related Carbon Dioxide Emissions, 1990-2011 95
Figure 6: Electricity Supply from Renewable Energies in Germany, 1990-2012 122
Figure 7: Brazilian Ethanol Production and Trade Balance, 2002–2011 189
Figure 8: Different Causal Links between Biofuels Production and Food Security 231
Figure 9: German Renewables-Based Final Energy Supply, 2012 249
Figure 10: Brazilian Consumption of Gasoline and Ethanol in Road Transport, 2002-2011 251
Figure 11: Brazilian Energy Supply, 2011 254

List of Abbreviations

AA	Federal Foreign Office (Auswärtiges Amt)
Abiove	Brazilan Association of Vegetable Oil Industries (Associação Brasileira das Indústrias de Óleos Vegetais)
ABNT	Brazilian Technical Standards Association (Associação Brasileira de Normas Técnicas)
ANFAVEA	National Association of Automobile Manufacturers (Associação Nacional dos Fabricantes de Veículos Automotores)
ANP	National Agency for Oil, Natural Gas and Biofuels (Agência Nacional do Petróleo, Gás Natural e Biocombustíveis)
BBE	German Bioenergy Industry Association (Bundesverband Bioenergie)
BEE	German Renewable Energy Federation (Bundesverband Erneuerbare Energien)
BMELV	Federal Ministry of Food, Agriculture and Consumer Protection (Bundesministerium für Ernährung, Landwirtschaft und Verbraucherschutz)
BMU	Federal Minstry for Environment, Nature Conservation and Nuclear Safety (Bundesministerium für Umwelt, Naturschutz und Reaktorsicherheit)
BMWi	Federal Ministry of Economics and Technology (Bundesministerium für Wirtschaft und Technolgie)
BMZ	Federal Ministry for Economic Cooperation and Development (Bundesministerium für wirtschaftliche Zusammenarbeit und Entwicklung)
BNDES	Brazilian Development Bank (Banco Nacional do Desenvolvimento)
BRICs	Brazil, Russia, India and China
CDES	Social and Economic Development Council (Conselho Nacional de Desenvolvimento Econômico e Social)
CDM	Clean Development Mechanism
CDU	Christian Democratic Union (Christlich Demokratische Union Deutschlands)
CEBRI	Brazilian Center for International Relations (Centro Brasileiro de Relações Internacionais)
CEM	Clean Energy Ministerial
CINDES	Centro de Estudos de Integração e Desenvolvimento

CPDA/UFRJ	Centre for Graduate Studies in Agricultural Development, Federal Rural University Rio de Janeiro (Centro de Pós-Graduação de Ciências Sociais em Desenvolvimento, Agricultura e Sociedade, Federal Rural University of Rio de Janeiro)
CSD	Commission on Sustainable Development
CSU	Christian Social Union (Christlich-Soziale Union in Bayern)
CTBE	Bioethanol Science and Technology Laboratory (Laboratório Nacional de Ciência e Tecnologia do Bioetanol)
CTC	Sugarcane Technology Center (Centro de Tecnologia Canavieira)
CTESS	WTO Committee on Trade and Environment in Special Sessions
DIN	German Institute for Standardization (Deutsches Institut für Normung)
ECOSOC	UN Economic and Social Council
ECOWAS	Economic Community of West African States
EEG	Renewable Energy Act (Erneuerbare-Energien-Gesetz)
Embrapa	Brazilian Agricultural Research Corporation (Empresa Brasileira de Pesquisa Agropecuária)
EU	European Union
FAO	Food and Agricultural Organization
FDP	Free Democratic Party (Freie Demokratische Partei)
FFLCH/USP	Faculty of Philosophy, Languages and Literature, and Human Sciences (Faculdade de Filosofia, Letras e Ciências Humanas da Universidade de São Paulo)
FGV	Getulio Vargas Foundation (Fundação Getulio Vargas)
G7 (G8, G20)	Group of Seven (Eight, Twenty)
GBEP	Global Bioenergy Partnership
GDP	gross domestic product
GIZ	Deutsche Gesellschaft für Internationale Zusammenarbeit
GTZ	Deutsche Gesellschaft für Technische Zusammenarbeit
IAEA	International Atomic Energy Agency
IBF	International Biofuels Forum
ICONE	Institute for International Trade Negotiations (Instituto de Estudos do Comércio e Negociações Internacionais)
IEA	International Energy Agency
IEE/USP	Institute of Energy and Environment of the University of São Paulo (Instituto de Energia e Ambiente da Universidade de São Paulo)

ILAC	Latin American and Caribbean Initiative on Sustainable Development (Iniciativa Latino-Americana e Caribenha sobre Deselvolvimento Sustantável)
iLUC	indirect land use changes
INMETRO	National Institute of Metrology (Instituto Nacional de Metrologia, Qualidade e Tecnologia)
IPEA	Institute for Applied Economic Research (Instituto de Pesquisa Econômica Aplicada)
IRENA	International Renewable Energy Agency
ISO	International Standardization Organization
KfW	Kreditanstalt für Wiederaufbau
MAPA	Ministry of Agriculture, Livestock and Food Supply (Ministério da Agricultura, Pecuária e Abastecimento)
MCTI	Ministry of Science, Technology and Innovation (Ministério de Ciência, Tecnologia e Inovação)
MDA	Ministry of Rural Development (Ministério do Desenvolvimento Agário)
MDGs	Millennium Development Goals
MDIC	Ministry of Development, Industry and Foreign Trade (Ministério do Desenvolvimento, Indústria e Comércio Exterior)
MMA	Ministry of the Environment (Ministério de Meio Ambiente)
MME	Ministry of Mines and Energy (Ministério de Minas e Energia)
MRE	Ministry of Foreign Relations (Ministério de Relações Exteriores)
NGO	Non-Governmental Organization
OAS	Organization of American States
OECD	Organization of Economic Cooperation and Development
OPEC	Organization of Petroleum Exporting Countries
PSDB	Brazilian Social Democracy Party (Partido da Social Democracia Brasileira)
PT	Workers' Party (Partido dos Trabalhadores)
REEEP	Renewable Energy and Energy Efficiency Partnership
REN21	Renewable Energy Policy Network for the 21st Century
RSB	Roundtable on Sustainable Biofuels
RSPO	Roundtable on Sustainable Palm Oil
SDGs	Sustainable Development Goals
SE4All	Sustainable Energy for All

SPD	Social Democratic Party (Sozialdemokratische Partei Deutschlands)
the Greens	Alliance 90/ The Greens Party (Bündnis 90/ Die Grünen)
UAE	United Arab Emirates
UBA	Federal Environment Agency (Umweltbundesamt)
UFRJ	Federal University of Rio de Janeiro (Universidade Federal do Rio de Janeiro)
UK	United Kingdom
UN	United Nations
UnB	University of Brasília (Universidade de Brasília)
UNCED	United Nations Conference on Environment and Development
UNCTAD	United Nations Conference on Trade and Development
UNDP	United Nations Development Program
UNEO	United Nations Environment Organization
UNEP	United Nations Environment Program
UNFCCC	United Nations Framework Convention on Climate Change
UNICA	Sugarcane Industry Association (União da Indústria de Cana-de-Açúcar)
UNIDO	United Nations Industrial Development Organization
USA	United States of America
USP	University of São Paulo (Universidade de São Paulo)
VDB	German Biofuels Association (Verband der deutschen Biokraftstoffindustrie)
WBGU	German Advisory Council on Climate Change (Wissenschaftliche Beirat der Bundesregierung Globale Umweltveränderungen)
WSSD	World Summit on Sustainable Development
WTO	World Trade Organization

1 Introduction

> "Sustainable development is the imperative of the 21st century. Protecting our planet, lifting people out of poverty, advancing economic growth – these are different aspects of the same fight. (…) We will not achieve any of these goals without energy" (Ki-moon 2011:3).

Energy is a vital component of sustainable development, as this statement by the United Nations (UN) Secretary-General Ban Ki-moon emphasizes. The availability of energy services is of utmost importance for the functioning of modern society. Most of our everyday activities require energy consumption, access to energy is central for fulfilling basic needs and today's economic processes strongly rely on energy input. At the same time, energy production and its use have a number of negative externalities, posing major threats to development achievements. Climate change serves as a notable example. More than two thirds of global greenhouse gas emissions originate from the energy sector (OECD/IEA 2013:79).

In order to ensure that the global energy system actually serves the purpose of sustainable development, it is crucial to increase the worldwide use of renewable energy. Renewables promotion helps to tackle the two major challenges faced by the global energy system today: satisfying a globally rising energy demand while mitigating the environmental damages caused by energy production and use. With a growing global population and worldwide efforts to climb the ladder of socio-economic development, global energy demand is rising sharply.[1] At the same time, the fossil fuel supply, which still dominates the worldwide energy system,[2] faces absolute limits. All around the world efforts are being undertaken to promote alternatives to fossil fuels. With rising energy prices and technological innovations, the exploitation of unconventional fossil fuels is becoming economically attractive. However, unconventional sources are even more environmentally harmful than conventional fossil fuels. Nuclear energy is also becoming less of an alternative. The Fukushima nuclear accident in March 2011 increased the perceived risks of nuclear energy, which not only undermined political acceptance but also augmented the costs of nuclear safety and insurance. Renewable energy not only causes less environmental damage than fossil fuels and nuclear energy, it is also becoming more and

---

[1] Between 2000 and 2011, world primary energy demand increased by one-third. The International Energy Agency (IEA) foresees a further increase of 45 percent until 2035 (compared to 2011) if the current policies are maintained (OECD/IEA 2013:57f.).

[2] According to OECD/IEA (2013:58), worldwide primary energy demand had a fossil fuel share of 82 percent in 2011.

more economically attractive as costs of renewable energy technologies are rapidly falling (see also Szulecki, Westphal 2014).[3]

Throughout the world, the renewable energy sector has gained momentum in recent years. Global renewable energy demand has increased by one third between 2000 and 2011 (OECD/IEA 2013:58). More and more governments promote renewable energy. Between 2005 and 2013, the number of countries with renewables support policies has far more than doubled from 48 to 127 (REN21 2013a, 2005). Global new investments in renewable energy have experienced a sixfold increase between 2004 and 2012. In 2012, investments in additional renewable energy capacities exceeded those in additional fossil-fuel generating capacities by more than $100 billion (Frankfurt School - UNEP Centre, Bloomberg New Energy Finance 2013). However, the challenge remains huge. Renewable energy sources provide only a small share of the global energy mix (13 percent in 2011) and this share has not risen between 2000 and 2011 (OECD/IEA 2013:58). In order to increase the contribution of renewables not only in absolute but also in relative terms, greater efforts are needed.

In global governance, renewable energy has started to move from the sidelines to the center stage. Since the turn of the millennium, a number of initiatives for transboundary cooperation on renewables have been launched. The most notable has been the creation of a new intergovernmental organization on renewable energy. The International Renewable Energy Agency (IRENA) was officially established in April 2011 and now counts more than 120 member states.[4] It shall act as the global voice and knowledge base for the worldwide use of renewable energy, advise its member states on renewables, and serve as a network hub for transboundary cooperation (IRENA 28.10.2011). In 2011, the UN Secretary-General launched Sustainable Energy for All (SE4All) which has been the major UN initiative on renewable energy so far. SE4All pursues the goal to double the share of renewables in global energy supply by 2030.[5] In September 2015, the UN General Assembly will adopt a set of Sustainable Development Goals which also cover the increased use of renewable energy.

Global governance can promote renewable energy in various ways. It can, for example, commit governments and other policy actors to increase the use of renewable energy within their spheres of influence, facilitate free trade of renewable energy technologies and services, transfer finance and technological know-how and

---

[3] On the environmental benefits of renewable energy and the increasing competitiveness see, for example, OECD/IEA (2013:197), REN21 (2013a), and Sustainable Energy for All, Renewable Energy Fact Sheed, http://www.se4all.org/wp-content/uploads/2013/09/RenewableEnergy.pdf (accessed March 21, 2014).
[4] IRENA, IRENA membership, http://www.irena.org/Menu/Index.aspx?mnu=Cat&PriMenuID=46&CatID=67 (accessed January 28, 2014).
[5] Sustainable Energy for All, our vision, http://www.se4all.org/our-vision/ (accessed March 20, 2014).

advise governments and other policy actors on how to promote renewables in an effective way. Creating and diffusing knowledge and social understanding are crucial building blocks of these undertakings, as encouraging new thinking can have a significant impact on the behavior of actors (Breitmeier, Young & Zürn 2006:55). At the same time, transboundary cooperation is impossible without a minimum of shared understanding (Haas 1992:29). Reaching such understanding is an important task in an emergent global governance area without a strong track record of transboundary cooperation. It is especially challenging in a global governance area such as renewable energy (or energy in general), which is repeatedly characterized as highly fragmented[6] and dominated by national policy processes which in turn are influenced by different resource endowments and geographic conditions.[7] Here, it is particularly likely that ideas on global governance are highly contested.

1.1 Research Purpose and Questions

From a theoretical point of view, the present study aims to advance global governance research by developing a Cognitivist approach to global governance analysis, focusing on the politics[8] behind the definition of global governance issues.

Global governance scholars have considerably enriched International Relations research by drawing attention to the political implications of an increasingly interdependent world (Mayntz 2008). Global governance research points to the transboundary nature of many policy problems and the multiple gains of transboundary cooperation.[9] Thus, a central argument of global governance research is that in light of transboundary interdependencies, transboundary cooperation can lead to better policy outcomes than unilateral action. Analyzing transboundary political steering in the absence of a supreme authority (Mayntz 2008:109, Rosenau 1992:7), it pays special attention to 'soft' non-hierarchical modes of political steering, involving state and non-state actors. As global governance research highlights the intertwined nature of different policy levels, it blurs the

---

[6] See, for example, Dubash, Florini 2011:11, Florini, Sovacool 2009:5239, Florini, Sovacool 2011:57f., Goldthau, Sovacool 2012:237, Newell, Phillips & Mulvaney 2011, Lesage, Van de Graaf & Westphal 2010:51, Westphal, Röhrkasten 2013:40.

[7] See, for example, Bastos Lima, Gupta 2013:58, Dubash 2011, Dubash, Florini 2011:13-16, Karlsson-Vinkhuyzen, Jollands & Staudt 2012:17, Kong 2011, Kuik, Bastos Lima & Gupta 2011:632.

[8] 'Politics' refers to the process of policy-making, involving political actors with conflicting preferences. It is one of the three main dimensions of policy-making, the others being 'policy' (referring to contents of policy-making) and 'polity' (referring to the institutional or formal aspects). See Schmidt 2004:535-537, 561.

[9] The term 'transboundary' is used here as it is common in global governance research. While the terms 'international' and 'transnational' can be understood as referring to relations between governments or non-state actors only, the term 'transboundary' is explicitly not limited to a certain type of actor but might include both state and non-state actors (Dingwerth, Pattberg 2006a:196). See Chapter 2.1 for further elaborations on this issue.

previous clear-cut division between foreign/international and domestic affairs (see, for example, Dingwerth, Pattberg 2006a, Finkelstein 1995, Rosenau 1999).

In contrast to the present study, most global governance approaches assume the existence of an unambiguously given reality and thus do not address the contested nature of ideas in global governance nor the political processes behind the establishment of social understandings (on this critique see Avant, Finnemore & Sell 2010a, Barnett, Duvall 2005b, Overbeek et al. 2010). They neglect that global governance ideas are contested in two ways: they involve different understandings and are subject to political struggles. This is problematic for two main reasons. First, they miss out one important step of political action at the global level. Global governance involves heterogeneous groups of actors with highly diverse backgrounds and views; the influence of actors decides whose ideas prevail in the end. Creating and structuring knowledge and social understandings is an important way of exercising power in global governance, as it influences what policy-makers and researchers consider 'real' or 'relevant'. Second, most global governance approaches fail to grasp what can be a subtle, but major, barrier to transboundary cooperation: different understandings of policy problems and solutions may complicate or even impede cooperation; yet, policy actors and researchers may take their own understandings for granted and therefore may not notice the underlying causes of cooperation problems.

To identify contested ideas in global renewable energy governance, the present study pays special attention to the governments of Germany and Brazil. Both have been leading actors in this area of transboundary policy-making. They regard themselves as international pioneers in the promotion of renewable energy and have undertaken several initiatives to shape transboundary policy-making on renewable energy. To the author's knowledge, the attention that global renewable energy governance receives in both administrations is exceptional by international comparison. The German government has been the "engine and brain" behind IRENA's creation (Röhrkasten, Westphal 2013:3). In Brazilian foreign policy, renewable energy was a priority issue for several years. The author's previous research on their bilateral cooperation (Röhrkasten 2009) suggests that their perspectives on global renewable energy governance differ considerably. The fact that Brazil is one of the few states that has not yet joined IRENA seems to reinforce the existence of major differences. Another factor makes their comparison especially appealing: as a member of the Organization of Economic Cooperation and Development (OECD), Germany belongs to the group of countries that in the past defined global challenges and how these should be addressed (Harris, Moore & Schmitz 2009:6f., Najam, Robins 2001:50, Williams 1993:24-28). Brazil, on the contrary, is typically regarded as part of the group of emerging powers that increasingly challenge the OECD-led world order.[10]

This study analyzes in depth how the ideas of the German and Brazilian governments on global renewable energy governance differ. It furthermore identifies possible reasons for these ideational differences. The main research question, therefore, is: *How and why do the ideas of the German and Brazilian governments on global renewable energy governance differ?* Four terms are central to the research question and require further clarification. Governments are regarded as composite actors, comprising the different policy actors of the national executive. To analyze each administration's ideas, government actors who make pronouncements on global renewable energy governance are taken into account. Ideas are defined as cognitive organizing concepts that give meaning to the material world (Adler 1997:319, Béland, Cox 2011:3f.). Ideas are products of cognition and connected to the material world only through interpretation. They make sense of the world, establishing how things relate to one another. Global governance, according to Finkelstein (1995:369), means "governing, without sovereign authority, relationships that transcend national frontiers". While Finkelstein argues for a wide-ranging approach to global governance analysis that not only covers the global scale but the whole range of transboundary relationships, including intra-regional or bilateral interactions as well, other global governance scholars limit their analysis to the global scale (see, for example, Kaul 2008:91, Schuppert 2008:17, Zürn, Koenig-Archibugi 2006:237-239). In order to narrow the field of research down, the present study concentrates on the global scale, as covering this is consensual among global governance researchers. Renewable energy derives from naturally regenerative or practically inexhaustible resources. As demonstrated in Table 1, it comprises energy that originates from direct and indirect solar radiation (such as photovoltaics, wind energy, hydropower and biomass), tidal energy and geothermics.

The time frame of the comparative empirical analysis begins with the World Summit on Sustainable Development (WSSD) in Johannesburg 2002 and its preparatory process in 2001/02, as this summit placed renewable energy on the international agenda and also induced both governments to engage in global renewable energy governance. The time frame ends with major initiatives by both governments to shape global governance on renewable energy in 2013: in the Brazilian case, this applies to the International Conference on Bioenergy organized by the government in April 2013; and in the German case, to the founding of a 'Renewables Club' by Environment Minister Peter Altmaier[11] in June 2013.

---

[10] See Alexandroff, Cooper 2010b, Armijo 2007, Cooper, Flemes 2013, Hirst 2006, Hurrell 2010, Hurrell 2000, Hurrell, Husar, Maihold 2010, Husar, Maihold 2009, Narlikar 2008, Nel 2010,Schirm 2010, Sengupta 2012, Soares de Lima, Tickner 2013.

[11] Altmaier was German Environment Minister from 2009 until December 2013. He is a member of the Christian Democratic Union (Christlich Demokratische Union Deutschlands, CDU).

Fossil fuels	Coal, petroleum, natural gas
Nuclear fuels	Uranium (nuclear fission), deuterium and lithium (nuclear fusion)
Renewable energy	Derived from direct and indirect solar radiation: solar heat, photovoltaics, biomass, wind power, hydropower, wave energy, sea current
	Tidal energy
	Geothermics

Table 1: Overview of Different Energy Sources[12]

1.2 State of the Art

A global governance perspective on renewable energy (and on energy in general) is rather new, both in policy-making and academic discussion. Since the second half of the 2000s, a genuine strand of academic literature on global (renewable) energy governance has been evolving,[13] facilitating a first mapping of this area of transboundary policy-making. Research foci have been laid on the historical evolution of global renewable energy governance, the global challenges addressed by transboundary policy-making, international organizations, partnerships and fora dealing with renewables and their governance activities. So far, there has been no encompassing and systematic assessment on contested ideas in global renewable energy governance. However, the research strand reveals scattered insights on contested issues in this area of transboundary policy-making.

Some authors identify cleavages between the group of industrialized countries and that of emerging and developing countries in global energy governance. Lesage, Van de Graaf & Westphal (2010:40,85f.) state that the former focus on environmental protection, while the latter prioritize access to affordable energy. In addition, industrialized countries tend to favor the expansion of global energy governance whereas emerging and developing countries hesitate due to sovereignty concerns. Kuik, Bastos Lima & Gupta (2011:627) add that sovereignty concerns weigh more strongly in developing countries than in industrialized countries because the former are more affected by global energy governance. When they expand their energy base, for example, they are confronted with environmental concerns raised by industrialized countries.

---

[12] Adapted version of Hirschl (2008:62).
[13] For analyses that provide overviews of global governance on renewable energy see, for example, Bastos Lima, Gupta 2013, Hirschl 2009, Lempp 2007, Suding, Steiner et al. 2006, Rowlands 2005. For overviews of global energy governance in general see, for example, Cherp, Jewell & Goldthau 2011, Colgan, Keohane & Van de Graaf 2011, Dubash, Florini 2011, Florini 2008, Florini, Goldthau 2011, Goldthau 2013a, Karlsson-Vinkhuyzen, Jollands & Staudt 2012, Lesage, Van de Graaf & Westphal 2010, Najam, Cleveland 2005, Sovacool 2009, Van de Graaf 2013b.

Research on the structural characteristics of global (renewable) energy governance points to fragmented and disconnected governance efforts which make communication between involved policy actors difficult. Cherp, Jewell & Goldthau (2011) claim that global efforts to enhance energy security, energy access and climate protection are not coordinated but instead build on three largely autonomous governance arenas which are characterized by different origins, paradigms and policy actors. Dubash & Florini (2011:11) argue that different understandings, languages and expertise lead to major communication problems between policy actors and spheres in global energy governance.

Analyses of UN conferences dealing with renewables mention issues that encountered controversial debates during conference processes. Biswas (1981:331), Odingo (1981:106) and Rowlands (2005:82) state that the modes and scope of North-South transfers, the role of multilateral corporations and sovereignty over national resources were controversial issues at a UN conference on renewable energy in 1981. Rowlands (2005:83f.) claims that during the Rio Conference in 1992, the Arab countries opposed the promotion of renewables as a whole and did not want the UN to engage in global renewable energy governance. Several authors underline that global targets for renewable promotion were among the most controversial issues at the WSSD in 2002 (Cleveland 2005:128, Corrêa do Lago 2009:162, IISD 2002:7, Karlsson-Vinkhuyzen 2010:184, Najam, Rowlands 2005:86-88).

Röhrkasten & Westphal (2013) shed light on contested issues that arose during the process of IRENA's creation. They emphasize that IRENA's creation in itself was contested, as several governments opposed it due to reservations about the promotion of renewables, sovereignty concerns or doubts on the value added of intergovernmental organizations. Concerning IRENA's institutional set-up and activities, they identify several issues that caused controversial debates. These included the questions as to whether IRENA should engage in standard-setting and regulation or rather build on the principle of voluntariness, if it should be formed by a small frontrunner coalition or instead strive for broad membership, and if its activities should cover all member countries or instead concentrate on developing countries only. In addition, the renewable energy sources that IRENA should promote were debated as well.

Several studies furthermore point to the competition between IRENA and the International Energy Agency (IEA). They interpret IRENA's creation as an effort to counterbalance IEA's influence in global energy governance: while the IEA is confronted with the criticism that it favors conventional energy, i.e. fossil fuels and nuclear energy, over renewables, IRENA provides 'renewables-friendly' research and policy advice (Lesage, Van de Graaf & Westphal 2010:69, Röhrkasten, Westphal 2013:12,14, Röhrkasten, Westphal 2012b:3, Van de Graaf 2012a, Van de Graaf 2012b:8).

The present study builds on the author's previous research on Germany and Brazil (Acosta, Zilla 2011, Röhrkasten 2009, Röhrkasten, Zilla 2012).[14] To the author's knowledge, these are the only comparative studies on Germany and Brazil that provide first insights on the ideational differences of both governments.

Röhrkasten (2009) is the most encompassing study so far. It assesses the potential for a strategic partnership between the German and Brazilian governments to shape 'global structural policy' on renewable energy. This assessment also includes a comparative analysis of both governments' preferences which shows that widely divergent perceptions and preferences prevent them from pursuing a joint strategy in 'global structural policy' (Röhrkasten 2009:57-81, 95-97). The German government emphasized the need to tackle market distortions that favor fossil fuels over renewable energy and to improve public awareness, information and know-how on renewable energy. It aimed to create IRENA to engage in policy advice, technology transfer and capacity-building. The Brazilian government, by contrast, criticized industrialized countries' trade barriers and prejudices against biofuels and aspired to create an international biofuels market. The comparative analysis furthermore demonstrates that the German government favored solar energy, wind energy and small hydropower, while the Brazilian administration concentrated on biofuels, particularly ethanol. In addition, it reveals that both governments advanced contrary assessments on the sustainability of export-led biofuels production in developing countries. While the Brazilian government underlined the benefits, its German counterpart issued warnings about it. Thus, this comparative analysis provides first insights on *how* ideas on global renewable energy differ between both governments. It furthermore gives one hint on *why* the ideas differ: it states that both governments favor renewable energy options that are important for their national exports (Röhrkasten 2009:95). The study gives recommendations for further research in order to enhance a better understanding of the diverging preferences and assessments of both governments, particularly with regard to biofuels: it suggests to analyze domestic policy processes, to differentiate between different government actors instead of treating governments as monoliths, and to integrate non-state actors into the analysis (Röhrkasten 2009:99). These recommendations are taken up by this PhD project.

Acosta & Zilla (2011) and Röhrkasten & Zilla (2012) focus on the governments of Germany and Brazil to investigate trade and value conflicts over biofuels. These studies state that the governments' assessments of the social and environmental implications of biofuels production differ widely. The administrations draw different conclusions on the relationsship between biofuels production and the scarcity of other natural resources and even set up diverging causalities between biofuels and food production. The studies emphasize that the different assessments

---

[14] One of the studies was published under the author's former last name Acosta.

are in line with the trade interests of each administration. While the Brazilian government wants to promote Brazilian biofuel exports and thus presents biofuels in very positive terms, the German administration raises social and ecological concerns to shield its domestic biofuels production. As such, these studies also point to economic interests as underlying causes for ideational differences.

1.3 Theoretical-Analytical Framework

The present study chooses a three-step approach to develop a Cognitivist approach to global governance analysis. As global governance research is the theoretical point of departure of this thesis, it first of all elaborates on global governance as a perspective on transboundary policy-making.

In a second step, it presents theoretical approaches that help to integrate contested ideas into global governance research. In order to shed light on the role of ideas in global governance, the thesis builds on insights from Constructivist research in International Relations which is based on the central argument that "international reality is socially constructed by cognitive structures that give meaning to the material world" (Adler 1997:319). In contrast to the Rational Choice assumption of an unambiguously given reality, Constructivist scholars emphasize that perspectives and understandings of reality vary across actors as these do not necessarily employ the same cognitive organizing concepts (Haas, Haas 2002:586). Moreover, the thesis pays special attention to the few global governance approaches that integrate contestation and power into their analysis, highlighting that global governance does not necessarily advance the benefits of all involved (see, for example, Avant, Finnemore & Sell 2010b, Barnett, Duvall 2005a, Barnett, Finnemore 1999, Hurrell 2005). Instead of speaking about global governance in passive terms only, they explicitly include policy actors. Thus, they do not only focus on global *governance*, but also on global *governors* – actors who exercise power across borders by creating issues, setting agendas, establishing and implementing rules or programs and evaluating and/or adjudicating outcomes (Avant, Finnemore & Sell 2010c). They furthermore question if global governance is *de facto* a global, or rather an OECD world concept. In line with this, they analyze what global power shifts due to the rise of new powers imply for global governance (see, for example, Alexandroff, Cooper 2010b, Cooper, Flemes 2013, Hurrell, Sengupta 2012, Najam, Robins 2001, Narlikar 2010, Williams 1993).

In a third step, the present study develops an actor-centered approach to ideational analysis. As it sheds light on the politics behind the definition of global governance issues, it focuses on policy actors as carriers and drafters of ideas. The basic behavioral assumptions build on Weak Cognitivism (Hasenclever, Mayer & Rittberger 2000:10-12, 1996:206-210). Weak Cognitivism assumes that goal-oriented political action and the social construction of reality go hand-in-hand. This variant of Constructivist research is compatible with the Rational Choice analyses that dominate global governance research: it focuses on those aspects that Rational Choice analyses take as given, analyzing how ideas shape perceived policy contexts and policy goals and how political action shapes ideas. As a consequence, Weak Cognitivism regards ideas not only as cause but also as result of political action. In line with Rational Choice, Weak Cognitivism also builds on interest-based behavior.[15] Yet, instead of assuming certain interests as given, it highlights that interests are defined by the policy actors themselves. Policy actors generate ideas self-consciously (Campbell 1998:378,381-383). If they are confronted with new issues, they tend to draw on known concepts (Breitmeier, Young & Zürn 2007:55). Actors can also frame ideas in a way that helps them achieve their political goals,[16] (over)emphasizing or concentrating on those aspects that are in line with their goals while suppressing those that are not. In addition, actors try to draft ideas in a consistent way in order to increase their persuasiveness (Mehta 2011:27). Some aspects might be more important for actors than others and thus influence how they draft less important aspects.

In order to analyze the contents of the governments' ideas on global renewable energy governance, the present study develops analytical categories that are grounded in both the theoretical-analytical framework and the empirical material under study (Dey 1993:96f.). The derivation of the categories starts out from Metha's analysis on the role of ideas in politics which states that ideas comprise both problem definitions and policy solutions (Mehta 2011:28-40). Applied to global governance, this means that global governance ideas specify why global governance is necessary or desirable, what direction and form it should take, and which global governors should be involved. Due to the findings of global governance research and salience in the empirical material under study, the present study furthermore considers 'meta-aspects' which reflect on the way global governance works or should work. Such meta-aspects, for example, specify if a policy actors views global governance in terms of advancing the mutual benefits of all involved or rather in terms of confrontation and power politics. In order to depict ideas on global *renewable energy* governance, further specifications are made. These not only ensure compatibility with academic discussions on global (renewable) energy governance but also mirror the thematic focus of the empirical material under study. As such, it is specified that problem definitions clarify why the promotion of renewables is a global issue, i.e. what global challenges it addresses, and why transboundary action is needed, that is, what factors hamper the worldwide spread of renewables.

---

[15] In that regard, it differs from Strong Cognitivism which focuses on ideas as social norms and regards political actors as role players following a logic of appropriateness. In contrast to Weak Cognitivism, Strong Cognitivism is incompatible with Rational Choice.
[16] See, for example, Adler, Haas 1992:374, Béland, Cox 2011:10, Blyth 2003:697, Schmidt 2011:52, Schmidt 2008:317.

Since 'renewable energy' is an umbrella term for different energy sources, it is furthermore important to indicate what kind of renewable energy sources the actors refer to when speaking about the worldwide promotion of renewables. The analysis of ideas on global renewable energy governance thus focuses on the following aspects: (1) global challenges addressed by the worldwide promotion of renewable energy, (2) specification of, and differentiation between, renewable energy options, (3) barriers to a worldwide promotion of renewables, (4) tasks for transboundary policy-making, (5) relevant global governors and (6) meta-aspects of/salient features in transboundary policy-making.

The search for possible reasons behind the ideational differences also starts out from research on ideas in policy-making (see, for example, Breitmeier, Young & Zürn 2007:53-57,191-225, Campbell 2002, Campbell 1998, Mehta 2011). More specifically, it is structured along the abovementioned elements which influence how policy actors generate ideas: the reliance on known concepts, the framing in line with political goals and the endeavor to draft ideas in a consistent way. Based on empirical research, the present study argues that the actors' policy contexts are important reasons behind ideational differences, as these influence what concepts are known to policy actors. It identifies three different policy contexts that influence the governments' ideas. As government actors who deal with global renewable energy governance act on two levels, the global and the domestic (Adler, Haas 1992:373f.), their ideas are shaped by the domestic policy context and their involvement and self-positioning in transboundary policy-making, which in turn influences their perception of the global policy context. The division of competencies is another important contextual factor, as global renewable energy governance involves government actors with different institutional backgrounds. Moreover, the self-interests of actors come in, i.e. goals to advance their conception of self-good (Hay 2011:79), as these direct the purposeful framing of global governance ideas. Due to the efforts of actors to draft ideas in a consistent way, some ideational differences are furthermore caused by other ideational differences. If policy actors, for example, define the barriers to a worldwide promotion of renewables differently, it is likely that their perspectives on the tasks for transboundary policy-making vary as well.

1.4 Methodological Approach

This thesis relies on a qualitative-interpretative research design. In Cognitivist analyses, interpretation is not only an important object of research but also an intrinsic part of how research is conducted (Adler 2002:95,101, Adler 1997:328, Houghton 2007:33, Price, Reus-Smit 1998:271-273, Sikkink, Finnemore 2001:395). An interpretative epistemology advances a different understanding of explanation and causalities than that of a positivist one. Instead of searching for deterministic laws and clear-cut causalities, it sheds light on the complex nature of relationships which often involve mutual constitution. According to an interpretative epistemology, explanation means understanding the constitution of things, i.e. understanding "how things are put together and how they occur" (Sikkink, Finnemore 2001:394). Inductive inquiry is of central importance for interpretative research: the starting point of interpretative research are not hypotheses and *a priori* assumptions but meanings that policy actors attribute to their reality. Contextual and historical interpretation objectifies subjective meanings. Thus, in order to understand social realities, tracing the political processes that led to the constitution of these realities is of utmost importance (Pouliot 2007).

The research design builds on a comparative analysis of two case studies, investigating the ideas advanced by the German and Brazilian governments on global renewable energy governance. The case studies shall facilitate a comprehensive understanding of the substantive contents of the ideas and the political processes behind their establishment. They rely on a case-sensitive analysis that aims at uncovering the 'within-case-logics'. Each case study comprises two main building blocks. The first part traces the process of government action on global renewable energy governance and thus helps to contextualize and historicize the ideas. More specifically, it analyzes (1) how and why the government engaged in global governance on renewable energy, (2) which policy actors were involved, and (3) what context factors and political processes at the domestic and global level influenced the administration's action. The process analysis primarily focuses on the governments, but also considers non-state actors that were involved in the policy processes under study. It thus uncovers what policy actors outside of government influenced the governments' action and ideas on global renewable energy governance. The second part of the case studies analyzes the ideas that the governments pronounce on global renewable energy governance. This part is based on a qualitative content analysis of government statements on the issue (Kracauer 1952, Mayring 2005). The data analysis relies on inductive coding (Mayring 2005:11f.) and is pre-structured along the six categories that help to analyze ideas on global renewable energy governance (for an overview, see also Table 5 on page 68): global challenges, renewable energy options, barriers, tasks, global governors and salient features. The same elements also structure the identification of ideational differences. The analysis of the reasons behind the ideational differences primarily builds on the process analyses within both case studies.

Three main sets of sources form the data corpus of this thesis: (1) official documents and statements by representatives of both governments, (2) qualitative interviews with government representatives and further policy actors involved in the processes under study, as well as with experts on the issue, and (3) academic literature. The official documents and statements are the primary sources for the analysis of each government's ideas. For the purpose of this study, publicly available and accessible documents and statements were assessed that deal with global renewable energy governance and that were issued during the time frame of analysis. These comprise policy papers, official submissions, press statements and website information of state bodies as well as speeches and articles by high-ranking government representatives. The principal function of the interviews is to contextualize and historicize the ideas of each administration.[17] Due to the scarcity of academic and grey literature on the issue, they are of central importance in tracing the policy processes under study and for revealing the factors and developments that influenced each government's reasoning and action. Overall, 75 interviews were conducted (39 interviews with representatives of policy actors and experts from Germany and 36 interviews with representatives of policy actors and experts from Brazil). The groups of interviewees in both countries comprise government officials, academics and policy consultants, as well as representatives from businesses and non-governmental organizations (NGOs). In the German case, they furthermore cover the German Parliament (Bundestag) and political foundations. Academic literature is the main source for the development of the theoretical-analytical framework and the analysis of global renewable energy governance. It furthermore serves as an additional source for the case study analyses.

1.5 Outline

Chapter 2 develops the theoretical-analytical framework of this thesis. It illustrates global governance as a perspective on transboundary policy-making and elaborates on how to integrate contested ideas into global governance research. It furthermore presents the actor-centered framework for the analysis of global governance ideas and provides further details on the methodological approach.

Chapter 3 enhances a sound empirical introduction to global renewable energy governance and related academic discussions. It traces the origins and evolution of global governance on renewable energy and characterizes this field of transboundary policy-making. To do so, it elaborates on the global challenges that global renewable energy governance addresses, illustrates structural characteristics of this area of transboundary policy-making and presents global governors and their governance activities.

Chapters 4 and 5 comprise the case studies of the German and Brazilian governments' ideas on global renewable energy governance. As outlined above, the two case studies contain two main building blocks: the first part traces the govern-

---

[17] The content analysis of the governments' ideas relies on official documents and statements instead of interviews as the former result of collective opinion-processes. Thus, they are better suited to depict the ideas of government actors than interview statements of individuals that might be biased, for example, due to individual perspectives or moments of abstraction during interview.

ment's action on global renewable energy governance, while the second part analyzes the contents of its ideas.

Chapter 6 contrasts the ideas of the German and Brazilian governments. It answers the main research question of this thesis, identifying the differences between the ideas of both governments and presenting the main reasons behind these.

The concluding chapter 7 resumes the main findings of this study, points to their implications for global renewable energy governance and gives suggestions for further research.

2 Theoretical-Analytical Framework: Contested Ideas in Global Governance

This chapter develops the theoretical-analytical lenses that will guide the empirical analysis. In a Cognitivist approach to global governance analysis, it is important to make the reader aware that theoretical approaches to political science analysis are always embedded in specific policy contexts:

> "There is no view from nowhere. All theories and concepts are bounded by time and place; they draw their relevance from the temporal sequences and particular contexts in which they are developed and deployed" (Hurrell 2011:149).

International Relations research is often influenced by perceptions of political relevance within the researcher's surroundings, and perceived changes in the political reality often lead to changes in analytical frames (Mayntz 2008:101, Tickner 2013:638). As such, in the following deliberations on theoretical approaches, policy contexts are explicitly taken into account.

The first section of this chapter presents global governance as a perspective on transboundary policy-making. To do so, it sketches the political and theoretical background of global governance research and lays out its theoretical foundations. The chapter then elaborates on theoretical approaches that help to integrate contested ideas into global governance research. Contrary to mainstream global governance research, these approaches shed light on the role of ideas, contestation and power in global governance, and also bring in policy actors. This section also touches theoretical discussions about the contestation around the concept of global governance itself and the accusation that global governance is not a global, but rather an OECD world, concept. The following section develops the actor-centered approach for the analysis of global governance ideas. It specifies basic assumptions about policy actors and their behavior, lays out how ideas and political action relate to one another and deduces analytical categories that guide empirical research. The chapter then presents the methodological approach of this thesis which relies on an interpretative research design and comparative case study analysis. It furthermore elaborates on the data source for the empirical research.

2.1 Global Governance As a Perspective on Transboundary Policy-Making

Global governance research focuses on the political implications of an increasingly interdependent world (Dingwerth, Pattberg 2006a:189-196, Dingwerth, Pattberg

2006b:395, Mayntz 2008:109). It analyzes how increasing interdependencies change the context of policy-making and how policy actors should respond to this changing context. Global governance research builds on the central argument that, in light of interdependencies, transboundary cooperation has the potential to achieve better policy outcomes than unilateral action. Thus, an important research focus is on the common challenges and shared goals that transboundary cooperation should address. Global governance research furthermore analyzes how to organize transboundary cooperation so that it can actually exploit its potential. In contrast to national policy-making which may rely on the formal power of governments to make and enforce binding decisions, a central challenge of transboundary political steering is the absence of a supreme authority.

Global governance, in the words of Finkelstein (1995:369), "is governing, without sovereign authority, relationships that transcend national frontiers". Whereas Finkelstein (1995:369) states that the concept of global governance is flexible in its reach, covering any interactions that transcend national frontiers while other authors, such as Zürn, Koenig-Archibugi (2006:237-239), Schuppert (2008:16) and Kaul (2008:91), limit it to the transcontinental scale. Rosenau (2009 [1995]:11) underlines that global governance does not build on a single organizing principle; rather, it should be understood as "the sum of myriad–literally millions of–control mechanisms driven by different histories, goals, structures, and processes". It comprises cooperation formats with different degrees of institutionalization.[18] Global governance scholars emphasize that different policy levels are closely interlinked; they thus challenge the mainstream assumption in International Relations of an international system comprising sovereign states that should be analyzed separately from domestic politics (Dingwerth, Pattberg 2006a: 189-193). For global governance scholars, states (or more specifically governments) are central but not the only actors in global governance. Thus, they analyze the interaction between a wide range of state and non-state actors and among different policy levels, comprising diverse steering mechanisms and spheres of authority (Dingwerth, Pattberg 2006a:196-198, Finkelstein 1995:369, Rosenau 1999:295).

---

[18] In that regard, it differs from international regime research which concentrates on cooperation formats with a relatively high degree of institutionalization. International regimes, according to the predominant definition by Krasner (1983b:2), comprise principles, norms, rules and decision-making procedures in a specific issue area (see also Hasenclever, Mayer & Rittberger 1996, Hasenclever, Mayer & Rittberger 2000). Global governance includes international regimes, but also other cooperation formats that do not meet the institutional requirements of international regimes.

2.1.1 Contextualizing Global Governance Research

Global governance research is rooted in the International Relations tradition of Liberal Institutionalism which explains international cooperation in terms of utilitarian reasoning by self-interested states.[19] The emergence of global governance research during the 1990s can be interpreted as a response to two developments: globalization as a process of increasing transboundary interactions and geopolitical changes due to the end of the Cold War (Mayntz 2008).

Globalization became one of the buzzwords of the 1990s. It denotes a process of increasing transboundary interaction which also challenges the problem-solving capacity of national governments (Held, McGrew 2003:190-193, Kaul, Grunberg & Stern 1999a:450f.).[20] Sometimes, the term 'globalization' is used to refer to economic interdependencies only, such as transboundary flows of goods, services and capital or the internationalization of related technology and know-how. However, transboundary interactions are not limited to the economic sphere, they comprise social, cultural and environmental phenomena as well (Castells 2005:10, Held et al. 1999:483f., Keohane, Nye 2000:105). Globalization therefore includes a wide variety of phenomena, ranging from the internationalization of communication systems and cultural transfers to transboundary pollution. Increasing transboundary interdependence was of course not an entirely new phenomenon of the 1990s. As early as the 15th century the 'discovery' of the New World resulted in substantial voluntary and forced migration movements, associated transfers of cultures and languages, and international trade (Held et al. 1999:484f.). Yet, the 1990s experienced a substantial increase in transboundary interaction – not only due to technological innovation with regard to information, communication, computing and transportation, but also due to political decisions such as the economic policies of deregulation and liberalization during the 1980s. It is commonly underlined that globalization during the 1990s differed not only in scale but also in qualitative terms from earlier globalizations, being characterized by higher extensity, intensity and velocity. This means that border-crossing interconnection encompassed larger distances, had a stronger influence on peoples' lives and changed more rapidly (see, for example, Held, McGrew 2003:186, Held et al. 1999:484, Hewson, Sinclair 1999:6f., Keohane, Nye 2000:108). As a result of these developments, governments were confronted with rapidly changing parameters of policy-making and their adaptive

---

[19] Central publications within Liberal Institutionalism are for example Keohane (1989), Keohane (1984), Keohane, Nye (1977) and Krasner (1983a). For a comprehensive overview see Hasenclever, Mayer & Rittberger 1996:183-196.

[20] It is important to note that globalization is not a linear process. Transboundary interaction may also decrease over time as, for example, economic interdependence did between 1914 and 1945. Neither does 'globalization' necessarily mean 'universalization'; the intensity of transboundary interaction may vary between different thematic areas and different countries or regions (Keohane, Nye 2000:107).

capacities often lacked behind the changing reality (Kaul et al. 2003a:21). Growing transboundary interdependence blurred the boundaries between domestic affairs and global matters (Held et al. 1999:483f.): policy results were increasingly influenced by factors outside of the national border; at the same time, policy decisions of national governments also caused transboundary repercussions. As a consequence, many scholars and policy-makers identified a need for more transboundary cooperation, integrating not only governments but also relevant private actors, such as businesses and non-governmental organizations.

At the same time, the end of Cold War raised the expectations on the scope and problem-solving capacity of international cooperation, especially within the (UN) which was no longer blocked by vetoes and distrust between the capitalist and communist world (Brühl, Rittberger 2001:2f., Hewson, Sinclair 1999:4, Knight 2009:173f.). Policy-makers and researchers alike hoped that international politics would no longer be based on exclusivity and hierarchy but instead organized in a more inclusive, participative and egalitarian manner. Challenging realist premises and advancing cosmopolitan thinking in International Relations research seemed to be easier than during the Cold War period: instead of focusing on the power politics between sovereign states in an anarchical international system, scholars pointed to the possibility of advancing common interests through transboundary cooperation (Barnett, Duvall 2005b:5f., Mayntz 2008:102f.). During the 1990s, the scope and agenda of international cooperation arrangements indeed expanded substantially (Knight 2009:173f.) – but many expectations on the effectiveness of these cooperation arrangements remained unfulfilled.

'Milestones' for the introduction and diffusion of global governance research in the discipline of International Relations were the publication of the volume "Governance without Government: Order and Change in World Politics" by Rosenau & Czempiel (1992) and the final report "Our global neighborhood" by the Commission on Global Governance (1995),[21] as well as the launch of the academic journal "Global Governance" in 1995 (see also Dingwerth, Pattberg 2006a:185f.). They all characterize the historical context that inspired their work in a similar way:

> "At a time when hegemons are declining, when boundaries (and the walls to seal them) are disappearing, (...) the prospects for global order and governance have become a transcendent issue. (...) One senses that the course of history is at a turning point, a juncture where the opportunities for movement toward peaceful cooperation, expanded human rights, and higher standards of living are hardly less conspicuous than the prospects for intensified group conflicts, deteriorating social systems, and worsening environmental conditions" (Rosenau 1992:1).

---

[21] The Commission on Global Governance dates back to an initiative of Willy Brandt after the fall of the Berlin Wall. With the backing of the then UN Secretary-General Boutros Boutros-Ghali the Commission started its work in 1992, encompassing twenty-eight distinguished personalities from different parts of the world. See Encyclopedia Britannica, Commission on Global Governance, http://www.britannica.com/EBchecked/topic/1917949/Commission-on-Global-Governance (accessed February 18, 2014).

"Time is not on the side of indecision. Important choices must be made now, because we are at the threshold of a new era. (...) We can, for example, go forward to a new era of security that responds to law and collective will and common responsibility by placing the security of people and of the planet at the centre. Or we can go backwards to the spirit and methods of what one of our members described as the 'sheriff's posse'–dressed up to masquerade as global action" (Commission on Global Governance 1995:xix).

"Ours is a time of great hope and great hopelessness, a time when ideological fault lines have disappeared, while the global rifts of wealth and power have widened. The emerging global unity of purpose has allowed many of us to imagine revitalized multilateral institutions forging cooperative responses to global problems; still, the reality of the UN system straining under so many new demands makes most of us questioning whether such a renaissance is possible" (Coate, Murphy 1995:1) – Editors of the Journal *Global Governance*.

All three quotations allude to the end of Cold War as a historical turning point that offered unprecedented opportunities for transboundary cooperation; at the same time they underline the great challenge of organizing transboundary policy-making in a way that enables the actual use of these new opportunities.

2.1.2 Blurring the Boundaries between Domestic and Global Affairs

Global governance approaches regard the domestic and global level of policy-making as being closely intertwined (see, for example Dingwerth, Pattberg 2006a:191-193). Rosenau emphasizes that spheres of political influence are not necessarily attached to territorial boundaries:

"The world is comprised of spheres of authority that are not necessarily consistent with the division of territorial space and are subject to considerable flux" (Rosenau 1999:295).

According to Rosenau, this is the central premise that the ontology of global governance research builds upon. Thus, global governance research does not restrict the exercise of authority to states. As global governance research overcomes the exclusive link between geography and political power, it deviates significantly from the Westphalian model of policy-making which is based on the principle of exclusive sovereign rule over a bounded territory (Held, McGrew 2003:185f., Held et al. 1999:487).

Global governance research underlines that increasing transboundary interdependencies alter both domestic and transboundary policy-making. This has important repercussions on sovereignty as a core political science concept. Whereas International Relations research used to focus on challenges to the external sovereignty of states only (i.e. prevention form outside interference), global governance scholars also shed light on the internal dimension of sovereignty which used to be the focus of domestic politics scholars only. Internal sovereignty, according to Reinicke (1998: 56f.), refers to the "ability of a government to formulate, implement and manage public policy" within the national territory. Thus, internal sovereignty also relates to domestic relationships between governments and non-governmental actors. It is important to note that this concept of sovereignty builds on an operational, and not a legal, understanding of sovereignty. Reinicke (1998: 70) identifies a "need for cooperation among governments on a scale and depth not yet witnessed". Governments risk their internal sovereignty if they do not cooperate with other administrations:

> "The only way for governments to achieve (...) internal sovereignty, is to pool, and thus share, internal sovereignty in those sectors in which globalization has undermined the effectiveness and efficiency of internal sovereignty at the national level" (Reinicke 1998: 71).

Kaul, Grunberg and Stern go in a similar direction, arguing that states need to cooperate if they wish to avoid losing their capacity for effective policy implementation at national level:

> "Nation states will witness continuing erosion of their capacities to implement national policy objectives unless they take further steps to cooperate in addressing international spillovers and systematic risks" (Kaul, Grunberg & Stern 1999a:451).

Held & McGrew (2003:191) argue that increasing interdependencies do not necessarily lead to a diminution but rather to a transformation of state power:

> "Contemporary globalization (...) has been accompanied by extraordinary growth in institutionalized arenas and networks of political mobilization, surveillance, decision making, and regulatory activity that transcend national political jurisdiction. This has expanded enormously the capacity for and scope of political activity and the exercise of political authority" (Held, McGrew 2003:191).

Whereas the power of private actors might increase in some issue areas, governments (and other policy actors) can also expand their political activities beyond national borders.

2.1.3 Governments: Central, but not the Only Actors

Most global governance researchers underline that governments remain important actors in global governance. In this regard, the use of the term 'governance without government' (Rosenau, Czempiel 1992) easily leads to misinterpretations since it suggests the absence of governments. Yet, Rosenau himself stated that this term does not exclude governments but rather refers to the absence of a supreme authority in global governance:

> "A focus on 'governance without government' does not require the exclusion of national or subnational governments from the analysis, it does necessitate inquiry that presumes the absence of some overarching governmental authority at the international level" (Rosenau 1992:7).

The relative importance of governments is contested in global governance research. Particularly in the early 1990s, global governance researchers commonly identified a withdrawal of the nation-state that other actors need to compensate (Mayntz 2008:103). In line with this, Castells (2005:10) even claims that "non-governmental actors become the voices and the movements that defend the needs, interests, and values of people at large". Yet, towards the end of the 1990s, this perspective changed and scholars commonly pointed at changing instead of waning political functions of states (Mayntz 2008:104). While some authors claim that there is no clear hierarchy between governmental and non-governmental actors (Dingwerth, Pattberg 2006a:191-193), others argue that states are the primary actors in global governance since they have the monopoly of legitimate coercive power. According to this view, states are not substituted but supplemented by other actors (see, for example, Commission on Global Governance 1995: 4, Keohane, Nye 2002 (2000):202, Koenig-Archibugi 2006:1).

Scholars agree that governments are not the only relevant actors in global governance. This openness to different policy actors is an important characteristic of global governance research:

> "The key distinction between international politics as it used to be and global governance as it currently manifests itself (...) is a proliferation of actors that dispose of at least some resources that are necessary to effectively steer the behavior of individuals and corporate actors across territorial boundaries" (Dingwerth, Pattberg 2009:42).

The use of the term '*transboundary* policy-making' mirrors this. While the terms 'international' and 'transnational' can be understood as referring to relations between governments or non-state actors only, the use of the term 'transboundary' explicitly includes both state and non-state actors (Dingwerth, Pattberg 2006a:196). According to Commission on Global Governance (1995:3), global governance must comprise all "actors who have the power to achieve results". Global governance scholars commonly underline that even the most powerful actors cannot solve transboundary problems unilaterally but rather need to engage in transboundary cooperation with other actors (Held, McGrew 2003:190,196). Besides governments, there are several groups of actors that are relevant for global governance; these include international organizations and bureaucracies, non-governmental organizations, hybrid organizations between the public and the private sphere such as the International Organization of Standardization (ISO), subnational governments, corporations and epistemic communities (Dingwerth, Pattberg 2009:44-48, Karns, Mingst 2010:14-21, Rosenau 2003:80-84, Simmons, Jonge Oudraat 2001:11f.). Held and McGrew point out that global governance increasingly involves actors with a transboundary institutional set up:

> "Most obvious is the rapid emergence of an increasingly dense web of intergovernmental organizations, international nongovernmental organizations (NGOs) and a wide variety of other transnational pressure groups and networks" (Held, McGrew 2003:188).

Zürn & Koenig-Archibugi (2006:237-239) and Schuppert (2008:16) distinguish three prototypes of global governance: governance by, with and without government. Intergovernmental governance (governance with government) includes intergovernmental networks, international regimes and international organizations. Supranational governance (governance by government) applies if rules are considered superior to national law or if officials in international organizations enjoy autonomy from national governments. Transnational governance (governance without government) involves only private actors. In reality it is common to find mixtures of these prototypes, as cooperation arrangements often comprise different kind of policy actors.

2.1.4 Value Added of Transboundary Cooperation

As global governance research builds on the central premise that in light of interdependencies, transboundary cooperation can lead to better policy outcomes than unilateral action, an important focus of global governance research is to highlight fields of action where transboundary cooperation can provide value added. Global governance researchers commonly point to global challenges that cannot be tackled by unilateral action. For example, when launching the journal *Global Governance*, the editors claimed that the journal should focus on "the entire range of global problems–economic development, peace and security, human rights, and the protection of the environment" (Coate, Murphy 1995:1).

Researchers associated with the United Nations Development Program (UNDP) (Kaul 2008, Kaul, Conceição 2006, Kaul, Grunberg & Stern 1999b, Kaul et al. 2003b) have developed the most comprehensive approach to identify areas for transboundary cooperation. They focus on the provision and use of global public goods, adapting insights from the public goods theory in microeconomics to the context of transboundary cooperation.[22] They point out that the provision of (global) public goods goes along with collective action problems: individually rational behavior leads to socially suboptimal outcomes. A precondition for collective action problems is interdependence: One actor's action (or non-action) influences the degree of another's goal achievement. Unilateral action in the presence of interdependence is unlikely to achieve the best outcome for all, as actors do not consider

---

[22] This approach is not entirely new. Russett & Sullivan (1971), Olson (1971) and Kindleberger (1986) chose similar approaches before. According to Coussy (2005:179), it is not theoretical novelty but rather a new demand for such a theoretical approach that led to the success of the UNDP approach to global public goods: "the current success of the notion of global public goods is due less to novel theoretical supply than it is reflective of a growing demand for a theory of public management of international relations".

the effects of their own acivities on others. Depending on the costs involved, cooperation may contribute to improving outcomes, thus realizing cooperation gains.

Class and type of global public good	Problem	Corresponding global bad
1. Natural global commons (stock variables, precede human activity)		
Ozone layer	Overuse	Depletion and increased radiation
Atmosphere	Overuse	Risk of global warming
2. Human-made global commons (stock variables: already produced)		
Universal rights and principles	Underuse (repression)	Human abuse and injustice
Knowledge	Underuse (lack of access)	Inequality
Internet (infrastructure)	Underuse (entry barriers)	Exclusion and disparities
3. Global conditions (flow variables: continued effort is required)		
Peace	Undersupply	War and conflict
Health	Undersupply	Disease
Financial stability	Undersupply	Financial crisis
Free trade	Undersupply	Fragmented markets
Freedom from poverty	Undersupply	Civil strife, crime and violence
Environmental sustainability	Undersupply	Unbalanced ecosystems
Equity and justice	Undersupply	Social tensions and conflict

Table 2: Typology of Global Public Goods[23]

Kaul et al. (2003c:8) define public goods as "the collection of things available for all people to access and to consume freely". They thus characterize public goods as goods that (should) belong in the public domain. *Global* public goods are defined as having costs and benefits that transcend national borders and geographic regions (Kaul 2008:91). According to Kaul, Grunberg & Stern (1999a:454f.) and Kaul & Mendoza (2003:100), global public goods comprise global natural commons, global human-made commons and global policy outcomes or conditions (see Table 2).

Every global public good has a corresponding global bad. Whereas natural global commons such as the ozone layer and the atmosphere are faced with over-

---

[23] Adapted version of Kaul, Grunberg & Stern (1999a:454f.).

use, human-made global commons, such as knowledge and universal rights, are confronted with the challenges of underuse and unequal distribution. Here, many people lack access to these goods. Global conditions such as free trade or environmental sustainability are commonly undersupplied. Out of the different global public goods, the International Task Force on Global Public Goods (2006) names six that should be the primary focus of transboundary cooperation: preventing the emergence and spread of infectious diseases, tackling climate change, enhancing international financial stability, strengthening the international trading system, achieving peace and security, and generating knowledge.

2.1.5 Transboundary Policy-Making in the Absence of a Supreme Authority

Whereas domestic policy-making can rely on governmental coercive power and thus on hierarchical steering (which does not mean that it must always rely on it), there is no world government with the ability to authoritatively decide on the common affairs of states and other policy actors. At global level, the possibilities of hierarchical steering are severely restricted. With the exception of a few supranational institutions, cross-border collective action depends on sovereign states or other policy actors opting voluntarily (at least in the formal sense) to cooperate (Young 1997).

Global governance builds on a variety of formal and informal, hierarchical and non-hierarchical steering mechanisms (Dingwerth, Pattberg 2006a:196, Rosenau 1992:4, Stoker 1998:17). Due to the absence of an overarching authority, 'soft', non-hierarchical modes of political steering are of central importance to global governance research. Whereas hierarchical modes of political steering are based on the authoritative allocation of values and rule enforcement, 'new modes of governance' do not build on formal hierarchies. They comprise softer elements, such as positive incentives, bargaining, learning and arguing (Börzel, Risse 2005:196-198). Finkelstein (1995:370f.) states that global governance comprises a wide range of activities, such as creating and exchanging information, formulating and implementing rules and principles, transferring resources, and mediating and resolving disputes. Rosenau (2009:3f.) underlines that a central challenge of global governance arrangements is to achieve compliance. He adds that governance structures mostly employ a combination of carrots and sticks to achieve compliance, ranging from financial incentives and persuasion to threat or coercion.

2.2 Integrating Contested Ideas into Global Governance Research

Theoretical global governance research often follows a functionalistic way of reasoning. Thus, it commonly assumes the existence of an unambiguously given reality and does not pay attention to different understandings of reality. The use of depoliticized concepts such as 'steering' and 'problem-solving' might even suggest that transboundary policy-making is mainly about management and not about political processes involving policy actors with different perspectives and interests – and also with different capacities to advance their perspectives and interests in transboundary policy-making. Only few scholars explicitly integrate politics and social construction into global governance research (see, for example, Avant, Finnemore & Sell 2010b, Barnett, Duvall 2005a, Barnett, Duvall 2005b:17, Barnett, Finnemore 1999:710-715). They highlight that global governance always involves contestation and that political processes decide on the definition of global governance issues.

2.2.1 The Role of Ideas in Global Governance

The scarce attention that global governance research pays to the role of ideas is not an isolated case. Mehta even claims that it has been a common feature in the whole discipline of political science to underestimate the role of ideas:

> "The role that ideas play in politics has long been appreciated more by the newspaper reader than by the average political scientist" (Mehta 2011:23).

Global governance research – like the discipline of International Relations in general – has been deeply influenced by the Rational Choice paradigm which leaves little or no space for ideas. As such, analyses inspired by the Rational Choice paradigm fail to grasp a central element of political action:

> "At the core of politics is the way ideas are packaged, disseminated, adopted, and embraced. The muddle of politics is the muddle of ideas" (Béland, Cox 2011:13).

Späth (2005:37-39) argues that global governance analyses mostly set in after common values and goals are defined, ignoring the political processes that lead to the definition of goals and objectives. Thus, they ignore one important step of political action at the global level. Dingwerth & Pattberg (2010:708f.) claim that global governance researchers should pay more attention to the social construction of global governance issues and ask how and why certain issues are considered as forming part of global governance and others not. In line with this, Avant, Finnemore and Sell emphasize that 'global problems' are not objectively given but socially constructed and that there is no objective 'best solution' to the identified problems:

> "Whether governance activities are 'a good thing' depends a great deal on answers to the questions 'good for whom?' and 'good for what purpose?'" (Avant, Finnemore & Sell 2010a:365-367).

Thus, global governance research should also ask *who* defines what is regarded as a global issue and object for transboundary policy-making.

Whereas the role of ideas is still underexposed in global governance research, there is a pronounced Constructivist research strand in the discipline of International Relations which emphasizes the importance of ideas and social construction in international policy-making. Its insights can enrich global governance research considerably. The central argument of the Constructivist research strand in International Relations is that international reality is based on cognitive structures:

> "International reality is socially constructed by cognitive structures that give meaning to the material world" (Adler 1997:319).

This statement illustrates that Constructivists do not negate the existence of a material world; yet, they argue that reality cannot be reduced to material conditions only (see also Adler 2002:100, Houghton 2007:27-30, Ruggie 1998:879). Constructivists pay particular attention to 'social facts' – such as sovereignty, rights and rules – which are based on collective knowledge and thus on human agreement (Adler 2002:100f., Searle 1995:1f., Sikkink, Finnemore 2001:303). In order to make sense of the world, actors use cognitive organizing concepts which rely not only on individual ideas, but to a large degree on widely shared or intersubjective ideas and understandings (Adler 2002:100, Sikkink, Finnemore 2001:392f.). As Adler & Haas (1992:386) underline, "the importance of these understandings lies not merely in being true but also in being shared". The "emphasis on the ontological reality of intersubjective knowledge", according to Adler (1997:322), forms Constructivism's main value added within the discipline of International Relations. Constructivists regard agents and structures to be mutually constitutive: social reality does not only shape human action, humans can also shape social reality (Adler 2002:100f., Houghton 2007:27-30). They emphasize that actors do not necessarily employ the same organizing concepts, and therefore do not have to share a common perspective or understanding of reality (Haas, Haas 2002:586). Thus, an important research focus of Constructivism is to analyze how actors derive meaning from a complex world and how they interpret the context of their actions (Haas 2002b:74, Katzenstein, Keohane & Krasner 1998:678-682).

In this sense, Constructivist ontology clearly contradicts the Rational Choice worldview which assumes the existence of an unambiguously given reality. Strict Rational Choice analyses assume parameters for action as objectively given (derived from material reality), leaving no room for distinctive interpretations. Sometimes, Rational Choice analyses allow for incomplete or asymmetric information about this reality, but they rule out different understandings of reality and the existence of different 'truths'. Constructivists, by contrast, underline that reality is shaped by both material and ideational factors (Ruggie 1998:879). While they do not negate the existence of naturally given facts, they point out that an important building block of reality is based on ideational structures (Houghton 2007:27-30).

While all Constructivists agree that ideas are a central building block of reality, they disagree on what kind of ideational factors matter most. Two major research strands can be identified, which (Hasenclever, Mayer & Rittberger 1996:205-217, 2000:10-12) call Strong and Weak Cognitivism. For Strong Cognitivsts, the ideational factors that matter most are social norms (see also Boekle, Rittberger & Wagner 2001:105). Weak Cognitivists, by contrast, focus on ideas as the "intellectual underpinnings" of international relations (Hasenclever, Mayer & Rittberger 2000:10). This thesis follows the Weak Cognitivist understanding of ideas because it is compatible with the Rational Choice model of action which has strongly influenced global governance research. Whereas Strong Cognitivism regards policy actors as role players that follow a logic of appropriateness,[24] Weak Cognitivism is compatible with the logic of consequentiality advanced by Rational Choice.

Weak Cognitivists highlight that intersubjective meanings are essential for transboundary cooperation, since the latter is impossible without a minimum of a shared understanding:

> "Before states can agree on whether and how to deal collectively with a specific problem, they must reach some consensus about the nature and the scope of the problem and also about the manner in which the problem relates to other concerns in the same and additional issue-area" (Haas 1992:29).

Before policy actors can engage in transboundary cooperation, they must agree on its goals and means. Thus, communicative action is a central component of transboundary cooperation:

> "Parties enter into debate in which they try to agree on the relevant features of the social situation and then advance reasons why certain behaviors should be chosen" (Hasenclever, Mayer & Rittberger 1996:213).

According to Weak Cognitivism, the creation and diffusion of social understanding is an important field of activity for transboundary policy-making. Transboundary policy-making is not only about solving transboundary problems, but also about encouraging new thinking about transboundary issues (Breitmeier, Young & Zürn 2006:191, 199-203). As the policy options that actors choose often conform to dominant knowledge systems, new thinking can have significant impacts on their behavior (Breitmeier, Young & Zürn 2007:55). Theories that assume a common problem understanding thus fail to recognize an important contribution of transboundary policy-making – which is to shape how policy actors define problems and which problem-solving strategies they consider (Breitmeier, Young & Zürn 2007:54).

---

[24] For examples of strong constructivist research see, for example, Boekle, Rittberger & Wagner 2001, Finnemore 1996, Finnemore, Sikkink 1998, Franck 1990 and Wendt 1992.

Breitmeier, Young & Zürn (2006:191) stress that an important purpose of international organizations is to increase consensual knowledge about the nature of the targeted problem and alternative policy options. Key to enhancing consensual knowledge are programmatic activities that create scientific knowledge, assess policies, monitor implementation, verify compliance and transfer financial means and technology (Breitmeier, Young & Zürn 2006:191f.). By changing both the content of existing knowledge and/or the degree of consensus regarding knowledge, international institutions can influence how policy actors frame issues, define problem-solving strategies and evaluate results (Breitmeier, Young & Zürn 2006:199f.,234). Haas (2002a:81-87) adds that international conferences, for example, rarely directly influence the behavior of policy actors but nevertheless produce important indirect effects: international conferences set the agenda in global governance, emphasizing new ideas or reframing existing ones. They provide a window of opportunity for raising consciousness and popularizing issues and generate new knowledge about problems, policy options and policy coalitions. Besides, international conferences help to identify new problems and threats, providing general alerts and early warnings. They furthermore might contribute to administrative reform within states and to the adoption of new international norms and standards.

2.2.2 Contestation and Power in Global Governance

It is often criticized that global governance research does not pay sufficient attention to power (Barnett, Duvall 2005b, Overbeek et al. 2010). This might be surprising, as power is a central concept in political science. In order to understand this 'power blindness', it is necessary to bear in mind the context that led to the emergence of global governance research. Such research can be understood as a countermovement to the Realist school of International Relations that almost exclusively focuses on power politics between sovereign states in an anarchical international system. According to the Realist framing, international politics is mainly about competition and confrontation.

In challenging Realist premises, global governance researchers highlight the possibilities of *cooperation*, focusing on the *advancement of common policy goals* (Barnett, Duvall 2005b:5, Mayntz 2008:102f.). It might be due to this focus that much global governance research tends to overestimate communalities while failing to grasp elements of contestation. According to the criticism by Barnett and Duvall, global governance scholars often present transboundary policy-making as a coordination mechanism that merely advances communalities and achieves better outcomes:

> "The prevailing definitions of global governance also have liberal undertones and mask the presence of power. Most definitions revolve around the coordination of people's activities in ways that achieve more desirable outcomes. Governance, in this view, is a matter of resolving

conflicts, finding common purpose, and/or overcoming inefficiencies between actors in situations of interdependent choice" (Barnett, Duvall 2005b:6).

Yet, global governance is not only about win-win options, and even if win-win outcomes are aspired to and/or attained, these may be distributed unequally. In addition, it is important to underline that cooperation is not equivalent to harmony (Keohane 1984). Cooperation always involves conflict. In harmonious situations, policy actors can automatically achieve their goals without needing to engage in cooperation. Only if conflicts in the pursuit of policy goals exist, is there a need for behavioral adjustment and thus for cooperation. Even if cooperation does not involve formal hierarchies, it is not free from power asymmetries.

Hurrell (2005:48) stresses that power is a central element in institutionalized transboundary cooperation. He argues that analyses of transboundary policy-making must shed light on the different capacities of policy actors to determine which issues are negotiated in what institutional arena, on the different capacities to set the agenda and institutional rules, on the different abilities to use carrots and sticks in the negotiation processes and, finally, the different capabilities to opt not to cooperate. Thus, there is no governance without power:

> "Governance and power are inextricably linked. Governance involves the rules, structures, and institutions that guide, regulate, and control social life, features that are fundamental elements of power" (Barnett, Duvall 2005b:2).

Barnett & Duvall (2005b:3,8) define power as the production of effects that form the capacities of actors to define and pursue their interests and ideals. This understanding of power is not limited to the exercise of power over others but also comprises "the normative structures and discourses that generate differential social capacities for actors to define and pursue their interests and ideals" (Barnett, Duvall 2005b:3). As a consequence, power does not necessarily involve expressed conflict. For the purpose of this thesis, the exercise of power through creating and structuring knowledge and social understanding is especially important. As Adler (2002:103) emphasizes, "the imposition of meanings on the material world is one of the ultimate forms of power, and thus is where constructivism's added value with regard to power lies".

Various dimensions of power must be taken into account in order to grasp its importance in transboundary politics (Barnett, Duvall 2005b). Power might work through direct relationships or diffuse connections. It can work through interaction, influencing the capability of actors to control their circumstances (power over), or through social relations that constitute actors as social beings with capacities and interests (power to). Distinguishing between the relational specificity (direct vs. diffuse) and the channels of power (interaction of specific actors vs. social relations of constitution), Barnett & Duvall (2005b:8-13) deduce four different types of power: compulsory, institutional, structural and productive (see Table 3).

Compulsory power denotes a relationship that enables one actor to directly control the actions or circumstances of another. This understanding of power is the most traditional, also represented in the following definitions of power by Weber and Dahl:

> "'Power' (*Macht*) is the probability that one actor within a social relationship will be in a position to carry out his own will despite resistance, regardless of the basis on which this possibility rests" (Weber 1964 [1947]:152).

> "*A* has power over *B* to the extent that he can get *B* to do something that *B* would not otherwise do" (Dahl 1957:202f.).

In global governance, actors exercise compulsory power in various ways. States, for example, might use economic sanctions to influence the policies of another state, multinational corporations and banks can shape the economic policies of states by threatening to move their capital to other states, and international organizations like the World Bank can control the development policies of borrowing states (Barnett, Duvall 2005b:3f., 13-15). Barnett & Finnemore (2005) add that international organizations may also use their authority to urge state and non-state actors to change their behavior.

		Relational specificity	
		Direct	Diffuse
Power works through	Interaction of specific actors	Compulsory	Institutional
	Social relations of constitution	Structural	Productive

Table 3: Types of Power in Global Governance[25]

Institutional power involves relationships of indirect interaction (Barnett, Duvall 2005b:3f., 15-17). It applies, for example, if actors design international institutions such that they represent their own interests and adversely affect the interests of others:

> "Global governance involves formal and informal institutional contexts that dispose actions in directions that advantage some while disadvantaging others. Understanding power in this way makes it much more difficult to approach global governance purely in terms of cooperation, coordination, consensus and/or normative progress; governance is also a matter of institutional or systemic bias, privilege, and unequal constraints of action" (Barnett, Duvall 2005b:17).

---

[25] Source: Barnett, Duvall 2005b:12.

If policy actors can enshrine their ideas within the set-up of institutions, they may achieve long-lasting effects on policy choices:

> "Although the institutionalization may reflect the power of some idea, its existence may also reflect the interest of the powerful. But even if this is the case, the interest that promoted some statute may fade over time while the ideas encased in that statute nevertheless continue to influence politics" (Goldstein, Keohane 1993:21).

Structural power means directly constituting the social capacities and interests of other actors (Barnett, Duvall 2005b:3f., 18-20). Contrary to institutional power, structural power focuses on the actor's social constitution instead of the action constraints of interest-seeking political actors. It concerns the internal structural positions that constitute actors. Structural power influences the conditions of their existence in two ways: it attributes differential social privileges to different positions (positions of super- and subordination) and not only constitutes the capacities of actors, but also their self-understanding and subjective interests. According to the authors, world-system theories build on such a conception of power. These theories argue that structures of production and economic exchange determine the identity of states – forming either part of the world system's core or its periphery.

Productive power also focuses on influencing the social constitution – the self-understanding and perceived interests – of other actors, but includes only indirect relationships. It works through systems of knowledge and discourses, involving "the socially diffuse production of subjectivity in systems of meaning and signification" (Barnett, Duvall 2005b:3f.). If problems in need of global governance are constructed, productive power comes in. With regard to productive power, it is important to analyze who defines what knowledge is regarded as authoritative and legitimate and to pay attention to the channels through which knowledge can gain influence (Barnett, Duvall 2005b:3f., 20-22).

2.2.3 Global Governors

Theoretical global governance approaches commonly focus on the structures and processes of transboundary policy-making whereas the actors involved receive less attention: "Global governance is something that happens; no one, apparently, does it" (Avant, Finnemore & Sell 2010c:1). Avant, Finnemore & Sell (2010b) focus their analysis on global governance actors who they call 'global governors'. Global governors are defined as follows:

> "Global governors are authorities who exercise power across border for purposes of affecting policy. Governors thus create issues, set agendas, establish and implement rules or programs, and evaluate and/or adjudicate outcomes" (Avant, Finnemore & Sell 2010c:2).

A wide range of actors can act as a global governor, such as governments, international organizations, corporations, professional associations and advocacy groups. In contrast to functionalistic approaches assuming easily identifiable issues in need of global governance, Avant, Finnemore & Sell (2010c:14-17) underline that there has to be a governor who defines something as a problem and is able to place it on the agenda.

According to Avant, Finnemore & Sell (2010c:3) the key to understanding global politics is to grasp how different policy actors relate to each other. They particularly point at two types of relationship: that between the global governors themselves and the relationship between governors and the governed. Since global governance arrangements are often not explicitly backed by force and coercion, it is crucial to understand the nature of authority that global governors enjoy and the channels through which the governed accept the authority of governors (Avant, Finnemore & Sell 2010c:8f.). Authority is not a commodity that exists in vacuum; it is a social relationship that builds on the recognition of others: it is the "ability to induce deference to others" (Avant, Finnemore & Sell 2010c:9). This deference may take various forms – subordinating, creating or changing preferences – and occur due to different reasons. Sources of authority both enable and constrain governors since their actions must be regarded as corresponding to their sources of authority (Avant, Finnemore & Sell 2010c:9-11).

Avant, Finnemore & Sell (2010c:11-13) distinguish five different bases of authority for global governors. Institution-based authority derives from offices in established organizational structures and is both defined and limited by the functions and rules of the respective institutions. Delegation-based authority roots itself in the delegation of another actor and is thus constrained by the preferences and aims of the delegators. Expertise-based authority builds on specialized knowledge and thus inheres in the actor. Principle-based authority derives from adherence to widely accepted principles or values; it can inhere in both actors and offices. Capacity-based authority is obtained due to perceived competence. The capacity for effective action may closely relate to institutional, delegated or expert authority. Whereas the expectation that a global governor will produce the desired outcomes can result in the deference of the governed, the lack of perceived capacity may also undermine authority based on other sources. To legitimize their action, global governors commonly appeal to a greater collective good (Avant, Finnemore & Sell 2010a:365-367). As a consequence of such proceedings, self-interests behind the action may be hidden.

Barnett & Finnemore (2005) analyze the exercise of power by international organizations. They underline that international organizations use their authority to exercise compulsory, institutional and productive power (Barnett, Finnemore 2005:175-181). International organizations directly influence the behavior of other actors in global governance, for example, by providing incentives and sanctions (compulsory power). They also have the power to structure situations and steer behavior towards certain outcomes – for example, by setting agendas or classifying and organizing information (institutional power). Besides, they shape underlying social meanings and thus contribute to the constitution of global governance (productive power). The authority of international organizations not only derives from their form as rational-legal bureaucracies; they also exercise principle-based authority, since they pursue goals that are widely seen as desirable (Barnett, Finnemore 2005:161f.). As such, they do much more than a functionalistic perspective would suggest. They not only help states and other actors to overcome collective action problems, but also construct the world in which cooperation takes place:

> "Bureaucracies, by definition, make rules, but in so doing they also create social knowledge. They define shared international tasks (like "development"), create and define new categories of actors (like "refugees"), create new interests for actors (like "promoting human rights"), and transfer models of political organization around the world (like markets and democracy)" (Barnett, Finnemore 1999:699).

Since international organizations have the power to create and structure knowledge and social understandings, they not only create categories of actors and action, but also fix meanings and diffuse norms and principles. Thus, they establish the boundaries of acceptable and legitimate action and grant authority to particular political actors (Barnett, Finnemore 1999:710-715).

2.2.4 Global Governance as a Concept of the OECD World?

It is repeatedly criticized that global governance is a concept of the OECD world. This criticism is directed at both transboundary policy-making and International Relations research. Gu, Humphrey and Messner underline that OECD countries commonly regard themselves as the central actors in global governance:

> "Until very recently, the OECD countries perceived themselves as the unchallenged centers of (…) world politics" (Gu, Humphrey & Messner 2008:274).

As a consequence, discussions on future global governance architectures are commonly based on the concept of a world order led by OECD countries. Within such a world order, OECD countries define global challenges and how these should be addressed, trying to show the rest of the world the 'right' way of problem-solving (Harris, Moore & Schmitz 2009:6f., Najam, Robins 2001:50, Williams 1993). Various authors argue that academic debates on global governance reproduce existing power asymmetries in transboundary policy-making and mirror the worldviews of only the most powerful states. They criticize that global governance approaches do not pay sufficient attention to the non-OECD world and thus falsely decontextualize and universalize the experiences and perspectives of industrialized countries (Conzelmann, Faust 2009, Humphrey, Messner 2008:201-203, Overbeek et al. 2010). Such OECD-centrism is not unique to global governance research. As Tickner (2013) argues, the whole discipline of International Relations is dominated by research coming from those countries which used to dominate international policy-making:

> "The center-periphery configuration of IR favors analytical categories and research programs that are defined by academic communities within the North while also reinforcing Northern dominance within international practice itself. The precariousness of the global South as both an agent of IR knowledge and a global actor seems directly to such self-referential practices in the field" (Tickner 2013:641).

As such, International Relations research not only mirrors but also reinforces existing power asymmetries in international policy-making.

Mayntz (2008:105-110) points out that the concept of global governance is particularly salient in International Relations research coming from Europe. To explain this salience, Mayntz refers to Europe's geopolitical context after the end of the Cold War:

> "Europe's relative weakness has understandably produced a powerful European interest in building a world where military strength and hard power matter less than economic and soft power, an international order where international law and international institutions matter more than the power of individual nations, where unilateral action by powerful states is forbidden, where all nations regardless of their strength have equal rights" (Mayntz 2008:106, referring to Kagan 2004:37).

According to Mayntz, it is no coincidence that researchers in European countries – which are relatively weak in terms of military or 'hard' power – use analytical approaches that favor collective problem-solving over unilateral action and that give more weight to economic and soft power than to military strength and hard power. Due to the institutional context of the European Union (EU), concepts such as multilevel governance and the pooling of sovereignty are particularly familiar for European researchers. In International Relations research coming from the United States, the concept of global governance is less salient. After the end of the Cold War, US American public discourse focused on an increasingly unipolar world order. International Relations scholars, influenced by the Liberal Institutional tradition, warned about the risks of power imbalances and underlined the benefits of multilateralism vis-à-vis unilateral action. Thus, US American International Relations research rather focused on 'multilateralism' than on 'global governance'.

The rise of 'new powers' such as China, India and Brazil questions the OECD-led world order. These countries were formerly associated with the 'powerless south' and are now seen as relevant actors for tackling global challenges. These 'emerging powers' blur the differences between the industrialized and the developing world, combining characteristics that were formerly associated either with industrialized or developing countries: low per capita income and the prevalence of poverty combined with high gross domestic product (GDP) and considerable capacity to influence international politics (Harris, Moore & Schmitz 2009:7). An important driver behind the academic discussions on emerging powers was a Goldman & Sachs publication pointing at the growing economic weight of four non-G7 members (Brazil, Russia, India and China, or "BRICs") and calling for the inclusion of these countries into the G7 (O'Neill 2001). Subsequently, the rise of new powers received a great deal of public attention and political science researchers began to develop different concepts to determine what new powers are and which countries should be counted to form part of this group.[26]

These power shifts will increasingly affect global governance arrangements and how global governance issues are defined. This will also cause repercussions in global governance research with politics and social construction being harder to ignore. The rise of new powers poses a central challenge to the current form of organizing global governance:

> "A key test of the 21st century order will be how the rising states relate to the organizational machinery of global governance" (Alexandroff, Cooper 2010a:6).

According to Alexandroff & Cooper (2010a:8), there is distrust between old and new powers. While the new powers are no longer willing to subordinate their interests and values, the old powers do not want to lose control of global governance arrangements that serve their interests and thus intend to "socialize" the new. Cooper & Flemes (2013) claim that it is a major puzzle as to whether emerging powers rely on established institutions in transboundary policy-making or build up parallel and/or competitive mechanisms. According to these authors, "the rise of new players has resulted in a fundamentally contested order of world politics and the meaning of essential concepts must be renegotiated" (Cooper, Flemes 2013:958). Parting from the assumption that new powers will increasingly influence global governance, various authors investigate the perspectives and preferences of new powers and ask how these relate to those of established powers.[27] Alexandroff & Cooper (2010a:3), for example, underlines that the G-20 lacks like-mindedness while the G-7 was characterized by "similar ways of looking at the global system". Narlikar (2008:97f.) points out that the negotiating behavior of rising powers differs from that of both established powers and smaller countries but insinuates that their motivations remain unclear. Lesage, Van de Graaf & Westphal (2010:85) affirm that "advanced industrialized and emerging economies often have completely different

---

[26] See, for example, "BRICSAM" (Cooper, Antkiewicz & Shaw 2007), "new leading powers" (Husar, Maihold & Mair 2010). "pivotal states" (Chase, Hill & Kennedy 1996), "regional leading powers" (Flemes 2010) and "anchor countries" (Stamm 2004).
[27] See, for example, Alexandroff, Cooper 2010b, Cooper, Flemes 2013, Gu, Humphrey & Messner 2008, Humphrey, Messner 2008, Hurrell, Sengupta 2012, Husar, Maihold & Mair 2009, Narlikar 2010, Wang, Rosenau 2009.

preferences". According to the authors, it is also visible in energy and climate policy: emerging countries give priority to economic growth rather than environmental protection and claim substantial financial compensations if they undertake actions to protect the climate. Narlikar (2008:99) and Hurrell (2006:1-3, 18) emphasize that emerging countries do not conform to existing values in international politics and that their ideas challenge the established international order. Cooper & Flemes (2013:948) even speak of "a psychological sense of the emerging states of being outsiders in the multilateral system, kept away from (...) privileges". According to these authors, the attitudes of emerging countries vis-à-vis the international system are marked by historical grievances and claims to serve as the representatives of developing countries, pursuing rapid socio-economic development and emphasizing the principles of sovereignty and non-intervention (Cooper, Flemes 2013:952). Hurrell & Sengupta (2012:465) also emphasize the self-perception of emerging powers as belonging to the 'South'. Despite their trajectories of economic growth, they are still confronted with severe internal development challenges and are only partially integrated into the "global economy whose ground rules have been set historically by the industrialized North" (Hurrell, Sengupta 2012:483).

The literature on emerging countries shows significant overlaps with the longer-lasting and more extensive academic discussions on North-South conflicts in international policy-making. This strand of literature makes it apparent that there are pronounced conflicts over the contents and forms of global governance arrangements and, by implication, over the definition of global governance issues too. Various authors assert that differences in living conditions and development levels between developing and developed countries result in diverse policy priorities and strategies in transboundary policy-making (see, for example, Biermann 2007, Chasek, Rajamani 2003, Grant 1995, Narlikar 2008, Roberts, Parks 2007 Williams 1993). They repeatedly point out that developing countries are less willing to give up sovereign rights and put a stronger focus on distributive issues. Williams (1993:12-17) and Grant (1995:570) state that developing countries have a shared sense of international distributive justice and that their vision of just international burden sharing clashes with that of the developed world. Referring to trade policy, Narlikar (2006:1006) points out that developing countries focus on equity in outcomes whereas their developed counterparts concentrate on equity of opportunity and process. In the literature on North-South conflicts in international environmental policy, Biermann (2007) emphasizes that developing countries prioritize economic development over environmental protection and that they regard industrialized countries as the historical culprits of global environmental problems. Subsequently, they demand that the developed world undertake actions to protect the environment and are only willing to engage in protection measures themselves if they receive financial compensation. In addition, they fear that industrialized countries could instrumentalize environmental policy in order to protect their own markets and hinder the economic catch-up process of developing countries. Roberts & Parks (2007) argue that a country's position in the global division of labor shapes the belief system of its decision-makers. According to these authors, global inequality hampers international cooperation by reinforcing structuralist worldviews and causal beliefs, polarizing preferences and eroding trust between North and South.

2.3 Actor-Centered Framework for the Analysis of Global Governance Ideas

This section develops the actor-centered framework that guides the analysis of global governance ideas. It elaborates on policy actors as drafters and carriers of ideas, lays out basic behavioral assumptions and specifies the interrelation between ideas and policy action. It then deduces the analytical categories that structure the comparative analysis of ideas on global renewable energy governance and the search for reasons behind ideational differences.

2.3.1 Policy Actors as Drafters and Carriers of Ideas

As this thesis builds on policy actors as drafters and carriers of ideas, it resembles the epistemic communities approach which introduced an actor-sensitive analysis of ideas in transboundary policy-making. The epistemic communities approach analyzes how political processes determine whose interpretations of reality are more viable in a particular historical context. It concentrates on networks of knowledge-based experts that specify cause-and-effect relationships, frame issues in public debates, propose policy options and highlight crucial aspects for negotiation (Haas 1992:1-7). Yet, as Adler (1997:343) stresses, other actors, be they governmental or non-governmental, can play a role similar to that of epistemic communities.

Although only individuals can think and act, it is possible to apply actor-centered concepts to units larger than the individual. Individuals often act in the name of a larger group or organization:

> "Though it remains true that in the final analysis only individuals are capable of purposive action, the world of empirical policy research is populated by collective actors of all types and in particular by corporate actors. The apparent contradiction is resolved by the fact that individuals will not always act on their own behalf but may act in a representative capacity and that they have the ability to identify with and to act from the perspective of larger units – a family, a group, a nation, and above all, organizations of all kinds, including firms, labor unions, political parties, government ministries, and the state" (Scharpf 1997:60f.).

Thus, individual behavior must be related to the unit of reference on whose behalf action is taken (Scharpf 1997:51-66).

The relevant policy actors for the purpose of this study are those that belong to the German or Brazilian government and make statements on global renewable energy governance. These pronouncements are not necessarily made in the name of the government as a whole. Often they are made in the name of specific government entities, such as its ministries. Thus, this study does not present governments as a monolith but rather regards them as composite actors. It differentiates between policy actors within the national executive and explicitly names the actors on whose behalf the statements are made.

2.3.2 Weak Cognitivism and its Basic Behavioral Assumptions

This analytical framework builds on Weak Cognitivism which emphasizes that goal-oriented behavior and the social construction of reality go hand in hand. As such, Weak Cognitivism can be understood as the 'middle way' between Rational Choice and Strong Cognitivism (see Table 4). Weak Cognitivism adopts the basic behavioral assumption of Rational Choice and thus regards policy actors as rational/goal-oriented players. In that sense, it clearly differs from Strong Cognitivism which assumes policy actors as role players following a logic of appropriateness. Yet, Weak Cognitivism contradicts the Rational Choice notion of an unambiguously given reality. In Rational Choice analyses, parameters of action are assumed as objectively given: policy actors are placed within an unambiguous context of action and pursue fixed and context-free policy goals. Weak Cognitivism, by contrast, emphasizes that reality is socially constructed and that social construction is both a part and result of goal-oriented behavior (Hasenclever, Mayer & Rittberger 1996:205-216, 2000:10-12)

	Reality	Basic behavioral assumption
Rational Choice	Objectively given	Rational/goal-oriented behavior
Strong Cognitivism	Socially constructed	Norms and role-driven behavior

▇ Weak Cognitivism

Table 4: Weak Cognitivism as a Middle Way between Rational Choice and Strong Cognitivism

According to Weak Cognitivism, policy choices always involve interpretation (Hasenclever, Mayer & Rittberger 1996:206-210). Before policy actors choose a course of action, they must assess their context of action, specify their goals and identify relationships between means and ends. Different policy actors may interpret their context of action differently; thus, Weak Cognitivists open their analyses for diverse understandings of reality and the existence of different 'truths'. Weak

Cognitivist research focuses on how actors derive meaning from reality, interpret their context of action and construct their policy goals in transboundary policy-making. It is important to note that interpretations are also the result of policy choices. This means that interpretations are not apolitical, but can also be employed for strategic purposes. While policy actors might take some interpretations for granted, they can also purposely draft ideas in a way that helps them to achieve policy goals.

The Weak Cognitivist notion of the policy goals pursued by actors differs significantly from Rational Choice analyses (Hasenclever, Mayer & Rittberger 2000:10-12). Rational Choice assumes that actors pursue objectively given, fixed and context-free self-interests. The content of the assumed self-interests varies between different schools of thought. Realist scholars of International Relations, for example, tend to focus on survival or power, whereas Liberal Institutionalists put the emphasis on wealth.[28] Weak Cognitivism does not assume specific policy goals, but rather stresses that policy actors themselves – not researchers – are the ones who have to define what policy goals they pursue. These policy goals can change over time and vary across policy actors. They are not limited to self interests but might also comprise other motivations. Thus, an important aspect of Weak Cognitivist research is to understand the processes by which policy goals originate and change (Sikkink, Finnemore 2001:394).

While Weak Cognitivism clearly contradicts the Rational Choice approach to interests, it does not reject the notion of interests *per se*:

> "Interest-based behavior certainly exists, but it involves *ideas about interests* that may encompass much more than strictly utilitarian concerns" (Schmidt 2008:318 highlithing S.R.).

Interests – the goals and strategies that policy actors pursue to improve their own well-being – are not assumed as being determined by objectively given material factors; instead, they are regarded as subjective responses to material and immaterial conditions (Béland, Cox 2011:11, Schmidt 2011:58). Thus, in contrast to Rational Choice, Weak Cognitivism analytically treats interests as contingent on how actors understand their context and define their preferences (Hasenclever, Mayer & Rittberger 1996:206). Interests are conceptions of self-good:

> "Constructions [of interests] are (...) subjective/intersubjective conceptions of self-good–of what it would advantage the individual to do or to have done either on his or her behalf or inadvertently by others" (Hay 2011:79).

And these conceptions of self-interests work as cognitive filters which help actors to orient themselves towards their environment (Béland, Cox 2011:10, Blyth

---

[28] For overviews of Realist and Liberal Institutionalist approaches see, for example, Baumann, Rittberger & Wagner 2001, Freund, Rittberger 2001 and Rittberger 2004.

2003:697, Schmidt 2008:317, Schmidt 2011:52). Weldes notes that the concept of interests *per se* is a product of social construction:

> "The very notion that 'interests' motivates action and so should be referred to in explanations of behavior and social outcomes is itself a relative novelty. It is within liberalism and the rise of capitalism that 'interests' first came to be understood as the motivating force driving the actions of individuals. That 'interest' as a general category, regardless of its content, should be of central importance to social analysis is thus itself a social construction rather than a natural fact" (Weldes 1996:306).

Weldes furthermore states that the concept of 'the national interest' is important for international policy-making for two reasons: it defines the state's foreign policy goals and works as a "rhetorical device through which the legitimacy of and political support for state action is generated" (Weldes 1996:276). Thus, the construction of legitimacy and interests is inextricably linked and should not be conceptualized as being two distinct processes. 'The national interest' provides the shared language in this process of interpretation. Its content results out of a process of meaning-creation through which policy-makers make sense of the international context and their place within it. It is thus based on the situation descriptions and problem definitions of policy-makers. Within this process of meaning-creation, particular understandings of the national interest become common sense, being perceived as a 'neutral' reflection of reality. Garrison (2007:105f.) adds that complex foreign policy situations easily lead to diverging problem definitions and consequently to different interpretations as to which policy alternatives conform to 'the national interest'.

In line with this, the epistemic communities approach highlights that domestic actors advance competing understandings of both the international context and national goals (Adler 1992:104). Drawing on Robert Putnam's understanding of international politics as two-level games (Putnam 1988), transboundary policy-making can be interpreted as containing both a domestic and international game (Adler, Haas 1992:373f.). The national game is about how ideas enter the political process and help define the national interest, serving as an input to the international game, which is characterized as follows:

> "In the international game, governments not only act out of concern for the domestic political environment, but they are also motivated by the need to solve international problems, whose interpretations and meanings are embedded in the national interest. As part of this game, governments transmit expectations and values that compete to become the basis of international behavior" (Adler, Haas 1992:374).

Thus, Adler and Haas emphasize that understandings of international problems are closely embedded in the national interest.

2.3.3 Ideas as Causes and Results of Political Action

Following Adler (1997:319) and Béland & Cox (2011:3f.), ideas are defined as cognitive organizing concepts that give meaning to the material world. They are products of cognition and connected to the material world only through interpretation. They make sense of the world, drawing connections between things and/or people. While they do not necessarily set up causalities (in the sense of 'x is responsible for y'), it is also possible that they establish how things relate to one another (for example, by identifying correlations). Ideas are causes of action as they help to think about how to understand and address policy problems:

> "Ideas are a primary source of political behavior. Ideas shape how we understand political problems, give definitions to our goals and strategies, and are the currency we use to communicate about politics. By giving definitions to our values and preferences, ideas provide us with interpretative frameworks that make us see some facts as important and others as less so" (Béland, Cox 2011:3).

As "subjective mental models" (Blyth 2003:697) ideas not only establish how policy actors construct the context of their actions, they also specify what actors want and what they consider to be appropriate or legitimate (Béland, Cox 2011:3f.). Ideas have both a cognitive and normative dimension. The cognitive dimension comprises empirical descriptions and theoretical analyses whereas the normative contains values and attitudes (Campbell 1998:384). Rein & Schoen (1977:243) and Mehta (2011:33) underline that normative and empirical descriptions are mutually reinforcing, as values for example influence what actors are willing to accept as fact. Thus, it is not possible to clearly separate the cognitive and normative dimensions of ideas.

Before policy actors act, they draft ideas, clarifying how they understand their context of action and how they shall respond to it. Thus, they specify what goals they want to pursue and which course of action they favor. In order to convince others of their favored course of action, they construct policy solutions that not only contain a specific course of policy action but also comprise symbols or concepts that legitimize and make it appealing and convincing to others (Campbell 1998:378, 381-383). Building on Kingdon (2003:131-139), Campbell presents four factors that may influence whether an idea affects policy decisions:

> "The probability that a programmatic idea will affect policy making varies in part according to the extent to which it provides clear and simple solutions to instrumental problems, fits existing paradigms, conforms to prevailing public sentiment, and is framed in socially appropriate ways" (Campbell 1998:399).

Paradigms are underlying theoretical and ontological assumptions held by experts and policy-makers about how the world works. When ideas as policy solutions fit the paradigm, they are regarded as "natural and familiar" (Campbell 1998:389f., 2002:22f.). Public sentiments – normative assumptions about what is desirable and appropriate – constrain the range of solutions that are considered legitimate

(Campbell 1998:392, 2002:23-25). Yet, public sentiment not only constrains the options for policy-makers, policy-makers may also "intentionally appropriate and manipulate public sentiments for their own purposes" (Campbell 1998:394).

The ideas that policy actors communicate might coincide with their actual perceptions of the context of their action and their policy goals; but policy actors can also purposely frame ideas in a way that helps them to achieve (communicated or not communicated) political goals. Thus, ideas that policy-makers communicate do not necessarily coincide one hundred percent with their actual thinking. As it is hard to verify what political decision-makers actually think, a sound political science analysis of ideas can only work with those ideas communicated by policy actors (Campbell 2002:26-28). Actors can self-consciously generate new ideas through a process of transposition and bricolage, recombining already existing and legitimate concepts (Campbell 1998:378, 381-383). When confronted with new problems, the policy actor uses known concepts. These concepts are only questioned if actors are confronted with clear inconsistencies (Breitmeier, Young & Zürn 2007:55). If it is not supposed that actors behave in a dishonest way, the ideas that they communicate do not contradict what they actually believe to be true. However, they may (over)emphasize or concentrate on those aspects that are in line with their political goals while suppressing others that are not. In order to achieve coherency between their ideas, and thus to increase the persuasiveness of their ideas, actors furthermore try to bind together the relevant features into coherent patterns (Mehta 2011:27).

Ideas as policy solutions are typically used by policy actors in a deliberate and self-conscious way (Campbell 1998:386). These ideas, which specify how to solve a problem and achieve previously established political objectives (Mehta 2011:27f.), are routinely contested in policy debates. Policy solutions are always attached to certain problem definitions. Ideas as problem definitions are not necessarily explicitly addressed in policy debates. Problem definitions are often taken for granted, entering political discussions only implicitly when policy solutions are based on different problem understandings (Mehta 2011:34f.). Policy actors may not even be aware of their participation in the process of problem definition and may not even notice that the definition itself is contested. Yet, it is also possible that political struggles explicitly include problem definitions when discussing political decisions. Mehta (2011:32) criticizes that political science researchers often merely concentrate on ideas as policy solutions, without paying attention to how the problems are defined. However, setting a problem is an essential element of political processes since problem understandings are crucial to further political development: "the questions we ask shape the answers we get" (Rein, Schoen 1977:236). Ideas as problem definitions do not only specify findings but also build the ground for policy action, permitting "the inquirer not only to explain the phenomena associated with its worries, but to set the directions of actions designed to reduce them" (Rein, Schoen 1977:240). Thus, problem definitions define the scope for possible policy solutions, excluding those that are not consistent with their understandings of reality. Nevertheless, problem definitions do not determine policy solutions, as many different solutions can be consistent with one particular problem definition (Mehta 2011:33). Of course, setting a problem does not always precede the construction of policy solutions. Policy actors might also be committed to certain policy solutions – for example, due to self-interests – and draft problem definitions in a way that these match to the previously established policy solutions.

2.3.4 Ideas on Global Renewable Energy Governance: Contents and Reasons Behind

To analyze the contents of the governments' ideas on global renewable energy governance, the present study derives analytical categories that are grounded in the theoretical-analytical framework and take specifics of the empirical material under study into account (Dey 1993:96f.). Starting point is the distinction by Mehta (2011:28-40) that ideas comprise both problem definitions and policy solutions. Applied to the empirical focus of this study, this means that global governance ideas establish why transboundary policy-making is necessary or desirable, what form and direction it should take and what global governors should be involved. Thus, global governance ideas construct the context of action in global governance and establish how transboundary policy-making should respond to this context. Building on insights of global governance research (and taking account of its salience in the empirical material under study), the present study furthermore considers 'meta-aspects', reflecting on how global governance works or should work. This analysis makes further specification in order to narrow down the field of research. Following Schuppert (2008:16), Zürn, Koenig-Archibugi (2006:237-239) and Kaul (2008:91), it only considers cooperation at a global scale.[29] Thus, it does not take ideas into account that merely relate to bilateral and intra-regional cooperation. In line with this, it considers only ideas on global governors with a transcontinental institutional set-up. It is important to note that transboundary policy-making is the primary point of reference for global governance ideas. Thus, global governance ideas only comprise of context factors and policy goals that are formulated with reference to transboundary policy-making. Ideas that primarily refer to policy actors themselves, such as self-positioning in global governance and self-interests in transboundary policy-making, might influence how actors construct global governance ideas, but they are not defined as global governance ideas themselves.

---

[29] As outlined in the introduction, covering the global scale is consensual among global governance scholars. Authors such as Finkelstein (1995) argue for a more wide-ranging approach to global governance analysis that covers the whole range of transboundary relations.

Further specifications are necessary in order to grasp global governance ideas in the realm of renewable energy. These enhance compatibility with academic research on global (renewable) energy governance and reflect the thematic focus of the empirical material under study. Global renewable energy governance comprises transboundary policy-making to promote renewable energy, i.e. to increase its worldwide production and use. Problem definitions specify why the promotion of renewable energy should be a subject for transboundary policy-making, thus presenting the need for, or the potential value added of transboundary policy-making. Such problem definitions contain three different aspects. First, they must specify why the promotion of renewable energy is a global issue: What are the global challenges addressed by the promotion of renewable energy? As it comprises very different energy options – such as wind and solar energy, hydropower and biomass – it is furthermore important to clarify what kind of renewable energy options the actors refer to when they speak about the worldwide promotion of renewables: what kind of renewables should be promoted? Is a distinction made between different renewable energy options? In addition, problem definitions must clarify why transboundary policy action is needed: what are the barriers that prevent the worldwide spread of renewables? With regard to policy solutions, two main aspects come in: the tasks of transboundary policy-making and the global governors involved. Thus, two main questions are relevant with regard to policy solutions: what are the main tasks for transboundary policy-making? Which global governors are relevant and why? Moreover, ideas on global renewable energy governance also contain 'meta-aspects' reflecting on how global governance works or should work: what features are salient in the way transboundary policy-making on renewable energy works (or should work)?

The analysis of reasons behind the ideational differences also takes research on ideas in policy-making as a starting point (see, for example, Breitmeier, Young & Zürn 2007:53-57,191-225, Campbell 2002, Campbell 1998, Mehta 2011). More specifically, it is structured along the abovementioned elements that influence the generation of ideas (see chapter 2.3.3): reliance on known concepts, framing in line with political goals and strive for coherency between ideas. Based on empirical research, the present study argues that the ideas formulated by policy actors often conform to prevailing knowledge systems of their policy contexts. If the ideas are not contested but widely shared within their policy context, the actors may even take these ideas for granted and perceive them as 'neutral' reflections of reality (Mehta 2011:34f., Weldes 1996). The government actors that deal with global renewable energy governance are embedded in different policy contexts. This analysis therefore differentiates between context factors on different levels. Following Putnam (1988) and Adler & Haas (1992:373f.), this study assumes that the government actors dealing with global renewable energy governance act on two levels, the domestic and the global. Thus, their ideas on global renewable energy governance can be shaped by both the domestic and global policy context. Since the promotion of renewable energy is a rather new area of transboundary policy-making, it is likely that the domestic policy context of renewable energy promotion has a particularly strong influence on how policy actors construct the context of action and policy solutions at the global level. The perception of the global policy context is influenced by the actors' own involvement and self-positioning in transboundary policy-making. A third context factor that comes in is the institutional background of the government actor in charge of global renewable energy governance. As the promotion of renewable energy is a cross-cutting issue, it can involve government actors with different institutional backgrounds and the division of competencies may vary between different administrations. The institutional background shapes selective perceptions (Scharpf 1997:39f.); thus, different ministries might consider different aspects as relevant.

As policy actors can purposely frame global governance ideas, two more factors come in. First of all, they can frame the ideas in a way that helps them to advance their conceptions of self-good (Hay 2011:79).[30] The self-interests that policy actors pursue can relate to both the global or domestic level: they might aim at improving their conditions within the global or the domestic context. The second factor is their endeavor to draft ideas in a consistent way (Mehta 2011:27). Policy actors might consider some aspects as more important and draft ideas in such a way that less important aspects conform to the more important ones.

2.4 Methodology

2.4.1 Interpretative Research Design

Interpretation is a central building block of a Constructivist epistemology. The Constructivist statement that interpretation and language shape objects of knowledge not only refers to the analyzed actors but also to researchers: interpretation is an intrinsic part of how research is conducted (Adler 1997:328, Adler 2002:95,101, Houghton 2007:33, Price, Reus-Smit 1998:271-273, Sikkink, Finnemore 2001:395,). As Constructivist research engages in interpretation, it "develops meanings about meanings" (Pouliot 2007:365). The core of a Constructivist approach are methods that aim at capturing intersubjective meanings (Price, Reus-Smit 1998, Sikkink, Finnemore 2001:395).

---

[30] On the framing of ideas in line with self-interests see also Adler, Haas 1992:374, Béland, Cox 2011:10, Blyth 2003:697, Schmidt 2008:317, Schmidt 2011:52.

An interpretative epistemology differs in several respects from a positivist one. It does not aim at establishing an all-encompassing truth, but instead focuses on contingent generalizations (see also Sikkink, Finnemore 2001:394):

> "These [contingent generalizations] do not freeze understanding of bring it to closure; rather, they open up our understanding of the social world" (Adler 2002:101).

As it aims at opening up understandings instead of freezing them, interpretative research offers a less parsimonious, rigorous and conceptually clear research design than a strict positivist epistemology does (Blyth 2003:701f.). Interpretative researchers furthermore advance a different understanding of explanation and causalities. As Adler (2002:101) emphasizes, an "interpretative practice of uncovering intersubjective meanings" is needed in order to explain causation. For interpretative scholars, deterministic laws – structural causal relations between entities and occurrences – can only be found in the physical world, not in the social realm (Adler 1997:329). Instead of trying to trace clear-cut causalities (in the sense of "x causes y"), interpretative research stresses the complex nature of relationships between variables that often involve mutual constitution ("coconstitution"). Whereas positivist researchers try to avoid overdetermination and interactive effects between independent variables, interpretative scholars regard these as necessary consequences of complexity which cannot be ruled out in empirical research (Haas, Haas 2002:588). To the contrary, a central task of constructivist research is to shed light on these complex relationships, trying to understand the *constitution* of things:

> "Understanding how things are put together and how they occur is not mere description. Understanding the constitution of things is essential in explaining how they behave and what causes political outcomes" (Sikkink, Finnemore 2001:394).

In order to draw causal or constitutive inferences, interpretative scholars rely on the historical reconstruction of social facts based on interpretative narratives, practices and discourses (Adler 2002:101, 109, Blyth 2003:697). Interpretative research focuses on understanding past and present developments; it does not claim to predict future developments:

> "A coherent constructivist methodological approach also means approaching research less as a predictive enterprise than as an effort to explain how past and present events, practices and interests became possible and why they occurred in time and space the way they did" (Adler 2002:109).

Pouliot (2007:379) emphasizes that from the viewpoint of Constructivists, there is not only one valid interpretation or theory. He suggests that researchers should concentrate on incisiveness in order to assess the relative validity of an interpretation. Thus, researchers should ask if the interpretation "sees further" and adds sense to previous interpretations.

Inductive inquiry is a central building block of an interpretative research program. As Pouliot (2007) highlights, induction needs to be the first step of any Constructivist research:

> "Induction is the primary mode of knowing because social facts constitute the essence of constructivism. Research must begin with what is that social agents, as opposed to analysts, believe to be real" (Pouliot 2007:364).

The starting point of interpretative analyses are not hypotheses that are tested later on or *a priori* assumptions, but the meanings that policy actors attribute to their reality. If scholars engage in inductive research, they aim to uncover this subjective (experience-near) knowledge while trying to avoid imposing their own perspectives on the objects of study (Pouliot 2007:359). An inductive proceeding furthermore implies that researchers try to remain as open as possible towards their research objects. Yet, an inductive proceeding does not mean 'category-free' research. In political science, it is always theory-guided and thus contains deductive elements. It is predominantly, though not completely, inductive (Zilla 2011a:58f.). The research for this thesis furthermore follows the procedure of abduction, moving back and forth between inductive and deductive steps. While a first version of the theoretical-analytical framework was developed before the inspection of empirical material, its final version was drafted after empirical research.

Next to uncovering subjective knowledge, interpretative research must also develop objectified knowledge. Such experience-distant knowledge is generated by 'standing back' from the object of analysis. To do so, Constructivist researchers engage in contextual and historical interpretation (Pouliot 2007:369). Pouliot suggests a three-step methodology for interpretative research that he calls "sobjectivism", combining induction, contextualization and historicization:

> "Subjectivism is a three-step methodology which moves along a continuum bordered at one end by subjective knowledge and at the other by objectified knowledge. One begins with the inductive recovery of agents' realities, then objectifies them through the interpretation of intersubjective contexts and thereafter pursues further objectification through historicization. In the spirit of grounded theory and abduction, however, these three steps should not be conceived as a linear pathway" (Pouliot 2007: 368).

According to Pouliot (2007: 368-374), Constructivist scholars should thus engage in a dynamic research process that recovers subjective meanings, puts them in context and sets them in motion by introducing time and history. To objectify subjective meanings, researchers must put them into a wider context of intersubjectivity (Pouliot 2007:366,370). Pouliot (2007:366f.) underlines that social realities are not naturally given, but are the results of social and political processes. To understand social realities it is necessary to shed light on the historical processes that led to their constitution. Several Constructivist researchers thus call for a process-centered research approach that analyzes political processes in their temporal dimension and traces political contestation around the social construction of reality (Adler

2002:109, Dessler 1999, Pouliot 2007, Ruggie 1995, Sikkink, Finnemore 2001:394). To trace the relevant political and social processes, researchers have to build a narrative: "a dynamic account that tells the story of a variety of historical processes as they unfold over time" (Pouliot 2007:367). In contrast to positivist causal analyses that search for constant antecedents, narrative-building focuses on contingent factors and practices that made a social fact possible (Ruggie 1995). "Narrative causality" analyzes how subjective and intersubjective meanings historically evolve and traces the historical constitution of social reality (Pouliot 2007:366f.). As Hurrell (2011:146) emphasizes, the goal of such a proceeding is "not so much to explain, but rather to 'make sense' of the big picture and to understand 'how things hang together'".

Thus, this analysis of global governance ideas not only focuses on the substantive content of the ideas, but also considers their contextual and historical embeddedness. This thesis thereby follows the claim by Schmidt (2008:305) that analyses of ideas should not only consider what is said (the "text"), but also where, when, how and why it was said (the "context"). In order to uncover the way in which ideas are produced, it explicitly considers the interactive processes by which actors generate and communicate ideas (Hardy, Harley & Phillips 2004:19, Schmidt 2008:306).

2.4.2 Comparative Case Study Analysis

This research design builds on a comparative analysis of two case studies. The focus on such a limited number enables intense observation and in-depth analysis, and thus 'advanced understanding':

> "The proximity to reality, which the case study entails, and the learning process that it generates for the researcher will often constitute a prerequisite of advanced understanding" (Flyvbjerg 2006:236).

In the first step, the two cases are analyzed separately. The case studies rely primarily on inductive inquiry, allowing for a case-sensitive analysis that uncovers the 'within-case logic'. The main purpose of the case studies is to facilitate a comprehensive understanding of the governmental ideas and to contextualize and historicize these ideas.

The first part of the case studies traces the process of government action in global renewable energy governance. To do so, it analyzes (1) how and why the governments engaged in global renewable energy governance, (2) which policy actors were involved and (3) what context factors and political processes at the domestic and global level influenced the administration's action. This analysis provides important contextual knowledge on where, when and for what purpose the government spoke about global renewable energy governance. In addition, it identifies which government actors are the carriers and drafters of the ideas under study and with whom they interact. Here, policy actors that do not belong to the governments but were involved in the policy processes under study also enter the analysis. To specify why the administrations engaged in global renewable energy governance, the analysis first and foremost relies on the purposes/aims and rationales formulated by government actors involved in the process. These purposes might relate to their self-interests or contain broader policy goals that are formulated with regard to transboundary policy-making.

The examination of each government's ideas on global renewable energy governance builds on a qualitative content analysis of government statements on the issue, which allows for a methodically controlled text analysis (Mayring 2005, Kracauer 1952). The texts are coded with conceptually and empirically grounded categories which are rooted in both the theoretical-analytical framework and the empirical material under study (Dey 1993:96f.). The use of qualitative research software (MAXqda) ensures a rigorous data analysis. The content analysis mainly follows the process of inductive coding presented by Mayring (2005:11f.). The coding is pre-structured along the general questions concerning the substantive content of global governance ideas. Thus, the following aspects pre-structure the coding of government statements: (1) global challenges addressed by transboundary policy-making on renewable energy, (2) different renewable energy options as objects for transboundary policy-making, (3) barriers to worldwide promotion of renewable energy, (4) tasks for transboundary policy-making, (5) global governors involved, and (6) salient features of transboundary policy-making on renewables. All subcodes are determined in an inductive way, thus facilitating proximity to the empirical material under study. Table 5 on the following page provides a short overview of the questions that guide the case study analyses.

After the two case studies follows the comparative analysis of government ideas, identifying the ideational differences between both governments. Again, this part of the study is structured along the elements that guide the identification of government ideas within both case studies. There is one important difference: while the analysis of the ideas within the case studies only takes those aspects into account that both governments communicate, the comparison also refers to aspects missed out by the governments. To allow for a comprehensive comparison, the analysis does not only identify ideational differences, but also considers similarities and congruent ideas. While an important aim of this analysis is to contrast the 'common ground' of the ideas that each government communicates on global renewable energy governance, ideational differences within the same government (either across time or between different government actors) are not concealed.

Tracing the government's action on global renewable energy governance
• How and why did the government engage in global renewable energy governance?   • Which actors were involved?   • Which context factors and political processes at the domestic and the global level influenced the government's action?
Identifying ideas on global renewable energy governance
Problem definitions   • What global challenges does the promotion of renewables address?   • What kind of renewables should be promoted? Is a distinction made between different sources of renewable energy?   • What are the main barriers to a worldwide promotion of renewables?   Policy solutions   • What are the main tasks for transboundary policy-making?   • Which global governors are relevant and why?   Meta-aspects   • What features are salient in the way transboundary policy-making on renewables works (or should work)?

Table 5: Guiding Questions for the Case Study Analyses

Based on the outcome of this inquiry, the question is posed as to which factors contribute to a better understanding of the identified ideational differences. According to the analytical framework presented above, this analysis focuses on context factors (domestic policy context, embeddedness in transboundary policy-making, division of competencies within government), the self-interests of both governments in global renewable energy governance, and the coherency between ideas. Again, this analysis is predominantly inductive. The analytical categories were developed after an intensive examination of the empirical material in order to capture the empirical phenomena under study in the best possible way while ensuring consistency within the analytical-theoretical framework.

2.4.3 Sources

Official documents and statements by representatives of the German and Brazilian governments are the primary sources for the analysis of both governments' ideas on global renewable energy governance. For the purpose of this study, publicly available and accessible documents and statements that deal with global renewable energy governance and were issued within the timeframe of analysis were examined – starting with the WSSD in Johannesburg 2002 and its preparatory process in 2001/02 and ending with each government's major initiative to shape global renewable energy governance in 2013 (the International Conference on Bioenergy in April 2013 in the Brazilian case and the founding of the 'Renewables Club' in June 2013 in the

German case). The official documents and statements comprise of policy papers, submissions, press statements, government declarations and programs, website information of government bodies as well speeches and articles by high-ranking government representatives. In the German case, the government bodies that made statements on global renewable energy governance are the Chancellery, the Federal Ministry for Environment, Nature Conservation and Nuclear Safety (Bundesministerium für Umwelt, Naturschutz und Reaktorsicherheit, BMU), the Federal Ministry for Economic Cooperation and Development (Bundesministerium für wirtschaftliche Zusammenarbeit und Entwicklung, BMZ), the Federal Foreign Office (Auswärtiges Amt, AA), the Federal Ministry of Economics and Technology (Bundesministerium für Wirtschaft und Technolgie, BMWi) and the Federal Ministry of Food, Agriculture and Consumer Protection (Bundesministerium für Ernährung, Landwirtschaft und Verbraucherschutz, BMELV). In the Brazilian case, the range of actors is less heterogeneous. Here, the Casa Civil, which is the Office of the Head of State and Government, and the Ministry of External Relations (commonly referred to as Itamaraty) are the only government bodies that publicly spoke about global renewable energy governance. High-ranking representatives from the German government who communicated on transboundary policy-making include Chancellors, Ministers and State Secretaries from the BMU, BMZ and AA. On the Brazilian side, this group comprises Presidents, Foreign Ministers and further leading staff within the Itamaraty. The Brazilian sources furthermore include monographs written by diplomats in the course of their diplomatic career and published by the Itamaraty. These monographs provide valuable first-hand insights and background information about Brazilian diplomacy in the realm of renewable energy (see Corrêa do Lago 2009, Feres 2010, Kloss 2012). Such sources are not available on the German side.

The empirical analysis furthermore relies on qualitative interviews with representatives of the German and Brazilian governments, representatives of further policy actors involved in the relevant policy processes and persons with expert knowledge on the issues under study. The primary function of these interviews is to facilitate a reconstructing analysis (Gläser, Laudel 2010), contextualizing and historicizing each government's ideas on global renewable energy governance. Due to the scarcity of academic and grey literature on the German and Brazilian involvement in global renewable energy governance, the interviews are of central importance in tracing the process of government action and revealing the context factors and policy processes that influenced each government's reasoning and action. The analysis of the substantive content of each administration's ideas, by contrast, primarily relies on the official statements and publications mentioned above; here, the interviews only serve as supplementary sources. As official statements and publications result of collective opinion-processes, they are better suited to depict the ideas of government actors than interview statements by individuals. The interviews were based on guidelines ("Leitfadeninterview"), trying to give the interviewees as much space as possible to develop their narratives and to bring in their specific expertise (Gläser, Laudel 2010:111-116). The interviews were conducted as systematizing expert interviews (Bogner, Menz 2002:37f.), aiming at the process and interpretative knowledge of the interviewees and focusing on the thematic comparability of the generated data.

For the purpose of this study, 75 interviews with a total of 86 people were conducted – 39 interviews with 41 representatives of policy actors and experts from Germany, and 36 interviews with 45 representatives of policy actors and experts from Brazil.[31] The interviews in Brazil were carried out during a two month research stay in Brazil in March/April 2013. Most of the interviews in Germany were conducted between August 2012 and February 2013, a second round took place in October/November 2013. One interview had been conducted in February 2010. From the Brazilian side, the two biggest groups of interviewees were government representatives (18 interviewees) and academics (17 interviewees). The interviewed government representatives covered the following state bodies: Itamaraty, Ministry of Mines and Energy (Ministério de Minas e Energia, MME), Casa Civil, Ministry of Agriculture, Livestock and Food Supply (Ministério da Agricultura, Pecuária e Abastecimento, MAPA), Environment Ministry (Ministério de Meio Ambiente, MMA), and the Ministry of Development, Industry and Foreign Trade (Ministério do Desenvolvimento, Indústria e Comércio Exterior, MDIC). The remaining interview partners came from businesses, NGOs and implementing agencies of the Brazilian government (see Appendix A2). From the German side, government representatives were the biggest group of interviewees (18 interviewees), covering BMU, AA,[32] BMZ, BMWi and BMELV. The remaining interview partners were drawn from implementing agencies of the German government, academia and consultancy, political foundations, the German Parliament (Bundestag), NGOs and businesses.[33] The interviews with German interview partners were conducted in German, the ones with Brazilian interview partners either in Portuguese or English.

Academic literature is the primary source for the development of the theoretical-analytical frame and the chapter that analyzes global renewable energy governance. It furthermore serves as an additional source for the case study analyses and the identification of reasons behind ideational differences.

---

[31] For confidentiality reasons, this study only mentions the function and institution of interviewees without revealing their names. Due to different approaches to confidentiality in both countries, the description of the function of Brazilian interview partners is more specific than in the German case.
[32] This includes the German Embassy in Brasília and the German Consulate General in Rio de Janeiro, Brazil.
[33] Please note that some of the interview partners from the German government and the implementing agencies were based in Brazil. In addition, all interview partners from German political foundations work in Brazil. In the Brazilian case, a diplomat who had worked at the Brazilian embassy in Germany was interviewed.

These sources are complemented by informal talks with representatives of relevant policy actors and experts as well as participant observations. The participant observations enabled the author to gain first-hand insights into discussions on global renewable energy governance in both countries. These comprised two meetings of the Global Bioenergy Partnership (GBEP) – the Bioenergy Week in Brasília, 18-23 March 2013, and the 4th meeting of the Working Group on Capacity Building in Berlin, 29-30 May 2013 – and two meetings of the German government on issues regarding IRENA – an expert discussion on IRENA's 2012 working program and its mid-term strategy in Berlin, 19 April 2012, and the presentation of IRENA's Global Renewable Energy Roadmap in Berlin, 8 August 2013. The research stay at two Brazilian institutions – the Centro Brasileiro de Relações Internacionais (Brazilian Center for International Relations, CEBRI) in Rio de Janeiro and the Division for International Studies, Political and Economic Relations at the Instituto de Pesquisa Econômica Aplicada (Institute for Applied Economic Research, IPEA) in Brasília – enabled a fruitful exchange of ideas with Brazilian researchers. And last but not least, working at the German Institute for International and Security Affairs (Stiftung Wissenschaft und Politik, Deutsches Institut für Internationale Politik und Sicherheit, SWP) from January 2010 until December 2013 facilitated a close and long-term interaction with German policy-makers involved in global renewable energy governance and renowned experts on these issues.

3 Global Governance on Renewable Energy

This chapter introduces the empirical part of the thesis, providing sound background information on transboundary policy-making on renewable energy. This context knowledge is crucial for the subsequent analysis of the ideas on global renewable energy governance advanced by the German and Brazilian governments, as it helps to understand how the ideas of both governments are embedded in the global policy context. As outlined in the introduction, a global governance perspective on renewable energy (and on energy in general) is rather new, both in policy-making and academic discussions. This chapter presents the state of art of these academic discussions.[34] Although the thesis explicitly focuses on renewable energy, it is important to note that the promotion of renewable energy cannot be considered in isolation but must take the broader context of global energy governance into account as policy actions concerning the different energy sources are closely interrelated.

The chapter begins with a comprehensive illustration of the origins and evolution of global renewable energy governance. As outlined in the theoretical-analytical framework, such a reconstruction of historical processes is crucial in the understanding of social realities. In the second section, the chapter focuses on the global challenges addressed by global renewable energy governance. This is an important research focus in academic discussions, as researchers who choose a global governance perspective typically refer to these challenges in order to explain why renewable energy should be an object of transboundary policy-making. The following subchapter presents two structural features that are often underlined when characterizing transboundary policy-making on energy: first, its relatively weak development and the dominance of national policy-making and second, its fragmentation. The chapter then elaborates on another research focus in the literature on global (renewable) energy governance: global governors and their governance activities. The concluding section shortly summarizes the insights of this chapter with

---

[34] For analyses that provide overviews of global governance on renewable energy see Bastos Lima, Gupta 2013, Hirschl 2009, Rowlands 2005, Steiner et al. 2006, Suding, Lempp 2007. For overviews of global energy governance in general see Cherp, Jewell & Goldthau 2011, Colgan, Keohane & Van de Graaf 2011, Dubash, Florini 2011, Florini 2008, Florini, Sovacool 2009, Goldthau 2011, Goldthau 2013a, Jollands & Staudt 2012, Karlsson-Vinkhuyzen 2010, Lesage, Van de Graaf & Westphal 2010, Najam, Cleveland 2005, Van de Graaf 2013b.

regard to contestation and social construction in global renewable energy governance.

3.1 Tracing the Origins and Evolution

In order to understand the current form of transboundary policy-making on renewable energy, a deep knowledge of its origins and evolution is of crucial importance. Tracing political processes helps to uncover path dependencies and draw causal and constitutive inferences (Adler 2002:101, 109, Blyth 2003:697). This subchapter analyzes the historical constitution of global renewable energy governance, starting from the origins of global energy governance which concentrated on fossil fuels and nuclear power, over initial attempts to promote renewable energy towards transboundary policy-making on renewable energy taking shape.

3.1.1 The Origins of Global Energy Governance

During the 20th century, transboundary cooperation on energy concentrated on energy security. Energy security concerns were clearly linked to fossil fuels and nuclear power; renewable energy received almost no attention. During this time, the International Atomic Energy Agency (IAEA), the Organization of Petroleum Exporting Countries (OPEC) and the IEA were created. They are still regarded as the most influential multilateral energy institutions (Lesage, Van de Graaf & Westphal 2010:182f.).

Nuclear energy was the first area of transboundary energy cooperation. In 1957, the IAEA was established as an independent international organization related to the UN system. Its aim is to promote "safe, secure and peaceful uses of nuclear science and technology."[35] Its creation was inspired by both concerns over the spread of nuclear weapons and an enthusiasm for the opportunities offered by this new energy source (Colgan, Keohane & Van de Graaf 2011). Its activities cover three main areas: nuclear technology for development, nuclear safety and security, and the threat of nuclear proliferation. The IAEA engages in research and technical cooperation, supports nuclear programs in its member countries, develops safety standards and implements safeguard agreements to prevent the further spread of nuclear weapons.[36] As such, it is the international organization with the most encompassing and wide-ranging mandate in global energy governance. With 159 member states, a regular budget of €338 million (2013) and 2300 people working at

---

[35] See IAEA, Our work, http://www.iaea.org/OurWork/index.html (accessed February 14, 2014).
[36] IAEA, IAEA Primer, Fact Sheets, March 2013, www.iaea.org/Publications/Factsheets/English/iaea-primer.pdf (accessed February 14, 2014).

its secretariat,[37] it is also the biggest international organization on energy. Yet, academic discussions on global energy governance rarely cover transboundary cooperation on atomic energy and elaborate on the IAEA.

Besides nuclear energy, transboundary energy cooperation was focused on fossil fuels, especially oil. It was organized around conflicting interests between states importing oil (IEA) and exporting it (OPEC). The main motivation behind the creation of both OPEC and the IEA was to build a counterweight to powerful market players. After the Second World War, seven international oil companies, primarily from the United States of America (USA) and the United Kingdom (UK), controlled most oil reserves. Oil prices were relatively low and the major powers USA, UK and Soviet Union did not worry about their energy supply as they enjoyed easy access to sufficient and cheap energy either domestically or in areas under their control (Karlsson-Vinkhuyzen 2010:180). Yet, oil-producing countries began to worry about their dependence on international oil companies and so in 1960 Iran, Iraq, Kuwait, Saudi Arabia and Venezuela created OPEC to improve their control over domestic oil reserves. Initially, OPEC concentrated on royalties and tax questions. In the subsequent years, other oil exporting countries, primarily from the Middle East and North Africa, decided to join OPEC. In the 1970s, OPEC members started to nationalize domestic oil industries and thus increase OPEC's control over worldwide oil production and pricing policies. Meanwhile, economic growth in industrialized countries fueled global oil demand and increased the dependency of OECD countries on oil imports from the Middle East. In 1973, OPEC's power became internationally apparent: Arab OPEC countries proclaimed an oil embargo against the US and the Netherlands in reaction to their involvement in the Israeli-Arab-War. This policy led to the first oil price shock of the 1970s, which is illustrated in Figure 1. The immediate reactions of the major oil importing countries of the OECD world were uncoordinated and even competitive, aggravating the overall economic burden they had to bear (Colgan, Keohane & Van de Graaf 2011, Karlsson-Vinkhuyzen 2010:180f., Lesage, Van de Graaf & Westphal 2010:57-61).

---

[37] IAEA, IAEA Primer, Fact Sheets, March 2013, www.iaea.org/Publications/Factsheets/English/iaea-primer.pdf (accessed February 14, 2014).

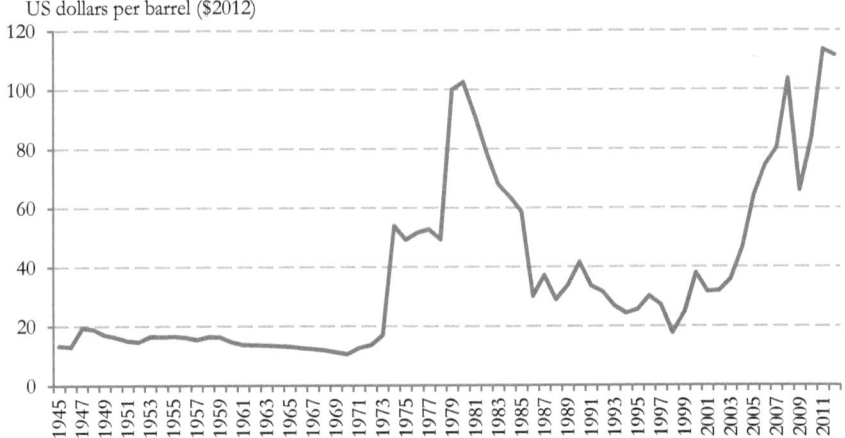

Figure 1: Crude Oil Prices, 1945–2012[38]

Today, OPEC's main focus is on setting collective volumes of oil production. It has 12 member countries,[39] accounting for 40 percent of the worldwide oil production and 75 percent of proven oil reserves (Lesage, Van de Graaf & Westphal 2010:58). Only countries with a substantial net export of crude oil can become OPEC members. Its mission is to coordinate the petroleum policies of its member states:

> "The mission of the Organization of the Petroleum Exporting Countries (OPEC) is to coordinate and unify the petroleum policies of its Member Countries and ensure the stabilization of oil markets in order to secure an efficient, economic and regular supply of petroleum to consumers, a steady income to producers and a fair return on capital for those investing in the petroleum industry."[40]

At its conferences, OPEC members decide on the volume of their collective oil production. OPEC also collects and disseminates data on basket prices, upstream investment, downstream capacity, oil taxes, oil reserves, production allocation and

---

[38] Source: illustration S.R. with data from BP, Historical data workbook, http://www.bp.com/content/dam/bp/excel/Energy-Economics/statistical_review_of_world_energy_2013_workbook.xlsx (accessed February 24, 2014)

[39] OPEC members are Algeria, Angola, Ecuador, Iran, Iraq, Kuwait, Libya, Nigeria; Qatar; Saudi Arabia; United Arab Emirates, Venezuela. See OPEC, Member countries, http://www.opec.org/opec_web/en/about_us/25.htm (accessed February 14, 2014).

[40] OPEC, Our mission, http://www.opec.org/opec_web/en/about_us/23.htm (accessed February 14, 2014).

further market indicators.⁴¹ The organization underlines that it aims to prevent oil price volatility but, in contrast to the period from the early 1970s to the mid-1980s, it does not set crude oil prices.⁴²

To counterweigh OPEC's power, the USA pushed for the creation of an international organization of oil importing countries. In 1974, OECD countries founded the IEA as an autonomous agency of the OECD (for comprehensive analyses on the IEA see also Colgan 2009, Florini, Sovacool 2009:5242f., Lesage, Van de Graaf & Westphal 2010:59-61, Leverett 2010:248-253, Van de Graaf 2012b). Its main purpose was to coordinate a collective response to major oil supply disruptions. As the main element of its activities, the IEA established an emergency system for sharing oil: members were obliged to hold oil emergency reserves equivalent to their net oil imports of 60 days (this amount was soon increased to an equivalent of 90 days) and to restrain oil demand in case of emergency. Thus, the IEA was endowed with the power to make legally binding decisions on its member states. Member states agreed that these decisions could be based on a majority vote and distributed voting rights according to the amount of net oil imports in 1973. However, in practice, consensus decisions have been the norm (Colgan 2009:8f.). Further objectives of the IEA included building an information system on the international oil market, developing alternative energy sources and increasing energy efficiency, as well as collaborating on energy technologies.⁴³ According to Leverett (2010), the USA also pushed for the establishment of the IEA because it wanted to mobilize support for market-based approaches to energy security in other OECD countries. During the 1980s the IEA in effect became "a multilateral forum for building support among industrialized energy consumers for upstream liberalization and market integration" (Leverett 2010:246f.). It encouraged member states to decontrol energy prices, reduce energy subsidies, deregulate national electricity markets and open domestic markets (Leverett 2010:252).

Over the course of time, the IEA has significantly expanded its activities and is now considered one of the most influential actors in global energy governance. Next to emergency measures to deal with oil supply disruptions, it serves "as the outstanding official organizational forum for information sharing and analysis on international energy matters" (Leverett 2010:252). Its activities do not only cover oil but all energy sources. Since the late 1970s, the IEA issues regular publications on energy topics, annually reviews the national energy policies of its member states and engages in dialogue with energy exporting countries and non-IEA energy consum-

---

⁴¹ OPEC, Data/Graphs, http://www.opec.org/opec_web/en/data_graphs/40.htm (accessed February 14, 2014).
⁴² According to OPEC, crude oil prices are set on the three major international petroleum exchanges in New York, London and Singapore. See OPEC, About OPEC, http://www.opec.org/opec_web/en/press_room/178.htm (accessed June 6, 2012).
⁴³ IEA, History, http://www.iea.org/aboutus/history/ (accessed February 14, 2014).

ers (Colgan 2009:14). IEA membership has expanded from 16 countries in 1974 to 28 at the present time.[44] Yet its overall weight in global energy consumption has decreased, as it excludes non-OECD countries such as China and India which are now major energy consumers. Its deceasing weight in global energy consumption challenges the IEA's capacity to influence global energy markets. Countries like China and India are unlikely to become IEA members as membership is restricted to OECD countries.[45] Moreover, IEA members must fulfill the exigencies of the emergency system and ensure that oil companies working in their jurisdiction comply with reporting requirements.[46] Despite several intents to reform, the IEA sticks to an anachronistic distribution of voting rights: the distribution of votes on its Governing Board is still based on net oil imports in 1973. A reallocation according to the current structure of energy imports would significantly shift voting rights, with the USA, UK and Canada losing a large share of votes and South Korea and Spain gaining (Colgan, Keohane & Van de Graaf 2011). As the IEA is a relevant actor in global renewable energy governance, its activities with regard to renewable energy will be analyzed more in detail in section 3.4.2.

The 1973 oil price shock also led to the creation of another influential multilateral forum of industrialized countries, the Group of Seven (G7). In reaction to the macroeconomic problems following the oil price shock, in 1975 the heads of states from France, Italy, Japan, UK, USA and West Germany agreed to organize annual meetings under a rotating presidency to discuss economic issues. Canada joined one year later and the forum became known as the G7. During the first few years, oil and energy supplies in general were important issues at G7 meetings. In 1979, for example, the G7 countries decided to restrict oil imports. With declining oil prices in the 1980s (see Figure 1 on page 76), energy issues lost importance. During the 1990s, G7 meetings focused on nuclear safety in the former Soviet Union. Yet the heads of states of the G7 had difficulties in working on other energy issues, as they could not reach common positions. Broader energy issues did not regain priority on G7's agenda until the 2000s (Colgan 2009:5, Colgan, Keohane & Van de Graaf 2011).

---

[44] IEA, member countries, http://www.iea.org/countries/membercountries/ (accessed March 6, 2014).
[45] In November 2013, the IEA signed a joint declaration with the governments of Brazil, China, India, Indonesia, Russia and South Africa „expressing mutual interest in pursuing an association", http://www.iea.org/newsroomandevents/ieaministerialmeeting2013/jointdeclaraction.pdf (accessed February 18, 2014).
[46] IEA member countries are: Australia, Austria, Belgium, Canada, Czech Republic, Denmark, Finland, France, Germany, Greece, Hungary, Ireland, Italy, Japan, Republic of Korea, Luxembourg, The Netherlands, New Zealand, Norway, Poland, Portugal, Slovak Republic, Spain, Sweden, Switzerland, Turkey, United Kingdom, and the United States. See IEA, member countries, http://www.iea.org/countries/membercountries/ (accessed February 14, 2014).

3.1.2 Initial Attempts to Promote Renewable Energy

Some initial attempts to address renewable energy in transboundary cooperation were already made during the Cold War Era. With the UN Conference on New Sources of Energy, the first major international meeting on renewable energy was held in Rome 1961 (Karlsson-Vinkhuyzen 2010:181f., Rowlands 2005:80, United Nations 1981:258f.). This expert conference was convened by the UN Economic and Social Council to exchange ideas and experiences on solar energy, wind power and geothermal energy. Special attention was paid to possible ways to increase the usage of these energy sources, especially in less developed areas. Conference participants stressed the need to explore geothermal resources and collect data on wind and solar radiation. They furthermore called for experimental stations for wind and solar energy and an adaption of existent technologies to the contexts of less developed countries (United Nations 1981:258f.). During the 1960s and the beginning of the 1970s, fossil fuel deployment was not yet an issue of global concern and it took almost twenty years before renewable energy re-emerged on the global agenda.

After the oil price shocks of the 1970s (see Figure 1 on page 76), policymakers around the world began to worry about fossil fuel dependence and consider alternative forms of energy. In 1980 the Report of the Independent Commission on International Development Issues, chaired by Willy Brandt, criticized the worldwide dependence on oil for industrial development. Highlighting the risk of depleting oil resources and high and volatile prices, it argued for an increased use of inexhaustible sources of energy. Among these, it counted solar energy, biomass, wind and tides, new forms of nuclear energy, hydropower and geothermal energy. It called for an international energy strategy, underlining the common interests of the international community for an energy transition:

> "All parties have an interest in the creation of an international framework and a political climate which can provide a confident and trusting collaboration of all countries to ensure long-term exploration and development of energy, and an orderly transition from a world economy and industry based on oil, to one that can be sustained through renewable sources of energy. They have a shared interest in conservation and in avoiding abrupt price changes. And all must be aware of the special need to protect the poorer oil-importing countries" (The Independent Commission on International Development Issues 1980:168f.).

The Brandt Report also claimed that energy prices should reflect long-term scarcities. To avoid economic disruption, the necessary price increases should be conducted in an orderly and predictable manner. It recommended the establishment of a global energy research center under auspices of the UN which should focus on renewable sources of energy, assist poorer developing countries with financial means to ensure their energy supplies matched demand, and increase financing for fossil fuel exploration and renewable energy development in the Third World (The Independent Commission on International Development Issues 1980:160-171).

In 1981, the UN convened an intergovernmental conference on renewable energy. The UN Conference on New and Renewable Sources of Energy in Nairobi discussed how to promote an orderly and peaceful transition away from fossil fuels.[47] The conference addressed energy sources widely considered as 'new renewable energy' such as hydropower, biomass, solar, wind and geothermal energy. Indeed, it also focused on energy sources that nowadays are not counted as 'new renewable energy', for example, fuel wood, charcoal, peat, oil shales, tar sands and draught animal power. According to Odingo (1981:104) it was decided to exclude coal, conventional oil and nuclear energy from the Conference to "avoid a serious political confrontation." The conference paid special attention to oil importing developing countries, as the oil price increases of the 1970s particularly hit this group of countries. It was feared that the lack of adequate and cheap energy would inhibit their continued economic development. During the preparatory process, the UN General Assembly established technical panels which analyzed, for each energy source considered, the availability of the energy source, availability and state of technologies, as well as the overall viability and environmental impacts. In addition, governments were asked to write national reports on new and renewable sources of energy. Odingo highlights the impact the request for national reports had on national energy policy-making:

> "This national effort was important for many reasons one of which was that for many developing countries it provided the first opportunity to make a hard look at the energy problem and begin the planning process which it is hoped will mature into the future after the main conference has passed. In some of these countries, new ministries of energy were hastily created for the first time in preparation for the energy conference" (Odingo 1981:104).

The Nairobi Plan of Action identifies the following policy areas in need of international support: energy assessment and planning; research, development and demonstration; transfer, adaptation and application of mature technologies; information flows, education and training; and mobilization of financial resources (Odingo 1981:106, United Nations 1981:689). North-South transfers of technological and financial resources were highly controversial issues at the conference. Many developing countries demanded firm commitments from industrialized countries, arguing for internationally agreed targets for financial transfers by industrialized countries and an unrestricted flow of technical information. In addition, they wanted to establish an international organization for monitoring purposes. These demands were strongly opposed by industrialized countries who called for a redistribution of existing financial resources instead of creating new mechanisms (Biswas 1981:331, Odingo 1981:106, Rowlands 2005:82, United Nations 1981:690). Further contentious issues included the role of multinational corporations (with the Eastern European countries against the involvement of private foreign capital) and sovereignty con-

---

[47] For a comprehensive documentation of the Conference see United Nations 1981:688-711.

cerns over national resources (with European Community members, Switzerland and the USA highlighting the primacy of international law) (Biswas 1981:332, United Nations 1981:689). The conference ended with a general agreement on the need for the international promotion of new and renewable energy, but without firm commitments or specific means of implementation. However, as an outcome of the conference, the UN General Assembly established the Committee on the Development and Utilization of New and Renewable Sources of Energy whose mandate was transferred to the ECOSOC in 1994 (Karlsson-Vinkhuyzen 2010:182). Despite its modest outcomes, the conference is widely regarded as a central step for transboundary cooperation on renewable energy (Hirschl 2009:4411, Karlsson-Vinkhuyzen 2010:182, Rowlands 2005:82f.,).

During the 1990s environmental concerns gained international attention while oil prices remained relatively stable (see Figure 1 on page 76). At the UN Conference on Environment and Development (UNCED) in Rio de Janeiro 1992 – one of the most influential UN conferences (if not the most influential) in the aftermath of the Cold War – renewable energy (and energy in general) played only a minor role (see, for example, Hirschl 2009:4411, Najam, Cleveland 2005:124f., Rowlands 2005:83f.). The documents adopted by the conference paid little or no attention to renewable energy. The final declaration of the summit, the Rio Declaration on Environment and Development, does not even mention energy; the Framework Convention on Climate Change – which some authors such as Karlsson-Vinkhuyzen (2010:193) and Najam & Cleveland (2005:128) highlight as the most influential indirect global governance arrangement for energy – does not refer to renewable energy and the action plan, Agenda 21, lacks a chapter on energy. However, the Agenda 21 chapters on changing consumption patterns (chapter 4), sustainable human settlement development (chapter 7), protecting the atmosphere (chapter 9), and sustainable agriculture and rural development (chapter 14) discuss the promotion of renewable energy – mostly in the context of environmental protection and with a clear focus on supporting renewable energy deployment in developing countries (United Nations Conference on Environment & Development 1992). Rowlands (2005:83f.) presents two reasons behind the slight attention paid to renewable energy within Agenda 21: first, declining oil prices during the 1980s reduced the economic pressure to develop alternatives to fossil fuels and second, the group of Arab countries was strongly opposed to promoting energy efficiency and renewable energy. Five years later, in 1997, the UN General Assembly acknowledged in its Rio+5 Review the missing attention paid to energy and emphasized it as one of the most important issues to be dealt with comprehensively (Karlsson-Vinkhuyzen 2010:183, Lesage, Van de Graaf & Westphal 2010:55f.).

3.1.3 Transboundary Policy-Making on Renewables Taking Shape

With the turn of the millennium, oil prices started to rise again (see Figure 1 on page 76) and the promotion of renewable energy gained more attention in transboundary policy-making. The G8,[48] which formerly focused on oil prices as a main energy issue, decided at its 2000 Summit in Okinawa (Japan) to set up a Task Force on Renewable Energy. The G8 leaders mandated the Task Force to make policy recommendations on how to support renewable energy in developing countries. In its final report, the Task Force recommended that the G8 take action in four areas: reducing technology costs by expanding markets, building a strong market environment, mobilizing finance and encouraging market-based mechanisms (G8 Global Renewable Energy Task Force 2001). The Task Force underlined that industrialized countries should support renewable energy within their own countries, as this would help reduce the cost of renewable energy and thus make it more accessible in developing countries. It furthermore advised industrialized countries to strengthen their support of renewable energy with technical and financial development assistance. With regard to market-based mechanisms, the Task Force recommended asking the IEA for further analysis on the competitiveness of renewable energy options, strengthen flexible mechanisms in the realm of climate policy and develop market mechanisms to address environmental externalities of energy technologies. As the USA and Canada did not approve the Task Force's conclusions, the G8 did not formally adopt its report. Nevertheless, the report influenced the further course of the international debate on how to promote renewable energy (Florini, Sovacool 2009:5243f., Hirschl 2009:4409, Rowlands 2005:85f.).

In 2001, the Ninth Session of the Commission on Sustainable Development (CSD) addressed energy as one of two sectoral themes. According to Karlsson-Vinkhuyzen, Jollands & Staudt (2012:14), this was the first time that sustainable energy was discussed at a high political level within the UN. The report of an intergovernmental group of experts on energy and sustainable development served as the point of departure for negotiations (United Nations Commission on Sustainable Development 2001b). In its final decision, the CSD emphasized the centrality of energy issues for sustainable development. Contrary to previous deliberations within the UN, the CSD did not concentrate on the promotion of renewable energy in developing countries but called for an increased use of renewable energy in both developing and industrialized countries:

> "The main challenge lies both for developed and developing countries in the development, utilization and dissemination of renewable energy technologies, such as solar, wind, ocean, wave, geothermal, biomass and hydro power, on a scale wide enough to significantly contribute

---

[48] G7 members are France, Germany, Italy, the United Kingdom, Japan, the United States, and Canada. The G8 comprises the G7 member states and Russia.

to energy for sustainable development. Despite some progress in promoting renewable energy applications in recent years (...) numerous constraints and barriers including costs continue to exist" (United Nations Commission on Sustainable Development 2001a:55f.).

Governments were called upon to build an enabling environment for the development and use of renewable energy sources. In line with previous UN agreements, the CSD-9 decisions emphasised the importance of North-South transfers for the worldwide promotion of renewable energy. However, the recommendations lacked the mechanisms of implementation (Karlsson-Vinkhuyzen 2010:183, Rowlands 2005:84f.).

At the WSSD in Johannesburg, South Africa, in 2002 energy issues were one of the top priorities (Annan 2002). According to Najam & Cleveland (2005:127), energy was one of the few areas where the Summit could advance the international agenda: the Johannesburg Declaration directly refers to energy and identifies it as a human need and the Johannesburg Plan of Implementation establishes energy as an issue of its own instead of treating it only as an element of other issues (United Nations 2002). During the preparatory process and the summit itself, policy-makers intensively discussed the promotion of renewable energy, focusing on three key issues: setting up internally agreed targets and timetables for the adoption of renewable energy, phasing out fossil fuel subsidies and transferring clean energy technologies to developing countries. Renewable energy and energy subsidies turned out to be among the most controversial issues at the conference (Corrêa do Lago 2009:162, IISD 2002:7, Karlsson-Vinkhuyzen 2010:184, Najam, Cleveland 2005:128, Rowlands 2005:86-88). Member states of the EU, Norway, New Zealand, Switzerland, Iceland, Tuvalu and Eastern European countries demanded time-bound targets for increasing the share of renewable energy in global energy supply. Brazil, on behalf of Latin American and Caribbean countries, presented a similar proposal (Corrêa do Lago 2009:162). The EU, Norway, New Zealand and Iceland furthermore proposed to set targets and timeframes for the phasing out of fossil fuel subsidies. Yet, the USA, Australia, Canada, Japan and many developing countries strongly opposed these proposals. G77/China feared that global targets and timetables for renewable energy would "divert attention away from the primary goal of ensuring universal access to energy services for the poor" (IISD 2002:7). The USA, Australia, Canada and Japan criticised the proposal as an inflexible one-fits-all approach. As policy-makers were unable to resolve these contentious issues, they did not adopt any substantial agreements but only broad principles for the promotion of renewable energy (IISD 2002, Röhrkasten, Westphal 2013:4). Alongside the Johannesburg Summit, the British government initiated the creation of the Renewable Energy and Energy Efficiency Partnership (REEEP), a multistakeholder partnership that brings together governments, businesses, banks and non-governmental organizations to strengthen markets for sustainable energy, primarily in emerging markets and developing countries. According to Florini & Sovacool (2009:5245), its creation was deeply influenced by the discussions of the G8 Renewable Energy Task Force set up in Okinawa two years earlier.

In 2004, the German government hosted the international conference Renewables 2004, the first of an international conference series on renewable energy.[49] Participants from 154 countries joined the conference, including representatives from governments, intergovernmental organizations, non-governmental organizations, the scientific community and the private sector. The conference discussed how to increase the share of renewable energy in both developing and industrialized countries. It focused on political frameworks, private and public financing, capacity building, and research and development (Expert Group on Renewable Energy 2005:51). An important outcome of the Renewables 2004 Conference was the creation of the multistakeholder Renewable Energy Policy Network for the 21st Century (REN21) which aims at promoting renewable energy in both developing and industrialized countries (see also Röhrkasten, Westphal 2013:6f.).[50]

At the G8 Summit in Gleneagles 2005 the G8 leaders adopted, together with Brazil, China, India, Mexico and South Africa, the Action Plan "Climate Change, Clean Energy and Sustainable Development"[51] which contains a declaration of intent to promote the development and commercialization of renewable energy (see also Florini, Sovacool 2009:5243f.). In the action plan, the G8+5 leaders declare their support for existing international cooperation initiatives such as REN21 and REEEP and for an expansion of IEA activities on renewable energy. They affirm their intention to support renewable energy in developing countries in terms of capacity-building, policy advice, and research and development. In Gleneagles, the G8 also agreed to begin a dialogue with emerging countries on energy issues. The G8 leaders assigned the IEA to develop alternative energy scenarios and advise on clean energy strategies, and the World Bank to create a framework for clean energy investment and financing. Furthermore, they initiated the launch of an International Bioenergy Partnership (GBEP) to support biomass and biofuel deployment, especially in developing countries. GBEP's activities focus on the measurement of

---

[49] Since then, four International Renewable Energy Conferences have been organized: he Beijing International Renewable Energy Conference (BIREC) in 2005, the Washington International Renewable Energy Conference (WIREC) in 2008, the Delhi International Renewable Energy Conference (DIREC) in 2010 and the Abu Dhabi International Renewable Energy Conference (ADIREC) in 2013. See REN21, IREC Selection Guidelines, http://www.ren21.net/REN21Activities/IRECs/IRECSelectionGuidelines.aspx (accessed February 14, 2014).
[50] Ren21, Mission and Concept, http://www.ren21.net/Portals/97/documents/Media%20Resources/REN21_Mission_and_Concept_060310.pdf (accessed February 14, 2014).
[51] G8, Action Plan "Climate Change, Clean Energy and Sustainable Development", http://www.g8.utoronto.ca/summit/2005gleneagles/climatechangeplan.pdf (accessed February 14, 2014).

greenhouse gas reductions from the use of bioenergy, the development of sustainability indicators for biofuels and capacity building for sustainable bioenergy.[52]

The 2006 and 2007 meetings of the UN CSD again focused on energy issues. Delegates emphasized the importance of energy for sustainable development, poverty eradication and achieving the Millennium Development Goals (MDGs). The conflicts that characterized the WSSD in Johannesburg 2002 arose again: Germany, on behalf of the EU, called for time-bound targets on energy efficiency, renewable energy and access to energy, whereas the G77 – particularly the Gulf States under the leadership of Saudi Arabia – blocked these proposals. At the end of the 2007 session, the EU even withheld its support to the compromise document presented by the chair of the session as it did not contain reference to time-bound targets on energy (IISD 2007, Röhrkasten, Westphal 2013:4).

In 2008, the German government initiated the creation of IRENA. It did so because of disappointments with UN processes on renewable energy. Instead of waiting for a UN-wide consensus, the German administration decided to push for the worldwide promotion of renewable energy by creating a new intergovernmental organization outside of the UN framework. After a number of bilateral consultations, the German government began the formal preparatory process and was soon joined by the governments of Spain and Denmark as major supporters of the initiative. In January 2009, IRENA's founding conference was held. IRENA was set-up as the first intergovernmental organization fully dedicated to the promotion of renewable energy. In contrast to the IEA which has its membership restricted to the OECD world, IRENA is open to all UN member states. Since 2011, IRENA is a fully-fledged intergovernmental organization. Two context factors facilitated IRENA's rapid creation in 2008/09 as they strengthened worldwide interest in the promotion of renewables: fossil fuel prices were rising tremendously (see Figure 1 on page 76) and the political attention for climate protection peaked in the wake of the Copenhagen climate summit in 2009 (Röhrkasten, Westphal 2013).

At the founding conference, IRENA's creation met widespread support. The statute was signed by 75 countries. However, a number of influential countries were missing among the signatories. At this time, creating IRENA did not meet the support of the emerging countries of Brazil, China, India, Indonesia, Mexico and South Africa nor the G8 countries Canada, Japan, Russia, UK and USA (Van de Graaf 2012a). Most of these countries decided to join afterwards. As of February 2014 only Brazil, Canada and Russia have refrained from doing so.[53] Some countries opposed IRENA's creation due to general reservations against renewable ener-

---

[52] G8, Action Plan "Climate Change, Clean Energy and Sustainable Development", http://www.g8.utoronto.ca/summit/2005gleneagles/climatechangeplan.pdf (accessed February 14, 2014).
[53] IRENA, IRENA membership, http://www.irena.org/Menu/Index.aspx?mnu=Cat&PriMenuID =46&CatID=67 (accessed February 14, 2014).

gy or the creation of new international organizations. Other countries, such as China and Brazil, raised sovereignty concerns. They feared IRENA would engage in standardization and impose restrictions on the energy policies of its member states. Some IEA member states did not want to create competition against the IEA. Reasons for supporting IRENA's creation varied as well. Next to the countries that were actually interested in the worldwide promotion of renewable energy, there were those that expected to gain access to financial means by joining IRENA. Other countries wanted to avoid the creation of a new intergovernmental organization without being able to influence it from within. The IEA member states of the UK, USA, Japan and Australia, for example, formed an alliance, trying to limit IRENA's budget and restrict its activities to developing countries – and as such trying to protect IEA's role as the central energy organization of the OECD states. IRENA's headquarters' location in Abu Dhabi in the United Arab Emirates (UAE) often meets surprise. As a member of the OPEC, UAE belong to the groups of countries that had opposed several attempts to promote renewable energy within the UN framework and that was expected to oppose IRENA's creation as well. Yet, the UAE and Nigeria expressed their support for the initiative from the beginning despite being OPEC members. This illustrates an important policy shift within OPEC. Today, several OPEC members no longer concentrate on fossil fuels only; in the search for alternative business models they also invest in renewable energy. For the UAE, gaining IRENA's headquarters was a question of national prestige: IRENA is the first intergovernmental organization located in the Arab world (Röhrkasten, Westphal 2013:8-10).

During IRENA's founding process, some controversial issues arose. The first relates to the degree of bindingsness of IRENA's activities: while some policy actors wanted IRENA to engage in standard-setting and regulation, it was finally agreed that IRENA would build on the principle of voluntariness. The scope of renewable energy sources that IRENA should cover was also discussed. Some policy actors raised austainability concerns with regard to biofuels and large hydropower. Here, agreement was reached to promote all kinds of renewable energy – if these are used in a sustainable way. The scope of membership was also discussed: should IRENA build on a small frontrunner coalition or on broad membership? Finally, it was decided to follow an UN-wide approach. Debates also concerned the geographical scope of IRENA's activities: should IRENA promote renewable energy at a global scale or concentrate its activities on developing countries? Here, IRENA has taken an ambiguous stance so far. Although its declared goal is to promote renewable energy globally, most of its activities focus on the developing world (Röhrkasten, Westphal 2013:8-10,13f.).

IRENA's creation has impacted how other global governors deal with renewable energy. Van de Graaf particularly points at the repercussions for the IEA:

"There are clear signs that the IEA's attitude towards renewable has changed in response to IRENA's creation. In September 2008, the IEA upgraded its renewable energy unit into a division, staffed by 9 full-time analysts. (...) In addition, the IEA has expressed itself unusually positive about solar energy in two recent reports published in mid-2010. (...) In February, the IEA announced that it would henceforth start issuing annual "medium term reports" on renewable energy" (Van de Graaf 2012a).

IRENA's Director-General Adnan Amin adds that due to IRENA's creation "not only the IEA, but also a number of other institutions, including in the UN framework, suddenly stepped up their game on renewable energy."[54]

At the climate summit in Copenhagen 2009, the US government initiated the founding of the Clean Energy Ministerial (CEM) – a ministerial forum that brings together government representatives from countries that are either major economies or leading in clean energy issues. In the same year, the G8 leaders strengthened IEA's work on low-carbon energies by establishing the International Low-Carbon Energy Technology Platform within the institutional structure of the IEA. Between 2009 and 2011, the Group of 20 (G20)[55] initiated four working groups on energy issues, one of those focusing on clean energy and energy efficiency (Van de Graaf, Westphal 2011:25f.).

The UN General Assembly declared 2012 as the International Year for Sustainable Energy for All, recognizing the growing importance of energy issues for sustainable development (see also Karlsson-Vinkhuyzen, Jollands & Staudt 2012:14). An important goal of this declaration was to raise awareness on energy issues and encourage action to promote sustainable energy.[56] In the run-up to the UN Year for Sustainable Development, the UN Secretary-General announced his initiative on SE4All which can be considered as the major UN initiative on renewable energy so far (Röhrkasten, Westphal 2013:4). SE4All pursues three main goals up to 2030: doubling the share of renewable energy in global energy supply, doubling the improvement rate for energy efficiency and ensuring that all people around the globe enjoy access to modern energy technologies.[57] While worldwide renewable energy supply has increased significantly since the start of this millennium, its share in the global energy mix has remained almost stable, at around 13 percent (see Figure 2). To guide action for goal achievement, the Secretary-

---

[54] Interview with Adnan Amin by Thijs van de Graaf, European Energy Review, 24 May 2012, http://www.europeanenergyreview.eu/site/pagina.php?id=3707 (accessed February 14, 2014).
[55] G20 members are Argentina, Australia, Brazil, Canada, China, France, Germany, India, Indonesia, Italy, Japan, Republic of Korea, Mexico, Russia, Saudi Arabia, South Africa, Turkey, the United Kingdom, the United States and the European Union.
[56] UNDP, Decade on Sustainable Energy for All, http://www.undp.org/content/undp/en/home/ourwork/environmentandenergy/focus_areas/sustainable-energy/2012-sustainable-energy-for-all/ (accessed November 19, 2013).
[57] Sustainable Energy for All, Objectives, http://www.sustainableenergyforall.org/objectives (accessed November 19, 2013).

General's High-Level Group on Sustainable Energy for All launched a Global Action Agenda which identifies effective courses of action.[58]

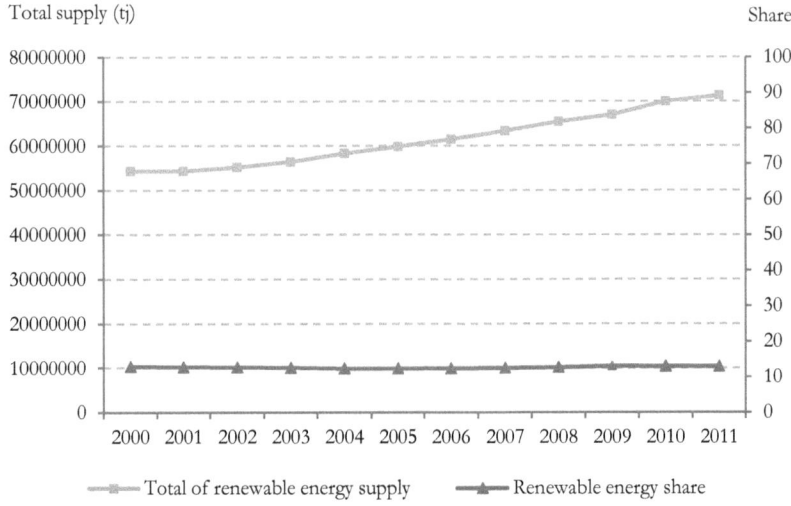

Figure 2: Worldwide Supply of Renewable Energy Sources, 2000-2011[59]

According to Bradshaw (2013:48), SE4All "is a somewhat belated acknowledgement that access to modern energy services was missing from the Millennium Development Goals." Despite the rising attention for energy issues within the UN, the 2012 United Nations Conference on Sustainable Development in Rio de Janeiro (Rio+20) did not achieve progress on energy issues. The final declaration again expresses its general support for promoting access to energy and renewable energy; yet it does not contain means of implementation. It does not even express its support for SE4All as it only "notes" the launch of the initiative.[60] Yet, the Rio+20 Summit had important indirect effects the worldwide promotion of renewable energy. The Summit endorsed the elaboration of Sustainable Development Goals

---

[58] Sustainable Energy for All, Actions and Commitments, http://www.sustainableenergyforall.org/actions-commitments (accessed November 19, 2013).
[59] Source: illustration S.R. with data from OECD iLibrary
[60] UNCED, The Future We Want, http://daccess-ods.un.org/access.nsf/Get?Open&DS=A/RES/66/288&Lang=E (accessed November 19, 2013), p.22.

(SDGs) which also cover the increased use of renewable energy.[61] The SDGs will be adopted by the UN General Assembly in September 2015.

3.2   Global Challenges Addressed by Renewable Energy Promotion

Energy policy-making is typically regarded as a national task. Scholars who analyze energy issues from a global governance perspective have a special need to explain why renewable energy should be the object of transboundary policy-making. Thus, research commonly focuses on the global challenges that global energy governance should address.[62]

Scholars explain the need or potential for global energy governance by pointing to transboundary interdependencies and policy problems that cannot be tackled by single policy actors alone. Transboundary interdependencies relate, for example, to trade in energy sources and energy technologies, and to transboundary externalities of energy production, transport and consumption. Some authors explicitly refer to the global public goods theory, highlighting collective action problems that go along with externalities and the provision of public goods. They underline that global energy governance is needed to address these collective action problems.[63] Cherp, Jewell & Goldthau (2011:76) introduce another aspect: They stress that most states face limited capacity to control their energy systems as their energy supply relies on foreign energy sources.

The promotion of renewable energy shall expand energy supply while reducing worldwide dependence on fossil fuels – and thus alleviate the risks that are associated with fossil fuels. Suding & Lempp (2007:4) argue that global renewable energy governance must address the "discrepancy between the global benefits of renewable energy and its continued under-exploitation": as a newcomer in energy markets, renewable energy requires public support that levels the playing field with fossil fuels, and transboundary cooperation might help national policymakers to fulfill this task. The relevance of renewable energy promotion for global governance is commonly underlined in relation to three global challenges: energy security, eradication of energy poverty and environmental sustainability (especially climate protection). It is often argued that these are challenges that no nation state can manage on

---

[61] UNCED, The Future We Want, http://daccess-ods.un.org/access.nsf/Get?Open&DS=A/RES/66/288&Lang=E (accessed November 19, 2013), p.42-44.
[62] See, for example, Cherp, Jewell & Goldthau 2011, Dubash, Florini 2011, Florini 2008, Florini, Sovacool 2011, Florini, Sovacool 2009, Goldthau 2012, Goldthau 2011, Lesage, Van de Graaf & Westphal 2010.
[63] See, for example, Cherp, Jewell & Goldthau 201176, Dubash, Florini 2011, Florini, Sovacool 2009:5141, Goldthau 2012, Karlsson-Vinkhuyzen, Jollands & Staudt 2012, Lesage, Van de Graaf & Westphal 2010:4.

its own and that transboundary cooperation offers the possibility of substantial mutual gains.[64] This section elaborates further on these global challenges.

3.2.1 Energy Security

Energy security is the first global challenge that the promotion of renewable energy shall address. Energy security is commonly defined as a reliable and affordable energy supply (Florini 2008:3, Florini, Dubash 2011:1f.).[65] In the realm of energy security, the increasing scarcity and geographic concentration of fossil fuels are identified as the main risks. It is important to note that the prevailing understanding of energy security as security of energy supply mirrors the view of energy importing countries. For energy exporting countries, energy security first and foremost relates to demand security, as they depend on export revenues and thus on a stable and predictable energy demand (Cherp, Jewell 2011:3, Karlsson-Vinkhuyzen 2010:186). A risk that affects both energy importers and energy exporters is price volatility which hinders the predictability of both import costs and export revenues. As such, it negatively affects investment decisions in exporting and importing states (Florini, Sovacool 2011:60-62, Kuik, Bastos Lima & Gupta 2011, Lesage, Van de Graaf & Westphal 2010:29f.).

Global energy demand is rising sharply, especially due to economic growth in emerging economies. Between 1990 and 2011, worldwide energy consumption increased by more than 40 percent (see Figure 3). At the same time the fossil fuel supply faces absolute limits. Fossil fuel reserves are exhaustible as they take hundreds of millions years to form. The extent of existing fossil fuel reserves and so the expected future availability of oil and gas is a subject of controversial debates – fueled by the lack of data on oil reserves and oil extraction. According to Lesage, Van de Graaf & Westphal (2010:19), oil extraction has exceeded oil discoveries every year since the late 1970s. Whereas some claim that the peak of oil extraction has already been reached, others argue that the world will never run out of fossil fuel supply as rising extraction costs will shift demand towards non-conventional fossil fuels and renewable energy. Next to biophysical limits, cyclical underinvestments might aggravate energy scarcity on the supply side (Lesage, Van de Graaf & Westphal 2010:2). Increasing competition over scarce resources is likely to lead to higher fossil fuel prices, with negative economic consequences for energy

---

[64] See, for example, Cherp, Jewell & Goldthau 2011, Florini, Sovacool 2011, Goldthau 2011, Karlsson-Vinkhuyzen 2010, Karlsson-Vinkhuyzen, Jollands & Staudt 2012, Kuik, Bastos Lima & Gupta 2011, Lesage, Van de Graaf & Westphal 2010, Najam, Cleveland 2005.
[65] Sometimes, 'energy security' is defined in broader terms, comprising access to energy and environmental sustainability as well (see, for example, Kuik, Bastos Lima & Gupta 2011). To avoid fuzzy concepts, this study sticks to the narrow definition of energy security.

consumers. As energy is a key driver of macroeconomic growth, rising energy prices may also slow down the world economy as a whole. This is especially true for oil prices: in the past, oil price hikes have caused worldwide recessions.

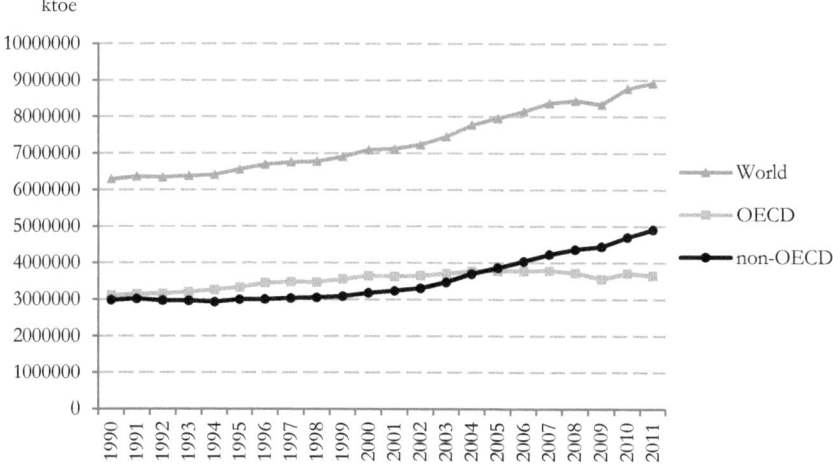

Figure 3: Development of Worldwide Energy Consumption, 1990-2011[66]

Besides, fossil fuel reserves are concentrated in a few countries that are perceived as highly volatile and instable (Florini, Dubash 2011:3f., Florini, Sovacool 2011:59f., Kuik, Bastos Lima & Gupta 2011:630, Lesage, Van de Graaf & Westphal 2010:20f.). Most of the remaining conventional hydrocarbon reserves are located in a handful of countries in the Middle East, the Caspian basin, central Asia and northwest Siberia. More than the half of the worldwide conventional gas reserves are located in three countries: Russia, Iran and Quatar (Lesage, Van de Graaf & Westphal 2010:20f.). Very few government-dominated firms control most oil supplies (Florini, Dubash 2011:3f.). The concentration of fossil fuel reserves does not only ease price rigging but also endows fuel extracting countries with considerable political power. Fossil fuel importing countries are vulnerable to supply disruptions and price shocks and so depend both economically and politically on fossil fuel exporting countries. The former also face the risk of depleting foreign exchange resources. Consequently, a common claim is to reduce dependence on energy im-

[66] Source: illustration S.R. with data from OECD iLibrary.

ports. Yet, as Florini and Dubash highlight, energy security should not be confused with national energy independence:

> "Energy independence is neither feasible for most countries nor particularly desirable as a goal in itself. Dependence on a world market that functions well is beneficial, not harmful" (Florini, Dubash 2011:3f.).

Well-functioning transboundary trade can enhance energy security and improve energy affordability (Yergin 2006:79). According to Dubash & Florini (2011:8f.), the main task of transboundary policy-making should be to improve the function of world energy markets.

Cherp & Jewell (2011) stress that academic debates on energy security comprise three distinctive perspectives: the political science perspective on sovereignty, the natural science perspective on robustness and the economic perspective on resilience. Whereas the sovereignty perspective identifies sabotage and terrorist attacks, political embargoes and malevolent exercise of market power as main threats, the robustness and resilience perspectives focus on risks such as failures of energy infrastructure, extreme natural events, demand outgrowing supply, resource depletion and market volatility. The proposed strategies to risk minimization vary concordantly: whereas the sovereignty perspective favors measures such as protecting infrastructure, controlling energy systems politically and economically, and switching to domestic fuels and trusted suppliers, the robustness and resilience perspectives propose measures like upgrading infrastructure, adopting safer technologies and switching to more abundant resources. Yet, all three perspectives agree on generic responses such as decreasing energy intensity and diversifying energy sources, for example by promoting regnewables.

The promotion of renewable energy can contribute to energy security in various ways. It can alleviate the increasing scarcity of global energy sources by expanding the energy supply. Therefore, it is seen as an instrument to alleviate energy price increases in the long run. Besides, the production of renewable energy is not limited to single world regions and so offers the possibility for decentralized production, therefore diversifying energy supply and decreasing the dependency on single – often politically instable – countries and regions. It also facilitates the domestic supply of energy and might consequently serve as an instrument to decrease import dependency – which is an important aspect of energy security according to the sovereignty perspective.

3.2.2 Access to Energy

The promotion of renewable energy also contributes to tackling energy poverty.[67] It is often underlined that the potential of renewable energy for small-scale and decentralized deployment is especially suited to improving access to energy. Energy does not only power the world economy, it is also a prerequisite for fulfilling human needs:

> "Energy *per se* is not a need, but it is absolutely essential to deliver adequate living conditions, food, water, health care, education, shelter and employment. For example, energy availability is a key determinant of how and how much food is grown, how food is cooked, the health impacts of how food is cooked or how living spaces are heated, the time required to 'procure' household energy, and so on" (Najam, Cleveland 2005:118).

Yet, one fifth of the world population has no access to electricity. As Figure 4 illustrates, electricity deprivation is especially widespread in Sub-Saharan Africa and South Asia. Dubash and Florini highlight that energy poverty is of central importance to global governance: it is globally prevalent, energy access is crucial for the success of the global anti-poverty agenda and there is huge potential for cross-border learning and mutual support.

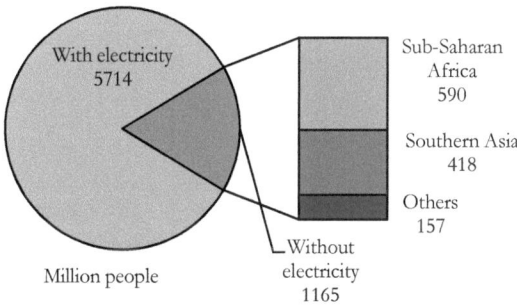

Figure 4: Electrification Access Deficit on a Global Scale, 2010[68]

Yet, as Dubash, Florini (2011:9) note, "although the eradication of poverty is a high-profile international cause, the poverty agenda is seldom directly linked to

---

[67] See, for example, Florini 2008:6, Florini, Sovacool 2011:66f., Karlsson-Vinkhuyzen, Jollands & Staudt 2012:14, Lesage, Van de Graaf & Westphal 2010:22f., Kuik, Bastos Lima & Gupta 2011, Najam, Cleveland 2005:118-120.
[68] Source: illustration S.R. with data from Sustainable Energy for All (2013:38).

energy." The MDGs, for example, do not include a goal on energy access, although energy is central for achieving all MDGs (for further elaboration of this aspect see Kuik, Bastos Lima & Gupta 2011:629). However, the SDGs will change this: they cover energy access as well.

Kuik, Bastos Lima & Gupta (2011) and Lesage, Van de Graaf & Westphal (2010:37f.) criticize that academic discussions on the strategic goals of energy policy-making often mirror the perspectives of developed countries and thus blind out energy poverty. The criticism by Lesage, Van de Graaf & Westphal (2010:37f.) focuses on the strategic triangle which is commonly used to describe the main objectives of energy policy-making and the tensions involved when pursuing all objectives at the same time. The strategic triangle comprises supply security, economic efficiency and environmental sustainability. The authors argue that in order to fully capture energy challenges in developing countries, access to energy must be integrated. Kuik, Bastos Lima & Gupta (2011) argue for an integration of energy access into the concept of energy security. Whereas developed countries do not face problems such as energy deprivation, providing poorer parts of their population with access to energy is a central governance task for developing countries. Kuik, Bastos Lima & Gupta (2011:630) state that energy security should therefore not relate to national security only, but also to human security. The latter comprises energy availability and affordability to the poorest instead of a country as a whole. Following such an understanding of energy security, it would not be enough to look at the national structure of energy availability and use. The quantitative and qualitative distribution of energy within a country/population would also come in (Najam, Cleveland 2005:120).

3.2.3 Environmental Sustainability

Environmental sustainability is the third challenge that the promotion of renewable energy addresses. Dubash and Florini emphasize the importance of environmental sustainability for global energy governance:

> "The rise of environmental sustainability as an important objective of global energy governance is arguably the single most dramatic shift in the global energy landscape over the last two decades" (Dubash, Florini 2011:10).

Bradshaw (2013:51) stresses that using any source of energy supply goes along with negative environmental externalities but the degree of damage caused varies between energy sources. Out of the different environmental sustainability concerns, most authors put special emphasis on climate change (Bradshaw 2013:51-56, Dubash, Florini 2011:10, Florini 2008:5f., Florini, Sovacool 2011:62f. Lesage, Van de Graaf & Westphal 2010:30-32). As Bradshaw (2013:51) emphasizes, climate change distinguishes itself from other environmental externalities since its impacts are not limited to geographic regions but affect the planet as a whole. Energy production and use contributes significantly to global warming; fossil fuel combustion is the principle source of global carbon dioxide emissions. According to OECD/IEA (2013:79), the energy sector is currently responsible for more than two thirds of global greenhouse gas emissions. Between 1990 and 2011, worldwide energy-related cabon dioxide (CO2) emissions increased by almost 50 percent (see Figure 5). Substituting fossil fuels with renewable energy offers the possibility of significantly reducing greenhouse gas emissions and thus contributes to climate change mitigation.

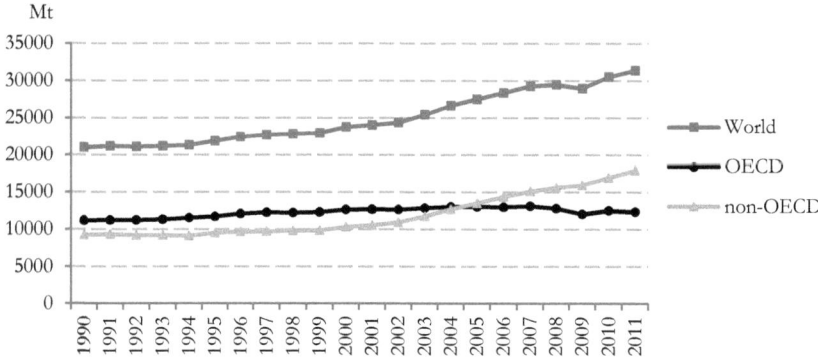

Figure 5: Global Energy-Related Carbon Dioxide Emissions, 1990-2011[69]

Environmental sustainability concerns do not only relate to climate change. Energy production and use have further negative externalities that do not respect borders: fossil fuel combustion leads to air pollution and acid rain which damage ecosystems and negatively affect human health. Extracting and transporting oil contaminates the marine environment – not only in times of crisis, for example during oil spills, but also through normal operations. Catastrophes such as nuclear meltdowns, coal mine collapse, natural gas explosions and dam breaches create huge and long-lasting environmental problems (Dubash, Florini 2011:10, Florini 20085f., Florini, Sovacool 2011:62f., Najam, Cleveland 2005:115f.). According to Florini and Sovacool, these negative environmental externalities make substituting fossil fuels an imperative:

---

[69] Source: illustration S.R. with data from OECD iLibrary.

"Yet even if climate change were not an issue, current fossil fuel dependence would still impose costly, often cross-border, environmental externalities. For some fuels such as coal and oil the negative externalities associated with their use are even greater than the existing market price for the energy they produce" (Florini, Sovacool 2011:62f.).

Thus, the promotion of renewable energy shall also reduce the negative environmental externalities of energy usage.

Nonetheless, the environmental sustainability of renewable energy is also contested. Particularly biofuels and large hydropower meet environmental criticism (Florini 2008:5f., Hirschl 2009:4407f., Lesage, Van de Graaf & Westphal 2010:41, Najam, Cleveland 2005:115f.). The production of biofuels might cause soil degradation and deforestation, or displace food production. The construction of large dams for hydropower generation can lead to serious interventions in the stability of ecosystems. And neither are solar and wind energy environmentally harmless: solar photovoltaic cells contain toxic substances and wind farms pose dangers to bird life (Florini 2008:5f., Lesage, Van de Graaf & Westphal 2010:41).

3.2.4 Trade-Offs Involved

There are substantial trade-offs between the three objectives of energy security, energy access and environmental sustainability. The energy source which is the most affordable in the short run, for example, is often not the one that causes less environmental externalities. This leads to trade-offs between energy security and energy access on the one hand and environmental sustainability on the other. Besides, improving energy access means increasing overall energy demand which has negative impacts on energy security and environmental sustainability (Newell, Phillips & Mulvaney 2011:2f.).

These trade-offs are important for global energy governance as not all objectives are necessarily achievable at the same time and different countries may prioritize their energy objectives differently. Lesage, Van de Graaf & Westphal (2010:40), for example, argue that for developing countries, access to affordable energy is typically more important than reducing greenhouse gases. Developed countries, in contrast, are more likely to focus on environmental sustainability. The authors base their argument on the pyramid of Maslow: humans first of all must fulfill their basic needs before thinking about higher ones. As energy poverty deprives basic needs, securing access to energy is the top priority of energy policymaking in every country. The uninterrupted availability of affordable energy – energy security – comes at the second step. Only then follows environmental sustainability because environmental protection tends to be regarded as a higher need. Newell, Phillips & Mulvaney (2011:68) point out that the current form of global energy governance is inclined to prioritize energy security while paying less attention to access to energy and environmental protection.

Newell, Phillips & Mulvaney (2011) bring another aspect in that strongly affects how global governance deals with the trade-offs in energy policy-making. They emphasize that the distribution of costs and the benefits of energy production and usage varies significantly between people and places and that the unequal distribution of economic, environmental and social outcomes is a result of power asymmetries in the decision-making processes at stake. They express "skepticism about the abundance of simple win-win solutions for tackling energy in a highly unequal world" (Newell, Phillips & Mulvaney 2011:21) and call for strong, participatory and transparent institutions to manage inevitable trade-offs in energy policy-making. According to the authors, decision-making processes at the global level are in need of major reform:

> "It is only through fundamentally changing *how* energy policy is made and *by* and *for whom* that we can expect some of the distributional inequities in access and exposure to the consequences of unsustainable energy production and use to be addressed" (Newell, Phillips & Mulvaney 2011:4).[70]

They emphasize that procedural justice is fundamental to achieve distributive justice.

Kuik, Bastos Lima & Gupta (2011:627) add that goal conflicts are especially severe in developing countries. The lack of adequate resources in these countries often imposes trade-offs in terms of who gets access to energy and in terms of considering the negative externalities of using specific energy sources. Besides, the trade-offs are also inherent in the international demands developing countries are confronted with when expanding their energy base:

> "Industrialized countries are increasingly implicitly questioning the right of developing countries to use fossil fuels because of its implications for climate change; or to build large dams because of ecological and social security concerns or expand nuclear energy because of its potential security implications" (Kuik, Bastos Lima & Gupta 2011:627).

At the same time, developing countries often lack the capital and expertise to invest in renewable energy (Cherp, Jewell & Goldthau 2011:76, Kuik, Bastos Lima & Gupta 2011:627). According to Karlsson-Vinkhuyzen, Jollands & Staudt (2012:16), the initial costs of developing and implementing renewable energy technologies form a major barrier to renewable energy deployment in poor countries. The author therefore claims that global governance activities should facilitate the transfer of knowledge and resources to these countries.

To a certain degree, trade-offs between the different objectives of energy policy-making can be altered as they are the results of technological conditions and political decisions (Dubash, Florini 2011:11). It is, for example, the failure to inter-

---

[70] The italics are part of the original quotation.

nalize full costs of energy usage which creates an apparent trade-off between affordability and environmental sustainability. If costs covered environmental externalities, polluting energy sources would not appear as low-cost alternatives. The promotion of renewable energy can contribute to ameliorating trade-offs in energy policy-making. It offers the potential to address the three challenges of energy security, energy access and environmental sustainability at the same time.

3.3 Structural Characteristics

In the academic literature on global energy governance, scholars commonly characterize transboundary policy-making on energy as being highly fragmented and relatively weakly developed, with national policy-making dominating. This section illustrates these two structural characteristics which affect global governance not only on renewable energy, but on energy in general.

3.3.1 Dominance of National Policy-Making

Traditionally, policy-making on energy issues has been regarded as merely a national task. Several authors emphasize that energy supply is central for economic development and military strength; as a consequence, energy is widely considered a strategic good that is crucial for the survival of a state and its political power in international relations (Gupta 2012:429,447, Hirschl 2009:4408f., Karlsson-Vinkhuyzen 2010:175, Karlsson-Vinkhuyzen, Jollands & Staudt 2012:11f., Lesage, Van de Graaf & Westphal 2010:183). It is this centrality of energy issues for the pursuit of national interests that leads to an unwillingness of governments to engage in transboundary cooperation on energy, as the following two statements point out:

> "There are great interests involved. All this implies that, generally, states do not tolerate that their energy sovereignty is compromised. States do not want to share power or even transfer the control over their national energy path" (Lesage, Van de Graaf & Westphal 2010:183).

> "Energy governance is an area where national security interests are more explicitly prioritized over global issues, and the nature of the explicit recognition of these interests has implied that countries are unwilling to promote and support a comprehensive global governance framework on energy" (Kuik, Bastos Lima & Gupta 2011:632).

Within International Relations research, energy issues have often been approached from a classical Realist perspective, concentrating on energy security and assuming a close relationship between fossil fuel supply, energy reserves and geopolitics (Cherp, Jewell & Goldthau 2011:76, Goldthau 2011:213). As a consequence, energy has been framed in the context of competition and power politics in zero-sum-games rather than in the context of transboundary cooperation and mutual gains. The predominant geopolitical view on energy in International Relations concentrates on fossil fuels, as these are still of primary importance to global energy supply. Issues addressed include battles over access to energy resources, nuclear energy and non-proliferation, the danger of terrorist attacks on energy infrastructure, the revival of energy mercantilism associated with China's engagement in Africa, and energy as a foreign policy tool for power politics (Cherp, Jewell 2011:1-3, Florini, Sovacool 2011:59-62, Goldthau 2011:214).

Global energy governance is weakly developed in comparison to related fields such as climate policy, development cooperation or trade.[71] According to Lesage, Van de Graaf & Westphal (2010:1) this applies even more to renewable energy sources than to fossil fuels as global energy governance has been mainly focused on fossil fuels. Karlsson-Vinkhuyzen (2010:175) even speaks of "almost a normative and institutional vacuum on energy" in global governance.

Political processes at the national level are an essential building block of global energy governance (Bastos Lima, Gupta 2013:58, Dubash 2011, Dubash, Florini 2011:13, Karlsson-Vinkhuyzen, Jollands & Staudt 2012:17, Kong 2011). It is often underlined that national processes dominate energy policy-making. According to Dubash & Florini (2011:16), national energy policy-making is not effectively linked to global institutions, resulting in an "extreme disconnect between national policy making and the transnational governors and mechanisms". The dominating national focus of energy policy-making prevents many policymakers from perceiving its transboundary dimension. Energy policymakers (beyond those that are directly engaged in transboundary governance) rarely notice the activities of international organizations and other global governors (Bastos Lima, Gupta 2013:58, Dubash, Florini 2011:16). Exceptions to this are the World Bank and other multilateral development banks which have considerably influenced national energy policy-making (Dubash, Florini 2011:14). According to Dubash & Florini (2011:13), "understanding global energy governance requires treating the boundary between domestic and international politics as porous" – even though in practice, the national and global levels are often perceived as disconnected. In order to grasp how domestic politics shape and are shaped by policy-making at the global level, it is crucial to study in depth the particularities around national policy-making. States face very different national contexts of energy policy-making that are strongly influenced by different resource endowments and geographic conditions (Dubash, Florini 2011:16, Kuik, Bastos Lima & Gupta 2011:632). Consequently, they are affected by transboundary relations in different ways: countries with a high energy import dependency have, for example, different concerns than energy exporting countries, and those with rich endowments of fossil fuels have different preferences than

---

[71] See, for example, Florini 2008:2, Florini, Sovacool 2011:69f., Gupta 2012:442, Karlsson-Vinkhuyzen, Jollands & Staudt 2012:11, Lesage, Van de Graaf & Westphal 2010:182f., Newell, Phillips & Mulvaney 2011, Van de Graaf, Westphal 2011:19f..

countries that need to develop renewable energy in order to use domestic energy sources.

In addition, transboundary policy-making needs to consider national sovereignty concerns, as some countries fear that multilateral processes will limit their freedom too much in deciding on their energy priorities (Karlsson-Vinkhuyzen 2010:183, Steiner et al. 2006:160). According to Kuik, Bastos Lima & Gupta (2011:627), this particularly applies to developing countries. These are most affected by global energy governance as they are, for example, confronted with the environmental concerns of industrialized countries when expanding their energy base. In line with this, Lesage, Van de Graaf & Westphal (2010:85f.) identify a major cleavage in global energy governance: while industrialized countries tend to favor the expansion of global governance, emerging and developing countries commonly raise sovereignty concerns.

3.3.2 Fragmentation

Energy is a cross-cutting issue of transboundary policy-making. It comprises very different policy problems which may appear as unrelated (Florini 2008:2). Besides, it interacts with different issue areas of global governance, such as security, environment, trade and trade-related issues, and development cooperation (Florini, Sovacool 2011:57f., Goldthau, Sovacool 2012:232). Within these issue areas, energy is often not regarded as a priority. Climate policy serves as a good example. Although international climate policy is an important driver behind the promotion of renewable energy, the support is only indirect: the central documents of international climate policy, the UN Framework Convention on Climate Change (UNFCCC) and the Kyoto Protocol, hardly even mention renewable energy (see also Hirschl 2009:4409-4411). Actors in related issue-areas might not even be aware of the side-effects their action has on energy issues. Therefore, decisions aiming at other goals often shape energy in an uncoordinated and incomplete way (Florini, Dubash 2011:1f.).

Global energy governance is repeatedly characterized as highly fragmented, with different aspects dealt with by different bureaucratic silos and analyzed by separate groups of experts and policy-makers. Several authors emphasize that the fragmented structure of global energy governance leads to an increased need for coordination.[72] Yet, different understandings, languages and expertise make com-

---

[72] See, for example, Dubash, Florini 2011:11, Florini, Sovacool 2009:5239, Florini, Sovacool 2011:57f., Goldthau, Sovacool 2012:237, Lesage, Van de Graaf & Westphal 2010:51, Newell, Phillips & Mulvaney 2011, Westphal, Röhrkasten 2013:40.

munication between different actors and governance spheres difficult (Dubash, Florini 2011:11).

Fragmentation occurs along two major lines: between different energy sources and between different issue areas/policy problems. Global governors and global governance institutions rarely cover the whole range of energy sources but instead concentrate on selected energy sources (see, for example, Dubash, Florini 2011:16, Florini, Dubash 2011:1f., Van de Graaf, Westphal 2011:19f., Westphal, Röhrkasten 2013:40). The intergovernmental organizations that exclusively deal with energy issues, for example, focus on a different subgroup of energy sources: IAEA concentrates on nuclear energy, OPEC on oil and IRENA on renewable energy. The only exception to this is the IEA which covers all energy sources but is often confronted with criticism that it over-represents fossil fuel interests (Lesage, Van de Graaf & Westphal 2010:69, Röhrkasten, Westphal 2013:12,14, Röhrkasten, Westphal 2012b:3, Van de Graaf 2012a, Van de Graaf 2012b:8). As a consequence, there is no holistic approach to energy-related issues in global governance:

> "To date, there is no single venue for international discussions on energy issues. There is no World Energy Organization. Instead, global energy governance is fragmented and dispersed into a patchwork of various overlapping and sometimes overtly competing organizations" (Lesage, Van de Graaf & Westphal 2010:51).

According to Florini & Sovacool (2011:69f.) and Dubash & Florini (2011:11), global governors are therefore failing to adequately address complex and interconnected energy challenges, including the trade-offs between different objectives of energy policy-making.

Cherp, Jewell & Goldthau (2011) state that global energy governance is characterized by largely autonomous governance arenas instead of a coherent framework for decision-making. They identify three governance arenas which each focus on one major challenge in energy policy-making: energy security, energy access and climate protection. Table 6 gives a comprehensive overview of these three arenas in global energy governance.

Energy security is the oldest – and still the dominating – arena in global energy governance. It focuses on a stable and secure global energy supply, and regards oil market stability as a primary goal. Its origin goes back to the oil crises of the 1970s. Within this arena, energy is primarily regarded as a national security concern for importing states while it is seen as a foreign policy issue for exporting states. The main paradigm in this governance arena is the sovereignty of nation states and the need to establish an arrangement that guarantees mutual energy security. The IEA and OPEC are central global governors in this arena. The G8 and the International Energy Forum are among the further relevant global governors. This governance arena weakly relates to issues in the realm of trade and security. Its main instruments include binding agreements on production quotas (OPEC) and on strategic petroleum stocks (IEA), as well as the gathering and the dissemination of oil data (Cherp, Jewell & Goldthau 2011:81).

	Energy security	Energy access	Climate protection
Central goal	Stable and secure global energy supply	Increase access to modern forms of energy	Reduce GHG emissions from energy usage
Historic origin	Oil crises of the 1970s	Development movement, 1960s-2000s	Environmental sustainability, 1990s-2000s
Main paradigm	Sovereign nation states acting in their self-interest establish arrangements guaranteeing mutual energy security	North-South assistance shall enhance development in developing countries	Concerted international action motivated by shared global goals can deal with 'common bads' such as climate change
Major actors	Major exporters and importers of energy (nation states) and their alliances (e.g. IEA, OPEC)	International development organisations (e.g. World Bank, UNDP), donor agencies, NGOs	Nation states (environmental ministries), IGOs (e.g. UNFCCC, IPCC, UNEP), NGOs, business networks
Interface with related arenas	Weakly related to trade and security	Strong link to poverty alleviation	Strong link to climate protection
Main instruments	Binding agreements on output control (OPEC), and on strategic petroleum stocks (IEA), oil data gathering and dissemination	Decentralized technical interventions, capacity building, financial and technical assistance, information exchange	Legally binding agreements on emissions reductions, financial and technical assistance, capacity building, formal information dissemination

Table 6: Arenas in Global Energy Governance[73]

The second arena in global energy governance concentrates on increased access to modern forms of energy in the developing world (on the intersection between global energy governance and development policy see also Carbonnier, Brugger 2013). Its roots are in the development movement that started during the 1960s. The basic paradigm of this governance arena is the necessity of North-South assistance to enhance development. International development organizations such as the

---

[73] Source: adapted version of Cherp, Jewell & Goldthau (2011:84).

World Bank and the UNDP are central global governors in this arena which strongly interfaces with transboundary cooperation on poverty alleviation. Other relevant actors include the aid agencies of donor countries and large transboundary NGOs. Whereas development cooperation on energy traditionally focused on large-scale infrastructure projects such as hydroelectric dams, it now concentrates on decentralized technical solutions, especially in the realm of access to electricity and modern cooking facilities. Its instruments cover transfers of financial and technical resources, capacity-building and information exchange (Cherp, Jewell & Goldthau 2011:81f.).

Climate protection is the third arena in global energy governance (on the intersection between climate protection and global energy governance see also Zelli et al. 2013). Its main goal is the reduction of energy-related greenhouse gas emissions. Its emergence goes back to the Rio Earth Summit in 1992 and the subsequent international climate protection efforts, especially under the UNFCCC. This governance arena builds on the post-Cold War paradigm that concerted transboundary action to address global bads such as climate change is both needed and feasible. Whereas energy is not the primary focus of this governance arena, "it is the arena with the clearest and most ambitious goal of systemic energy transition" (Cherp, Jewell & Goldthau 2011:82). The main actors in this governance arena are states and international organizations such as the UNFCCC, the Intergovernmental Penal on Climate Change and the United Nations Environment Program (UNEP). Large transboundary NGOs and private sector networks also play an important role. It is worthwhile to note that actors in this arena usually have a stronger focus on environmental issues than on energy issues. The UNFCCC focal points in member states are for example located in environment ministries, not in energy ministries. This governance arena encompasses a variety of governance instruments, such as legally binding commitments to reduce greenhouse gases, financial and technical assistance, capacity building and the formal dissemination of information (Cherp, Jewell & Goldthau 2011:82f.).

Cherp, Jewell & Goldthau (2011:85) emphasize that the complexity of global governance arrangements in the realm of energy is not a problem but rather a necessary condition for success – if the arrangements fit the problems they address. Yet, global governors and global governance arrangements in all three arenas address energy challenges in isolation, despite inherent links:

> "The most fundamental problem is that there are surprisingly few links between the three governance arenas. They are virtually isolated from each other, each with its own distinct goals, actors and mechanisms. Due to the lack of interconnection, the existing global energy governance fails to address the major energy challenges in an integrated manner" (Cherp, Jewell & Goldthau 2011:83).

Whereas the focus on short-term market stability in the realm of energy security fails to address long-term challenges such as rising global energy demand and increasing concentration of fossil fuel production, global governance efforts in the realm of climate protection do not effectively influence the policy agenda of key energy actors. Governance in the realm of energy access remains marginal in both the energy sector and in development cooperation.

3.4 Global Governors and their Governance Activities

Global governors take up much space in the research on global energy governance.[74] This stands in a clear contrast to theoretical global governance research which often builds on structure-centered approaches without paying much attention to the actors involved. As outlined in the theoretical part of my thesis, global governors steer behavior and affect social understandings across borders. They may do so in different ways: creating issues, setting agendas, establishing and implementing rules and programs, evaluating outcomes and reacting to non-compliance (see, for example, Avant, Finnemore & Sell 2010c:2). The following section presents the main global governors in transboundary policy-making on renewable energy and briefly illustrates their institutional set-up and governance activities. The literature on global energy governance typically focuses on intergovernmental organizations as the main global governors (see also Goldthau 2013b:2). However, the following deliberations also include further transboundary organizations with a transcontinental scope that are relevant actors in global renewable energy governance. These organizations involve nation states as members, but are not limited to them.

3.4.1 IRENA

IRENA is a global governor of primary importance in global renewable energy governance as it is the only intergovernmental organization that exclusively focuses on the promotion of renewable energy. It has met with increasing interest in academic discussions.[75]

IRENA's main objective is to "promote the widespread and increased adoption and the sustainable use of all forms of renewable energy" (Conference on the Establishment of the International Renewable Energy Agency 2009:Art.II). According to IRENA's mid-term strategy, being "neutral/unbiased among all renewable

---

[74] Some studies provide a comprehensive mapping on global governors. For global renewable energy governance see, for example, Suding, Lempp 2007 and Steiner et al. 2006, and for global energy governance in general (Florini 2008, Florini, Sovacool 2009, Lesage, Van de Graaf & Westphal 2010:51-72. Others scholars focus on individual global governors.
[75] See, for example, Cherp, Jewell & Goldthau 2011:82, Hirschl 2009:4413f., Karlsson-Vinkhuyzen, Jollands & Staudt 2012:16, Lesage, Van de Graaf & Westphal 2010:68f., Najam 28.10.2010, Röhrkasten, Westphal 2013, Röhrkasten, Westphal 2012b, Van de Graaf 2013a, Van de Graaf 2013b, Van de Graaf 2012a.

technologies" is a fundamental value of its work (IRENA 28.10.2011:13). IRENA's activities have a global scope; it explicitly aims at supporting both developing and developed countries (IRENA 28.10.2011:13). 124 states and the EU are IRENA members. Further 40 states have signed IRENA's statutes but not yet ratified them.[76] In 2013, IRENA annual budget was of $29.7 million, comprising a core budget of $18 million and voluntary contributions.[77] Voluntary contributions mainly come from IRENA's host countries, the UAE and Germany. The former hosts IRENA's headquarters while the IRENA Innovation and Technology Centre is located in Germany. IRENA's staff comprises around 70 people.[78] Its institutional set-up resembles UN structures.

IRENA seeks to serve three main functions in global renewable energy governance: to be the global voice and knowledge base for the worldwide use of renewable energy, advise its member states on renewable energy, and serve as a network hub for transboundary cooperation (IRENA 28.10.2011).[79] IRENA builds on the principle of voluntariness. It does not work on binding commitments or political declarations but offers knowledge services and policy advice on a voluntary basis.[80] Although IRENA's activities cover both developing and developed countries, there has been a stronger focus on developing countries, as IRENA identifies the highest need for support in these countries (IRENA 28.10.2011:16f.). Despite IRENA's declared goal to be neutral between different renewable energy technologies, its activities have so far privileged the power sector.[81] IRENA's Director-General, Adnan Amin, furthermore acknowledges that biomass has been an "underdeveloped" area of IRENA's work program.[82] IRENA's representatives repeatedly underline that it does not serve as a financing or implementing agency, but concentrates on policy advice and knowledge services.[83] Yet, with regard to financ-

---

[76] IRENA, IRENA membership, http://www.irena.org/Menu/Index.aspx?mnu=Cat&PriMenuID=46&CatID=67 (accessed February 14, 2014).
[77] IRENA, Work Programme and Budget for 2013. Report of the Director-General, http://www.irena.org/DocumentDownloads/WP2013.pdf (accessed February 18, 2014).
[78] IRENA, Annual Report 2012, <http://www.irena.org/DocumentDownloads/4thCouncil/C_4_2_Annual%20Report%202012.pdf> (accessed November 18, 2013).
[79] Presentation by Francisco Boshell, IRENA, at the conference "International Governance for Renewable Energy and the Role of Emerging Powers", Berlin, June 21, 2012; interview with executive staff, IRENA, August 10, 2012.
[80] Presentation by Francisco Boshell, IRENA, at the conference "International Governance for Renewable Energy and the Role of Emerging Powers", Berlin, June 21, 2012; interview with executive staff, IRENA, August 10, 2012.
[81] Interview with executive staff, IRENA, August 10, 2012.
[82] Interview with Adnan Amin by Thijs van de Graaf, European Energy Review, 24 May 2012, http://www.europeanenergyreview.eu/site/pagina.php?id=3707 (accessed February 14, 2014).
[83] Presentation by Francisco Boshell, IRENA, at the conference "International Governance for Renewable Energy and the Role of Emerging Powers", Berlin, June 21, 2012; interview with executive staff, IRENA, August 10, 2012.

ing, IRENA has so far taken an ambiguous stance as it cooperates with the Abu Dhabi Fund for Development in the financing of renewable energy projects in developing countries.[84]

IRENA's activities are structured around three areas. The Knowledge, Policy and Finance Centre serves as "a global knowledge repository and center of excellence for renewable policy and finance issues".[85] It engages in statistical work and analyses renewable energy policies, investment frameworks and the impacts of renewable energy deployment. The Country Support and Partnership Division is responsible for policy advice. Among its core activities are the Renewables Readiness Assessments which aim at improving the framework conditions for renewable energy deployment in developing countries.[86] The IRENA Innovation and Technology Centre analyzes renewable energy technologies and innovations. It publishes cost data relating to different renewable energy technologies and develops technology roadmaps.[87]

3.4.2 IEA

The IEA aims at ensuring "reliable, affordable and clean energy for its 28 member countries and beyond".[88] It is widely considered a central actor in global energy governance and also receives much attention in academic discussions.[89] Its activities cover the whole range of different energy sources. As such, it works on renewable energy as well as on conventional sources of energy. The IEA comprises 27 member states which all belong to the OECD. It also cooperates with non-IEA member states. The voting rights of member states in IEA's Governing Board are distributed according to the countries' net oil imports in 1973. Thus, the USA is – by far – endowed with most voting rights; its voting rights amount to more than one fourth of the total.[90] The organization has an annual budget of around €27 million (2012) and around 260 people work at its secretariat in Paris.[91]

---

[84] IRENA, "IRENA/ADFD Project Facility", http://www.irena.org/menu/index.aspx?mnu=Subcat&PriMenuID=35&CatID=109&SubcatID=159 (accessed July 11, 2013).
[85] IRENA, Knowledge, Policy and Finance Centre, http://www.irena.org/menu/index.aspx?mnu=cat&PriMenuID=35&CatID=109 (accessed March 06, 2014).
[86] IRENA, Country Support and Partnerships, http://www.irena.org/menu/index.aspx?mnu=cat&PriMenuID=35&CatID=110 (accessed March 06, 2014).
[87] IRENA, IRENA Innovation and Technology Centre, http://www.irena.org/menu/index.aspx?mnu=cat&PriMenuID=35&CatID=112 (accessed March 06, 2014).
[88] IEA, About us, http://www.iea.org/aboutus/ (accessed November 14, 2013).
[89] On IEA see, for example, Colgan 2009, Colgan, Van de Graaf, 2014, Figueroa 2012, Florini, Sovacool 2009:5242f., Jacoby 2009, Lesage, Van de Graaf 2013:86-90, Leverett 2010, Van de Graaf 2012b).
[90] Own calculations based on Colgan, Keohane & Van de Graaf (2011). After the USA (which counts with 46 votes) follow Japan (17 votes), Germany (11 votes), France (9 votes), Italy and the UK (with 8

The IEA considers itself to be "at the heart of global dialogue on energy, providing authoritative and unbiased research, statistics, analysis and recommendations."[92] Whereas IEA's emergency mechanisms concentrate on fossil fuel shortages, its activities in the realm of research and policy advice include conventional energy sources as well as renewable energy. IEA publications such as the World Energy Outlook and Key World Energy Statistics are worldwide renowned as the leading informational sources on world energy markets. When elaborating on IEA's contribution to global energy governance, Newell, Phillips & Mulvaney (2011:35) particularly highlight IEA's provision of energy data. The IEA regularly conducts peer-reviews of the energy policies of its member states and traces the implementing progress of its policy recommendations. From time to time, the energy policies of non-member countries are reviewed as well (Colgan 2009).

While the IEA claims that it provides unbiased research and analysis, it is sometimes criticized for favoring fossil fuels in its research and policy advice (Lesage, Van de Graaf & Westphal 2010:69, Röhrkasten, Westphal 2013:12,14, Röhrkasten, Westphal 2012b:3, Van de Graaf 2012a, Van de Graaf 2012b:8). IEA scenarios lack transparency concerning modelling and underlying assumptions. Its findings have to be approved by its member states and among the IEA members are some governments that have clear interests in fossil fuels (Röhrkasten, Westphal 2013:12,14, Röhrkasten, Westphal 2012b:3). In line with this, Lesage, Van de Graaf & Westphal (2010:69) point out that IEA's role in the worldwide promotion of renewable energy is seen as ambivalent, as some fear that the organization might hinder a rapid expansion of renewable energy due to its overrepresentation of fossil fuel interests.

IEA's Renewable Energy Division, which is part of the Energy Market and Security Directorate, collects renewable energy data and provides analysis and reporting on renewable energy technologies, policy developments and issues of market integration.[93] It develops roadmaps on different renewable energy technologies and since 2012 publishes a yearly Medium Term Renewables Energy Market Report.[94] This report tracks deployment trends of renewable energy and forecasts

---

votes each) and Canada (7 votes). All other IEA members are endowed with three to five votes (Belgium and Luxembourg form one party). See Colgan, Keohane & Van de Graaf 2011.
[91] IEA FAQs: Organization and Structure, http://www.iea.org/aboutus/faqs/organisationandstructure/ (accessed June 6, 2012).
[92] IEA, What we do, http://www.iea.org/aboutus/whatwedo/ (accessed February 14, 2014).
[93] See IEA, Working Party on Renewable Energy Technologies, Strategic Plan 2013-2015, http://www.iea.org/media/aboutus/standinggroups/rewp_strategy.pdf (accessed November 14, 2013) and IEA, Working Party on Renewable Energy Technologies, Mandate 2013-2015, http://www.iea.org/media/aboutus/standinggroups/rewp_mandate.pdf (accessed November 14, 2013).
[94] According to interviews with BMU staff, this report can be seen as an IEA-internal rival to the World Energy Outlook, as its forecasts contain more ambitious projections on renewable energy expansion than the World Energy Outlook.

market developments over a period of five years. The Renewable Energy Working Party – made up by officials of member country governments and the European Commission – advises IEA bodies on renewable energy issues and guides the work of the Renewable Energy Division, also reviewing and assessing its studies and reports. The Renewables Energy Division and the Renewable Energy Working Party oversee the work of the IEA, implementing agreements that deal with renewable energy. The IEA implementing agreements facilitate cooperation between interested member and non-member countries, international organizations, private sector representatives and NGOs, focusing on energy research, deployment and demonstration. Implementing agreements on renewable energy cover bioenergy, geothermal energy, hydropower, ocean energy, photovoltaic power, solar heating and cooling, solar power and chemical energy systems, as well as wind energy.[95] The Implementing Agreement on Renewable Energy Technologies addresses crosscutting issues that influence renewable energy deployment. The Renewable Energy Division also contributes to IEA's International Low-Carbon Energy Technology Platform which, according to the IEA, is its "chief tool for multilateral engagement on clean technologies between its member and partner countries, the business community and international organizations."[96] The platform enhances the sharing of best practice and creates guidance on technology roadmaps. The IEA furthermore counts with a Renewable Industry Advisory Board, representing the renewable energy industry.[97]

3.4.3 REEEP

The multistakeholder partnership REEEP aims at strengthening markets for renewable energy and energy efficiency.[98] It was initiated by the British government in Johannesburg 2002. REEEP counts 385 partners, comprising governments, NGOs and private sector representatives. Among the REEEP partners are 45 governments.[99] Consequently, the vast majority of REEEP partners are non-governmental

---

[95] IEA, Implementing Agreements, http://www.iea.org/countries/membercountries/australiatestpage/implementingagreements/ (accessed November 14, 2013).
[96] IEA, International Low-Carbon Energy Technology Platform, http://www.iea.org/aboutus/affiliatedgroups/low-carbonenergytechnologyplatform/ (accessed November 14, 2013).
[97] See IEA, Working Party on Renewable Energy Technologies, Strategic Plan 2013-2015, http://www.iea.org/media/aboutus/standinggroups/rewp_strategy.pdf (accessed November 14, 2013) and IEA, Working Party on Renewable Energy Technologies, Mandate 2013-2015, http://www.iea.org/media/aboutus/standinggroups/rewp_mandate.pdf (accessed November 14, 2013).
[98] On REEEP see, for example, Florini 2008:15, Florini, Sovacool 2009:5245f., Newell, Phillips & Mulvaney 2011:37f., Steiner et al. 2006, Suding, Lempp 2007, Szulecki, Pattberg & Biermann 2010.
[99] REEEP, REEEP Partners, http://www.reeep.org/partners (accessed February 18, 2014).

actors. Next to REEEP's international secretariat located in Vienna, it has four regional secretariats covering Southern Africa, Latin America and the Caribbean, South Asia and East Asia and a focal point in West Africa.[100] It is funded by voluntary contributions from governments and industry associations.[101] Most funding comes from Austria, Germany, IRENA, Norway, OPEC Fund for International Development (OFID), Switzerland and the UK. In 2012/13, REEEP received donations in the amount of €4.5 million.[102]

REEEP aims at "accelerating clean energy in developing countries and emerging markets."[103] In the past, REEEP initiated and funded projects with high potential for mass replication, supporting governments in establishing favorable regulatory and policy frameworks as well as encouraging innovative finance and business models.[104] Nowadays, REEEP concentrates on private sector development. It identifies business opportunities to increase access to clean energy and cooperates with actors in the public and private sector to implement these opportunities. It mobilizes public and private funding for energy projects and provides expertise on information and knowledge management to facilitate exchange of climate and energy data. Besides this, REEEP supports policy-maker networks such as the Sustainable Energy Regulation Network (SERN) and Renewable Energy and International Law (REIL).[105]

3.4.4 REN21

REN21 is a multistakeholder policy network that aims at promoting favorable policies for renewable energy (on REN21 see Florini 2008:15, Hirschl 2009, Newell, Phillips & Mulvaney 2011:38f., Suding, Lempp 2007, Szuleki, Pattberg & Biermann 2010). It was initiated by the German government at the Renewables 2004 Conference in Bonn. Its steering committee, which is the central decision-making body of

---

[100] REEEP, Regional Secretariats, http://www.reeep.org/regional-secretariats (accessed February 18, 2014).
[101] The REEEP website lists the following donors: Australian Clean Energy Council (CEC), Climate and Development Knowledge Network (CDKN), the governments of Australia, Germany, Norway, Switzerland and the UK as well as the OPEC Fund for International Development (OFID). REEEP, Donors, http://www.reeep.org/donors (accessed February 18, 2014).
[102] REEEP, Annual Report 2012/13, http://www.reeep.org/reeep-annual-report-2012-13 (accessed February 18, 2014), page 28.
[103] REEEP, REEEP leaflet, http://www.reeep.org/reeep-leaflet-a5-format (accessed February 18, 2014).
[104] REEEP, Executive Summary "REEEP at a Glance", http://www.reeep.org/48/about-reeep.htm (accessed September 24, 2012).
[105] REEEP, Annual Report 2012/13, http://www.reeep.org/reeep-annual-report-2012-13 (accessed February 18, 2014); REEEP, REEEP leaflet, http://www.reeep.org/reeep-leaflet-a5-format (accessed February 18, 2014).

REN21, comprises representatives from governments, intergovernmental organizations, industry, science and academia and NGOs.[106] According to Newell, Phillips & Mulvaney (2011:39), the annual budget of REN21 comprises around $1 million. It is predominantly financed by the German government.[107] A small secretariat, located in Paris and provided by the Deutsche Gesellschaft für Internationale Zusammenarbeit (GIZ) and UNEP, is responsible for daily work.[108]

REN21 aims at "providing policy-relevant information and research-based analysis on renewable energy to decision makers, multipliers and the public to catalyse policy change".[109] It furthermore serves at a platform to connect various stakeholders working on renewable energy. Its main policy instrument is the annual publication of the Renewables Global Status Report. REN21 characterizes the background of this publication as follows:

> "It grew out of an effort to comprehensively capture, for the first time, the full status of renewable energy worldwide. The report also aimed to align perceptions with the reality that renewables were playing a growing role in mainstream energy markets and in economic development" (REN21 2012:6).

Further publications comprise regional status reports on China, India and the Middle East and North Africa[110] and the 'Renewables Global Futures Report' (REN21 2013b) which explores long-term trends of renewable energy deployment. REN21 also provides a number of web-based tools.[111] In addition, it co-organizes, together with the host countries, the International Renewable Energy Conferences (IREC).[112]

---

[106] REN21, The Steering Committee, http://www.ren21.net/AboutREN21/SteeringCommittee/tabid/5018/Default.aspx (accessed September 24, 2012).
[107] Interview with former staff member, Division International and European Affairs of Renewable Energy, BMU, November 2012.
[108] REN21, Secretariat, http://www.ren21.net/AboutREN21/Secretariat/tabid/5020/Default.aspx (accessed September 24, 2012).
[109] REN21, About REN21, http://www.ren21.net/AboutREN21.aspx (accessed February 18, 2014).
[110] REN21, Regional Status Reports, http://www.ren21.net/REN21Activities/RegionalStatusReports.aspx (accessed February 18, 2014).
[111] The Renewables Interactive Map tracks the worldwide development of renewable energy technology and policy (http://map.ren21.net/, accessed February 19, 2014). Together with REEEP, REN21 provides the Information Gateway for Renewable Energy and Energy Efficiency reegle, covering policies, regulation and financing on renewable energy and energy efficiency (http://www.reegle.info/, accessed February 19, 2014).
[112] REN21, IREC Selection Guidelines, http://www.ren21.net/REN21Activities/IRECs/IRECSelectionGuidelines.aspx (accessed February 19, 2014).

3.4.5 GBEP

GBEP was initiated by the G8+5 countries in Gleneagles 2005 to facilitate transboundary cooperation on bioenergy (on GBEP see Bastos Lima, Gupta 2013:52,56, Lesage, Van de Graaf 2013:88, Scarlat, Dallemand 2011:1637, Suding, Lempp 2007). Since this date, the G8 has continuously declared its support for GBEP and expanded its mandate. 23 countries and 14 international organizations and institutions are GBEP partners, a further 26 countries and 11 international organizations serve as observers.[113] GBEP is registered as a CSD partnership.[114] The GBEP secretariat, which is responsible for daily work, is hosted by the Food and Agricultural Organization (FAO) and mainly funded by the Italian government. Additional funding is provided by Brazil, Germany, the Netherlands, the UK, USA and FAO.[115] There is no publicly-available information as to the amount of its funding.

GBEP's purpose is "to organize, coordinate and implement targeted international research, development, demonstration and commercial activities related to production, delivery, conversion and use of biomass for energy, with a focus on developing countries."[116] It facilitates high-level policy dialogues at global level, supports policy-making and market development at the national and regional level, encourages exchange of information, skills and technologies, and facilitates market integration by tackling barriers within the supply chain.[117] GBEP's activities are organized around two taskforces and a working group. The taskforce on sustainability has developed voluntary sustainability indicators for bioenergy,[118] the taskforce on greenhouse gas methodologies covers lifecycle analyses of transport biofuels and solid biomass,[119] and the working group on capacity building for sustainable bioenergy is responsible for disseminating and implementing the outcomes of both taskforces. It organizes workshops, study tours and public forums.[120] In addition,

---

[113] GBEP, GBEP Partners, http://www.globalbioenergy.org/aboutgbep/partners-membership/partners00/en/ (accessed February 19, 2014).
[114] GBEP, History, http://www.globalbioenergy.org/aboutgbep/history/en/ (accessed February 19, 2014).
[115] GBEP, Secretariat, http://www.globalbioenergy.org/contact-us/en/ (accessed on February 19, 2014).
[116] GBEP, purpose and functions, http://www.globalbioenergy.org/aboutgbep/purpose0/en/ (accessed on February 19, 2014).
[117] GBEP, purpose and functions, http://www.globalbioenergy.org/aboutgbep/purpose0/en/ (accessed on February 19, 2014).
[118] GBEP, Task Force on Sustainability, http://www.globalbioenergy.org/programmeofwork/taskforce-on-sustainability/en/ (accessed February 19, 2014).
[119] GBEP, GBEP Task Force on GHG Methodologies, http://www.globalbioenergy.org/programmeofwork/ghg/en/ (accessed February 19, 2014).
[120] GBEP, Working Group on Capacity Building for Sustainable Bioenergy, http://www.globalbioenergy.org/programmeofwork/working-group-on-capacity-building-for-sustainable-bioenergy/fi/ (accessed February 19, 2014).

GBEP conducts awareness-raising and information management activities on bioenergy.[121]

3.4.6 CEM

CEM focuses on advancing clean energy technologies.[122] It was launched by the US government at the UNFCCC conference in Copenhagen 2009 as a ministerial forum, involving those countries that are either among the major economies or leading in clean energy issues. It comprises 23 participating countries. CEM integrates representatives of businesses and NGOs through public-private roundtables, side events and public fora. The cornerstones of CEM's activities are 13 initiatives which focus on different aspects in the realm of energy efficiency, clean energy supply and cross-cutting issues. CEM members are not expected to participate in all CEM activities but rather to choose those in which they are most interested. Next to this work, CEM holds annual ministerial meetings. There is no publicly-available information about the amount and the sources of its funding.

CEM's aim is "to promote policies and programs that advance clean energy technology, to share lessons learned and best practices, and to encourage the transition to a global clean energy economy."[123] CEM initiatives aim to enhance intergovernmental coordination and provide energy decision-makers with the necessary information and tools to improve policy frameworks. The four CEM initiatives in the realm of clean energy supply focus on bioenergy, carbon capture, hydropower and solar and wind energy. The working group on bioenergy works on a global bioenergy atlas and capacity-building strategy. The working group is chaired by Brazil and Italy; further participants include Sweden and Denmark. It closely cooperates with GBEP.[124] The initiative on the sustainable development of hydropower shares expertise and best practice in relation to sustainability and financing issues. Besides this, it seeks to convince multilateral development and financing initiatives to support sustainable hydropower in developing countries. The governments of Brazil, France, Mexico, Norway and the USA as well as the IEA participate in this

---

[121] GBEP, Raise awareness and facilitate information exchange on bioenergy, http://www.globalbioenergy.org/programmeofwork/raise-awareness-and-information-exchange/en/ (accessed February 19, 2014).
[122] CEM, About the Clean Energy Ministerial, http://www.cleanenergyministerial.org/about/index.html (accessed September 25, 2013); CEM, Our work, http://www.cleanenergyministerial.org/OurWork.aspx (accessed December 4, 2013).
[123] CEM, About the Clean Energy Ministerial, http://www.cleanenergyministerial.org/about/index.html (accessed September 25, 2013).
[124] CEM, Fact Sheet: Bioenergy Working Group, prepared for the CEM Meeting in New Delhi, 17-18 April 2013, http://www.cleanenergyministerial.org/Portals/2/pdfs/factsheets/FS_BIO_April2013.pdf (accessed February 19, 2014).

initiative.[125] The Multilateral Solar and Wind Working Group has developed a global atlas for solar and wind energy. It furthermore works on a joint capacity-building strategy and assesses economic value creation by solar and wind industries. The working group is chaired by Denmark, Germany and Spain. Further member countries are Australia, Brazil, France, India, Japan, Korea, Mexico, Norway, South Africa, UAE, UK and US. In addition, the European Commission, IEA, IRENA, REEEP, REN21 and UNEP participate in the initiative.[126]

3.4.7 UN Bodies and Agencies

Several UN bodies and specialized agencies deal with renewable energy (see, for example, Karlsson-Vinkhuyzen 2010, Karlsson-Vinkhuyzen, Jollands & Staudt 2012:14, Najam, Cleveland 2005, Schubert, Gupta 2013). Yet, UN work on energy is widely perceived as dispersed and weakly coordinated (Westphal, Röhrkasten 2013:41). Through their agenda-setting, UN Conferences and the UN CSD have made significant contributions to global governance on renewable energy (see sections 3.1.2 and 3.1.3). However, they failed to undertake further action to promote renewable energy. Among the UN bodies dealing with renewable energy are the UNDP, UNEP, the United Nations Industrial Development Organization (UNIDO) and the FAO. Although the UNFCCC does not explicitly aim at the promotion of renewable energy, its efforts to mitigate climate change can be considered as a major driver behind the expansion of renewable energy: mitigating climate change implies reducing the production and use of fossil fuels which are major emitters of greenhouse gas emissions (Gupta 2012:434). In the wake of the WSSD in Johannesburg 2002, the UN established UN Energy as an inter-agency mechanism to coordinate its work on energy issues. Around 20 UN agencies, programs and organizations participate at UN Energy. In spite of this, it does not have a strong role within the UN system (Newell, Phillips & Mulvaney 2011:35, Schubert, Gupta 2013, Westphal, Röhrkasten 2013:41). With his SE4All initiative, the UN Secretary-General has recently undertaken important steps to promote renewable energy. The initiative aims at doubling the global share of renewable energy by 2030 and highlights various activities for goal-achievement. Yet, it did not attain the support of the UNCED 2012.

---

[125] CEM, Fact Sheet: Sustainable Development of Hydropower, prepared for the CEM Meeting in New Delhi, 17-18 April 2013, http://www.cleanenergyministerial.org/Portals/2/pdfs/factsheets/ FS_Hydro_April2013.pdf (accessed February 19, 2014).
[126] CEM, Fact Sheet: Multilateral Solar and Wind Working Group, prepared for the CEM Meeting in New Delhi, 17-18 April 2013, http://www.cleanenergyministerial.org/Portals/2/pdfs/factsheets/ FS_MSWWG_April2013.pdf (accessed February 19, 2014).

The World Bank, as an independent specialized agency of the UN, also impacts the worldwide promotion of renewable energy. It finances energy infrastructures in developing countries and advises developing countries on energy issues (on the World Bank see, for example, Expert Group on Renewable Energy 2005:53, Goldthau, Sovacool 2012:237, Nakhooda 2011, Newell, Phillips & Mulvaney 2011:37). Nakhooda (2011:124f.) particularly highlights the World Bank's contribution to spreading new ideas about energy governance and the scale of its resources in the realm of knowledge development, research and policy analysis. World Bank activities do not concentrate on renewable energy, but cover the whole range of energy sources. However, it has specific programs aimed at the promotion of renewable energy. It for example implements a program that helps developing countries to map their renewable energy potential.[127]

3.4.8 G8 and G20

The G8 and G20 are seen as further key actors for the promotion of renewable energy.[128] In global energy governance, the G8 and G20 have used their power to set the international agenda, steer existing global governors and create new ones (see sections 3.1.2 and 3.1.3). The G8, for example, created GBEP, and GBEP's mandate still depends on G8 support. Colgan and Van de Graaf highlight the influence that the G8 exercises over the IEA:

> "Formally speaking, the governing board is the IEA's highest decision-making organ, but the agency has in the past been tasked by another, outside body: the G7, and later the G8" (Colgan, Van de Graaf 2014:13).

They claim that the G8 has strongly influenced the IEA's agenda. This influence does not stem from formal links between the G8 and the IEA but rather from the political power that G8 member states enjoy within the organization. With regard to renewable energy, the authors particularly refer to the impact of the G8 summit in Gleneagles 2005, as the G8 called upon the IEA to produce new studies and reports on climate change and clean energy. According to Lesage & Van de Graaf (2013:89), the G8's focus on energy policy since the Gleneagles Summit has significantly strengthened IEA's work on clean energy technologies. In 2010 the G8 for

---

[127] Worldbank, Mapping the Renewable Energy Revolution, http://www.worldbank.org/en/news/feature/2013/06/17/mapping-the-energy-revolution (accessed November 19, 2013).
[128] See, for example, Colgan, Van de Graaf 2014, Ebinger, Avasarala 2013, Florini, Sovacool 2009:5243f., Lesage, Van de Graaf 2013:89f., Lesage, Van de Graaf & Westphal 2009, Newell, Phillips & Mulvaney 2011:63, Van de Graaf, Westphal 2011.

example created, together with IEA ministers, the International Low-Carbon Energy Technology Platform within the institutional structure of the IEA.[129]

The G20 also impacts global renewable energy governance through agenda-setting and the steering of other global governors. At the 2009 summit in Pittsburg, for example, the G20 established energy and climate change as two pillars of its plan for international economic recovery. It furthermore commissioned the OECD, IEA, OPEC and the World Bank to jointly study ways on how to phase out inefficient fossil fuel subsidies (Ebinger, Avasarala 2013:194, Lesage, Van de Graaf 2013:89f.). The phasing out of fossil fuel subsidies would have important indirect effects on the promotion of renewable energy. Yet, as Ebinger & Avasarala (2013:194) underline, after 2009 energy issues only played a marginal role at the G20 summits.

3.5 A Snapshot at Contestation and Social Construction

While transboundary policy-making on energy for a long time focused almost exclusively on conventional energy sources, renewable energy has gained center stage in global governance. Major drivers behind transboundary cooperation on renewables have been energy security concerns in times of rising oil prices and global efforts to mitigate climate change. Besides, efforts to improve worldwide access to energy also came in. Global environmental concerns have been primarily voiced by industrialized countries, while developing and emerging countries have tended to prioritize access to affordable energy. The UN framework has been of crucial importance for transboundary agenda-setting, yet reaching UN-wide consensus on the promotion of renewable energy has been a difficult (and oftentimes impossible) undertaking. For a long time, countries with vested interests in the production of conventional energy, particularly fossil fuel exporting countries, tried to obstruct transboundary cooperation on renewable energy. This has changed in recent times as the localization of IRENA's headquarters in the UAE illustrates. Further countries, particularly from the developing world, feared that global action could interfere with the priorities of their domestic energy policies.

For the global governors dealing with renewables, the creation and diffusion of knowledge and social understandings are important fields of activity. All the analyzed global governors engage in this undertaking. This also involves struggles over the prerogative of interpretation, as the relationship between IEA and IRENA reveals. The IEA provides worldwide-renowned data and analysis on world energy markets and engages in policy advice. Its publications must be approved by its members who all belong to the OECD world. As the IEA is confronted with the criticism that its research and policy advice favors conventional energy over renew-

---

[129] IEA, International Low-Carbon Energy Technology Platform, http://www.iea.org/aboutus/affiliatedgroups/low-carbonenergytechnologyplatform/ (accessed February 19, 2014).

ables, IRENA was created to counterweigh its influence in global energy governance. The IRENA also concentrates on knowledge services and policy advice. With its exclusive focus on renewables and its worldwide membership, it aims to establish itself as the global voice and knowledge base for the worldwide use of renewables. Out of the further global governors dealing with renewable energy, the G8 stands out. It has used its power to steer existing global governors, particularly the IEA, and has created new ones such as the GBEP.

4 German Ideas on Global Renewable Energy Governance

This chapter enhances a sound understanding of the German government's action and ideas on global renewable energy governance. German policies to promote renewable energy received worldwide attention in the aftermath of the Fukushima nuclear accident in 2011, when the German government decided to phase out nuclear power and expand its investments in renewable energy. However, the German 'Energiewende' did not start in 2011 but rather dates back to the first term of the government coalition of the Social Democratic Party (Sozialdemokratische Partei Deutschlands, SPD) and the Alliance 90/The Greens Party (Bündnis 90/ Die Grünen, the Greens) between 1998 and 2002 (see also Röhrkasten, Westphal 2012a:332f., Schreurs 2013). Since then, the German government has also undertaken several initiatives to strengthen transboundary cooperation on renewable energy. On the sidelines of public attention, it has successfully pushed for the creation of a new international organization for renewable energy, the IRENA (Röhrkasten, Westphal 2013:3).

The first section of this chapter traces how the German government has engaged in transboundary policy-making on renewable energy. It focuses on its three main initiatives to shape global governance on renewable energy: first, the organization of the International Conference Renewables 2004 and the establishment of the multi-stakeholder network REN21; second, the German initiative to found an IRENA; and third, the launch of the Renewables Club in 2013. As outlined in chapters 1 and 2, the process tracing not only presents the key aspects of the three initiatives, but also illustrates which policy actors were involved, what policy goals the government pursued and what context factors and political processes at the domestic and global level influenced the German administration's action.

The second section illustrates the German government's ideas on global renewable energy governance. In line with the theoretical-analytical framework, it focuses on the following elements (see also Table 5 on page 68): global challenges addressed by renewable energy promotion, specification of, and differentiation between, renewable energy options, barriers to a worldwide promotion of renewables, tasks for transboundary policy-making, relevant global governors and salient features in transboundary policy-making.

4.1 Tracing the Government's Action on Global Renewable Energy Governance

4.1.1 The Renewables 2004 Conference and the Establishment of REN21

The German government started to engage in global governance on renewable energy in the context of the WSSD 2002. At the summit, the German government strongly supported the EU proposal to increase the worldwide share of renewable energy to at least 15 percent of primary energy supply by 2010, and to adopt an action plan to enhance goal achievement.[130] Yet, the EU proposal did not succeed. As a consequence, the German government established, together with other EU member states and Small Island Developing States, the Johannesburg Renewable Energy Coalition – a loose international coalition of like-minded countries aiming at a more ambitious cooperation on renewable energy (Hirschl 2009:4411, Röhrkasten, Westphal 2013:5f., Rowlands 2005:88, Suding, Lempp 2007:5). In their Coalition Declaration, they emphasized the importance of time-bound targets for the development of renewable energy markets at the national and global level (Johannesburg Renewable Energy Coalition 2005). Besides, the German Chancellor Gerhard Schröder (SPD, 1998–2005) announced that Germany would convene an international conference on renewable energy (Schröder 2002c).

In 2004, Germany hosted the Renewables 2004 Conference in Bonn (see also Röhrkasten, Westphal 2013:6f.). The conference was formally co-organized by the Federal Ministry for the Environment, Nature Conservation and Nuclear Safety (BMU), which by then was in charge of renewable energy promotion in the German government, and the Federal Ministry for Economic Cooperation and Development (BMZ) (BMU 2013b).[131] Yet, according to Hirschl (2008:415), the BMU was *de facto* leading the conference process.[132] During the preparation process of the conference, the German government closely cooperated with the Johannesburg Renewable Energy Coalition. They jointly organized four regional preparatory conferences, covering Latin America and the Caribbean, Africa, Europe and Asia (Johannesburg Renewable Energy Coalition 2005).

The Renewables 2004 Conference was convened as a multi-stakeholder conference, with 3600 participants from 154 countries. Among the participants were

---

[130] International Institute for Sustainable Development (IISD), WSSD Info News No8, http://www.iisd.ca/wssd/infonews8.html (accessed July 04, 2013).
[131] Interview with executive staff (#1), Division International and European Affairs of Renewable Energy, BMU, December 2012 and interview with board member, German Renewable Energy Federation, December 2012.
[132] This aspect was reinforced in the interview with the board member of the German Renewable Energy Federation in December 2012.

representatives from governments, intergovernmental organizations, NGOs, the scientific community and the private sector. The German government was represented at a high level. Chancellor Gerhard Schröder, Development Minister Heidemarie Wieczorek-Zeul (SPD, 1998–2005) and Environment Minister Jürgen Trittin (the Greens, 1998–2005) joined the conference. At the conference, participants discussed how to increase the share of renewable energy in both developing and industrialized countries. The conference particularly focused on political framework conditions, financing, capacity building and on research and development.[133]

In their speeches, Schröder and Trittin explicitly referred to the unsatisfactory result of the WSSD 2002 (Schröder 2004, Trittin 2004b). According to Schröder, the Renewables 2004 conference should provide what the WSSD 2002 had not achieved, namely a way forward for increasing the worldwide share of renewable energy:

> "We are not starting from scratch with this project for the future. We are continuing what we started together in Rio in 1992 and in Johannesburg in 2002. There, at the World Summit on Sustainable Development, the participating governments declared that there is a need to increase the use of renewable energies. (…) It was because we didn't get around to defining specific targets in Johannesburg that I issued an invitation to this conference. I am very grateful to you for attending. The presence of delegates from more than a hundred countries shows that there is a growing willingness in the international community to assume responsibility for a sustainable energy policy" (Schröder 2004).

At the conference, Schröder criticized that the UN framework did not pay sufficient attention to the promotion of renewable energy. He called for new partnerships and networks between public and private actors at the national and global level in order to strengthen transboundary cooperation on renewable energy:

> "Most of the "global challenges" we face are not yet being addressed by us globally. In the United Nations framework, for example, renewable energies continue to play a subordinate role. There are lots of organizations that include them in their agendas but I still don't see a source of impetus that would steadily move things forward towards achieving this objective on a global scale. On an issue like energy supplies we need to develop new partnerships and networks in which governmental, non-governmental, national, and international players can work together continuously" (Schröder 2004).

In line with this, Heidemarie Wieczorek-Zeul (2004b) underlined the need for a "global architecture" on renewable energy.

At the Renewables 2004 Conference, three outcomes were adopted: a political declaration on a joint vision for a sustainable energy future,[134] policy recommendations for governments, intergovernmental organizations, local authorities, the

---

[133] REN21, Renewables 2004 Conference, http://www.ren21.net/REN21Activities/IRECs/ Renewables2004.aspx (accessed December 10, 2013).
[134] Renewables 2004, Political Declaration, http://www.renewables2004.de/pdf/Political_declaration_final.pdf (accessed July 18, 2012).

private sector and civil society,[135] and an international action program containing voluntary commitments by conference participants.[136] In her closing speech, Heidemarie Wieczorek-Zeul summarized the vision endorsed in the political declaration as follows:

> "We have endorsed a vision in which renewable energies will become a most important and widely available source of energy. And let us be clear: because of the huge demand in the developing world, it will need more energy, much more energy than the world is using today. So the expansion needed in renewable energies is enormous. This vision is not wishful thinking. It is possible, if we all mobilize the political will. Internationally respected institutions and bodies have stated that this is possible. But it won't happen by itself. "Business as usual" will not be enough" (Wieczorek-Zeul 2004a).

She referred to the action program as "a major contribution" towards the fulfillment of the commitments made in the Johannesburg Plan of Implementation and affirmed that the voluntary commitments of the action program would be monitored on a voluntary basis (Wieczorek-Zeul 2004a). Trittin added that the conference outcomes helped to combine voluntary action with UN structures:

> "Renewables 2004 is something new. Here in Bonn, we want to combine voluntary initiatives for increasing the use of renewable energies with the United Nations sustainable development structures. This is the inner link between the Declaration, the Policy Recommendations and the International Action Programme" (Trittin 2004a).

Parallel to the Renewables 2004 Conference, an International Parliamentary Forum for Renewable Energies took place in Bonn, chaired by the SPD Parliamentarian Hermann Scheer (EUROSOLAR, World Council for Renewable Energy 2009a, Röhrkasten, Westphal 2013:6). More than 300 members of parliament from 70 countries attended this forum. In his report to the delegates of the Renewables 2004 conference, Hermann Scheer presented the International Parliamentary Forum's call for the establishment of an IRENA:

> "The International Parliamentary Forum calls for an International Renewable Energy Agency (IRENA) as the most important institutional measure, set up as an intergovernmental organization. Membership should be voluntary, and all governments should have the opportunity to join at any time. Its aim should be to advise authorities in the policy development phase, to promote non-commercial technology transfer, to provide human capacity building, transparent certifications and standardizations" (Scheer 2009 [2004]:85).

According to Scheer's presentation, IRENA should have the institutional structure of an intergovernmental organization with voluntary membership, cooperating closely with UN organizations. It should focus on policy advice, technology transfer, capacity building, certification and standardization. Scheer furthermore referred

---

[135] Renewables 2004, Policy Recommendations, http://www.renewables2004.de/pdf/ policy_recommendations_final.pdf (accessed July 18, 2012).
[136] Renewables 2004, International Action Programme, http://www.renewables2004.de/pdf/ International_Action_Programme.pdf (accessed July 18, 2012).

to "double standards in the field of international energy institutions", claiming that the mere existence of the International Atomic Energy Agency (IAEA) justified the creation of an international organization for renewable energy. According to Scheer, IRENA should promote renewable energy the way the IAEA had done with nuclear energy (Scheer 2009 [2004]:85). Yet, the participants of the Renewables 2004 Conference did not follow the International Parliamentary Forum's call.

At the Renewables 2004 Conference, the German government initiated the multistakeholder network REN21 (Röhrkasten, Westphal 2013:6). In her closing address, Heidemarie Wieczorek-Zeul outlined the basic structure of the planned multi-stakeholder network:

> "The conference has called for a global network on renewables, a network of those actors in governments and international organizations who are highly committed to a sustainable energy future. A network that also reaches out to other stakeholders, who have been such a prominent part of this conference: to the private sector, civil society, the professional community and the local authorities" (Wieczorek-Zeul 2004a).

She proposed that the network should focus on the three main topics that the Renewables 2004 conference had addressed:

> "This network may well organize its work around the three basic topics which have proved useful for this conference: policy frameworks, financing options, research and capacity development. Around each topic, a cluster could be established of institutions and actors with comparative advantages in dealing with these topics in greater detail. However, the structure should be kept to a minimum" (Wieczorek-Zeul 2004a).

In January 2005, REN21 started its work. It unites stakeholders from governments, international organizations, industry associations, civil society and science and academia. Financed by the German government,[137] REN21 aims at facilitating "knowledge exchange, policy development and joint action towards a rapid global transition to renewable energy."[138] An important focus of REN21 is to provide policy-relevant analysis to catalyze policy change, in both industrialized and developing countries. In addition, it tracks the implementation of the Renewables 2004 action program and co-organizes follow-up conferences (BMZ 2008:23).

These initial steps for strengthening transboundary cooperation on renewables were made against the background of domestic policies to support renewable energy. The domestic promotion of renewables was a political priority of Schröder's government (1998–2005), which was based on a coalition of the SPD and the Greens. In the agreement for the first term of government (1998 – 2002), the coalition parties had concurred to transform the German energy system by promoting renewable energy, reducing energy usage and phasing out nuclear power (SPD,

---

[137] Interview with a former staff member, Division International and European Affairs of Renewable Energy, BMU, November 2012.
[138] REN21, About REN21, http://www.ren21.net/AboutREN21.aspx (accessed February 14, 2014).

Bündnis 90/Die Grünen 1998:17-20). Environmental concerns were the main motivation behind this step. The Greens originate from protests against nuclear power; and as the name of this party unequivocally suggests, ecology is a defining building block of its identity. Their ecological approach from the beginning was strongly focused on phasing out nuclear power and on mitigating climate change; and they regarded energy policy as a decisive field for policy action. Consequently, the call for an ecological energy transition was an important building block of their election program in 1998 (Bündnis 90/Die Grünen 1998:14, 23-25, 42f.).[139] But it was not only the Greens who pushed for a transformation of the German energy system. In its 1998 election program, the SPD also called for the phasing out of nuclear energy, the taxation of environmentally damaging energy usage, and the promotion of renewable energy, particularly solar energy (SPD 1998:32,36-38). In its first term, the Schröder government undertook several crucial steps to promote renewable energy within Germany. In 2000, it officially announced its commitment to phase out nuclear energy and issued the Renewable Energy Act (Erneuerbare-Energien-Gesetz, EEG) to establish feed-in tariffs for electricity from renewable sources.

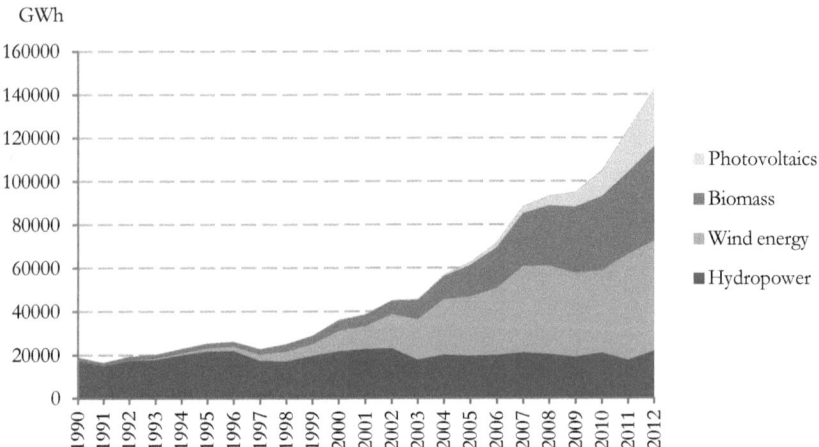

Figure 6: Electricity Supply from Renewable Energies in Germany, 1990-2012[140]

The feed-in tariff gave renewable energy priority access to the grid and provided an important boost for the German renewable energy industry, particularly for solar,

---

[139] In their election program 2002, the Greens presented the phasing out of nuclear energy, the energy transition and climate protection as the main achievements of their government participation 1998 – 2002 (Bündnis 90/Die Grünen 2002:7).
[140] Illustration S.R. with data from BMU (2013c:18).

wind and bioenergy (Hirschl 2008:189). Figure 6 illustrates the development of the German renewables-based electricity supply. It shows that, since the first Schröder government, the contribution of wind energy and biomass has increased significantly, while the contribution of hydropower has remained relatively stable. Photovoltaics has also experienced growth, supplying significant amounts of energy since 2004.

According to some policy actors, these ambitious domestic policy measures endowed the German government with a special legitimacy to push for the promotion of renewables at the global level.[141] In addition to the domestic promotion of renewable energy, the governing coalition of SPD and the Greens increasingly promoted renewable energy within the German development cooperation (Bündnis 90/Die Grünen 2002:80). In 2002, the ruling coalition made another step that would later on prove to affect not only the domestic promotion of renewables, but also the government's engagement in global renewable energy governance. It transferred the competency for renewable energy from the Federal Ministry of Economics and Technology (BMWi) to the BMU (Röhrkasten, Westphal 2013:6). This was a tribute to the Green Party which in the 2002 elections had gained votes while the SPD had lost votes. The Greens headed the BMU and called for expanded competencies. The BMWi was widely criticized for not providing sufficient support for renewables. It had argued against introducing non-market measures to promote renewable energy. Consequently, it had also opposed the EEG (Hirschl 2008:128f., Jacobsson, Lauber 2006:272, Laird, Stefes 2009:2622). The BMU's focus on nuclear safety and climate protection, by contrast, endowed it with an institutional interest in the promotion of renewables.[142]

The German government's initiatives on global renewable energy governance were strongly influenced by the SPD Parliamentarian Hermann Scheer who had been lobbying for the creation of an international organization for renewables since 1990 (see also Röhrkasten, Westphal 2013:4-8). In 1988, Scheer founded the European Association for Renewable Energy Eurosolar, an NGO aiming at the complete substitution of nuclear and fossil energy with renewable forms of energy.[143] Prior to the Earth Summit in Rio de Janeiro in 1992, Scheer published a Memorandum for the Establishment of an International Solar Energy Agency

---

[141] Interview with executive staff, German NGO Forum on Environment and Development, October 2013; Interview of Franz Alt with Hermann Scheer, July 2009, "Die größten Hindernisse gegen IRENA waren mentale", http://www.hermannscheer.de/de/index.php?option=com_content&task=view&id=697&Itemid=172 (accessed December 13, 2013).

[142] The BMU had been founded after the nuclear reactor disaster in Chernobyl. Climate protection became one of its core competencies. See BMU, 25 Jahre Bundesumweltministerium – die umweltpolitischen Meilensteine (May 2011), http://www.bmu.de/bmu/chronologie/ (accessed July 05, 2013).

[143] Eurosolar, Who we are, http://www.eurosolar.de/en/index.php?option=com_content&task=view&id=150&Itemid=52 (accessed July 05, 2013).

(ISEA) within the UN (Scheer 2009 [1990]). In the memorandum, he underlined the importance of renewable energy for global environmental protection, referring to environmental damages caused by fossil fuels and the risks of nuclear energy. According to Scheer (2009 [1990]:9), an ISEA should focus on "unconstrained international technology transfer in the field of direct and indirect solar energies (in other words: renewable energy sources)." Scheer presented his idea at the UN headquarters and convinced the Austrian government, which was led by social democrats, to suggest the founding of an ISEA at the 1990 meeting of the UN General Assembly. Despite these efforts, the preparatory committee of the Rio Conference did not put the creation of an ISEA on the conference agenda. According to Scheer, this was due to several reasons. Next to a general reluctance to create a new international organization, confidence was missing that renewable energy could play a significant role in the worldwide energy supply. In addition, there was strong opposition against pushing alternatives to conventional forms of energy (Scheer 2009 [2000]:27f.).[144] In the run-up to the WSSD in Johannesburg in 2002, Scheer again tried to put the creation of an international organization for renewables on the agenda – yet without success. According to Mez & Brunnengräber (2011:183), oil producing countries and countries with a large share of conventional sources in their domestic energy supply strongly opposed Scheer's proposal. In 2000, he published the Memorandum for the Establishment of an International Renewable Energy Agency (IRENA) (Scheer 2009 [2000]). This time, he proposed to create an organization outside of the UN system. He doubted that an UN-wide consensus was possible and therefore preferred a coalition of the willing to take the lead. In addition, he changed the agency's denomination. The former name had suggested that the organization would only cover direct solar energy, but Scheer envisaged an organization covering all renewable energy options. In 2001, Eurosolar convened an international impulse conference for the creation of IRENA. At this conference, the German Development Minister Heidemarie Wieczorek-Zeul (SPD, 1998 – 2009) expressed her political support for initiating IRENA. At the conference, Scheer argued that international cooperation should no longer disadvantage renewable energy vis-à-vis nuclear energy. States should not only create an IRENA, but also adopt an "International Treaty for the Proliferation of Renewable Energy":

> "In the case of renewable energy, there is as yet no adequate treaty framework and no binding international, institutional and financial initiatives comparable to those applying in the field of nuclear energy. (…)
>
> The same international effort that has been put into the development of nuclear energy is required for the development of renewable forms of energy, particularly so in view of the fact that the world's energy problems are more acute than ever. Anyone who believes that a specialised

---

[144] Interview of Franz Alt with Hermann Scheer, July 2009, "Die größten Hindernisse gegen IRENA waren mentale", http://www.hermannscheer.de/de/index.php?option=com_content&task=view&id=697&Itemid=172 (accessed December 13, 2013).

agency for renewable energy is unnecessary ought to be consistent and stop providing funds for the IAEA's technology transfer activities.

(...) It is imperative that renewable energy receive equal treatment with nuclear energy within the system of international treaties and institutions. Consequently, we require not only an IRENA, which should have at least the same institutional and financial resources as the IEAE – more in fact.

Also needed is a supplementary protocol to the Non Proliferation Treaty (...). It should take the form of an International Treaty for the Proliferation of Renewable Energy" (Scheer 2009 [2001]).

According to Scheer, this Eurosolar conference gave the necessary impulse for the German government to organize the Renewables 2004 Conference.[145] Hermann Scheer proposed to Schröder that he invite governments to an international conference on renewable energy and initiate IRENA's creation at that conference (EUROSOLAR, World Council for Renewable Energy 2009b:6).

In 2002 Scheer, together with the Green Parliamentarian Hans-Josef Fell, achieved the inclusion of a call for an international renewable energy agency into the coalition agreement for the second term of the Schröder government (SPD, Bündnis 90/Die Grünen 2002:36f., see also Röhrkasten, Westphal 2013:6). Despite the declaration in the coalition agreement, the German government was reluctant to undertake concrete steps for its realization. The Green Environment Minister, Jürgen Trittin, did not want to promote an initiative of an SPD parliamentarian.[146] BMU staff aimed at creating an UN Environment Organization (UNEO) and were reluctant to campaign for the founding of two international organizations at the same time.[147] In 2003 the German Bundestag adopted a resolution, drafted by Hermann Scheer, which called upon the German government to take concrete measures for the founding of IRENA (Deutscher Bundestag 2003). In an internal paper reacting to the Bundestag resolution, BMU and BMZ expressed their concerns about creating it, doubting that important industrialized and developing countries would support the initiative. In addition, they claimed that placing such an agency's creation on the agenda of the Renewables 2004 Conference could bear the risk that other countries would get the impression that the German government was pursuing a hidden agenda, trying to use the conference to become the host of a new international organization. The BMU commissioned Adelphi Research, a German non-profit institution for applied environmental research and policy analysis, with

---

[145] Die größten Hindernisse gegen IRENA waren mentale", Interview by Franz Alt with Hermann Scheer, July 2009, http://www.hermannscheer.de/de/index.php?option=com_content&task=view&id =697&Itemid=172 (accessed July 05, 2013)
[146] Interview with a former Member of the Bundestag, the Greens, November 2012. This information was confirmed by BMU staff.
[147] Interview with former staff member, MdB Hermann Scheer and SPD Parliamentary Group, German Bundestag, November 2012.

an extensive study to assess different institutional options for such a body, ranging from an NGO to an intergovernmental organization.[148] The study presented the founding of an intergovernmental organization as the best option for effective promotion of renewable energy on a global scale (Pfahl et al. 2005:76,78). Yet, it questioned the political viability of creating a new international organization and therefore concluded that the formation of a partnership was the best politically feasible option (Pfahl et al. 2005:XIII).[149] In line with these recommendations, the German government did not raise the issue of the agency's creation at the conference but instead founded the multistakeholder partnership REN21. The Schröder government did not take further steps to initiate IRENA.

4.1.2 Founding IRENA

In 2007, the German government finally started its main initiative on global renewable energy governance: the creation of an intergovernmental organization with global scope that dedicates itself to the worldwide promotion of renewable energy (see also Röhrkasten, Westphal 2013:7f.). It started to push for IRENA's creation five years after the intentions to do so were announced in the coalition agreement. Two factors were central for this policy change: the election of a new government and a further disappointment with the UN processes on renewable energy (Röhrkasten, Westphal 2013:7). After the federal elections of 2005, the composition of the German government changed. The 'grand coalition' of the Christian Democratic Union (Christlich Demokratische Union Deutschlands, CDU), its Bavarian sister party Christian Social Union (Christlich-Soziale Union in Bayern, CSU) and the SPD replaced the coalition of SPD and the Greens. As three years before, Scheer managed to include the plea for an international renewable energy agency into the coalition agreement (CDU, CDU & SPD 2005:42). The government change led to a more favorable context for IRENA's creation. This might be surprising, as the Greens were replaced by CDU/CSU. The Greens had been strong advocates of renewable energy, while CDU/CSU had favored nuclear energy over renewables. This time, Scheer's party headed all ministries that were relevant for starting the initiative for IRENA's creation – the BMU, BMZ and the Federal Foreign Office (Auswärtiges Amt, AA).[150] This facilitated his access to the relevant ministers. The Development Minister, Heidemarie Wieczorek-Zeul, had already communicated her

---

[148] Internal document of the German Bundestag.
[149] The study was concluded before the Renewables 2004 Conference but was published after the conference in 2005. According to a former staff member of the German Bundestag, the summary of the study was not drafted by the authors of the study, but by BMU staff. It is therefore likely that this statement mirrors the political priorities of the BMU and not the opinion of the official authors of the study.
[150] Interviews with former IRENA special envoys (#1 and #2), AA, November and December 2012.

support for Scheer's initiative at the International Impulse Conference for the Creation of an IRENA 2001 (Wieczorek-Zeul 2009 [2001]). The newly appointed Environment Minister, Sigmar Gabriel (SPD, 2005-2009), had expressed his support at a Eurosolar conference shortly before taking office (Gabriel 2009). As such, Gabriel changed the internal position of his ministry which so far had not supported IRENA's creation.[151] It was an incident at global level that gave the German government the final push for IRENA's creation. The 2007 session of the UN CSD proved to be a serious blowback for the German strategy to promote renewable energy on a global scale. It wanted the UN CSD to approve time-bound targets for increasing the global share of renewable energy. Yet, the UN CSD ended without any agreement. After this experience, the BMU decided to change its strategy and push for the worldwide promotion of renewable energy outside the UN framework, initiating the creation of an international renewable energy agency. The BMU became the driving force behind the German initiative. It was joined by the BMZ and the AA. CDU/CSU politicians and their ministries did not take an active stance towards IRENA's creation. Chancellor Merkel (CDU, since 2005) expressed her support for the initiative without actively engaging with it.[152] The BMWi and its Minister, Michael Glos (CSU, 2005–2009), did not support IRENA's creation but neither obstructed it.[153]

In 2007, the German government began bilateral talks to explore the political viability of creating a new international organization for renewable energy (see Röhrkasten, Westphal 2013:7f.). The AA appointed three former ambassadors as special envoys to hold bilateral talks with governments in Asia, North America, Africa and Latin America. Within these world regions, the special envoys visited countries that the German government regarded as influential and/or like-minded – either due to strong diplomatic ties or the countries' interest in the promotion of renewable energy.[154]

In 2008, the German government officially started the preparatory process for IRENA's creation (see Röhrkasten, Westphal 2013:8f.). At the Washington

---

[151] Interview with former staff member, MdB Hermann Scheer and SPD Parliamentary Group, German Bundestag, November 2012.
[152] Interviews with a former Member of the Bundestag, the Greens, and former staff member, MdB Hermann Scheer and SPD Parliamentary Group, German Bundestag, November 2012.
[153] Interviews with a former Member of the Bundestag, the Greens, German Bundestag, November 2012, former IRENA special envoy (#2), AA, November 2012 and a staff member, Division International Energy Policy, AA, February 2010.
[154] The following countries were visited: Argentina, Brazil, Canada, Chile, China, Colombia, Costa Rica, Ethiopia, Ghana, India, Indonesia, Japan, Kenya, Malaysia, Mali, Mexico, Nigeria, Senegal, Singapore, South Africa, South Korea and USA. Out of the countries visited, the German government considered the support of the emerging powers of China, India, Brazil, Mexico and South Africa as especially important. Interview with staff member, Division International Energy Policy, AA, February 2010 and internal document of the AA "Ortez zur IRENA-Gründungskonferenz vom 04.02.2009".

International Renewable Energy Conference WIREC 2008, it formally presented to the participating governments a concrete proposal for an international renewable energy agency.[155] In April 2008, it held the First Preparatory Conference for IRENA in Berlin to discuss first activities and a possible structure for IRENA (The Government of the Federal Republic of Germany, Initiative for an IRENA 2008:19). The German delegation to the conference comprised above all AA and BMU staff; BMZ and BMELV sent staff as well. In addition, Scheer participated with two of his employees, and one representative of the Federal Environment Agency (Umweltbundesamt, UBA) was also present.[156] After this conference, the German lobbying activities for IRENA were joined by Denmark and Spain.[157] Both countries had a strong record in renewable energy promotion. In addition, Denmark sought to position itself as a frontrunner in climate protection prior to the Copenhagen Climate Summit 2009. In summer 2008, the German government organized two preparatory workshops in Berlin, discussing IRENA's initial activities, its statutes and financing (The Government of the Federal Republic of Germany, Initiative for an IRENA 2008:19). In October 2008, the final preparatory conference was held in Madrid. Here, discussions on IRENA's statutes were finalized. In addition, agreement was reached on how to proceed with the initial steps of institution-building (The Government of the Federal Republic of Germany, Initiative for an IRENA 2008:19). Germany invited governments it considered to be politically influential and/or interested in the promotion of renewable energy to the preparatory process.[158] In addition to the governments consulted by the IRENA special envoys, the German government invited G8 states, European countries and governments that had expressed their political support for creating IRENA, such as Egypt, Jordan, Morocco and the UAE. Between 40 and 60 governments participated at the preparatory conferences and workshops.[159] Moreover, the German government consulted international organizations during the preparatory process. Among these were the UNFCCC, UNEP, UNIDO, IEA, REN21 and REEEP. Additional consultations were held with industry associations and NGOs at the

---

[155] AA, IRENA: Promoting renewable energy worldwide, http://www.auswaertiges-amt.de/EN/Aussenpolitik/GlobaleFragen/Energie/IRENA-Gruendung_node.html (accessed July 18, 2013)

[156] Preparatory Conference for the Foundation of the International Renewable Energy Agency (IRENA), Berlin, 10-11.04.2008, List of participants.

[157] According to BMU staff, Denmark and Spain only joined the initiative in a formal sense. *De facto*, they did not play a major role in IRENA's creation.

[158] Interview with executive staff, International Cooperation, BMU, November 2012.

[159] IRENA, First Preparatory Conference, http://www.irena.org/menu/index.aspx?mnu=Subcat&PriMenuID=13&CatID=30&SubcatID=57 (accessed February 14, 2014); IRENA, Workshops I + II, http://www.irena.org/menu/index.aspx?mnu=Subcat&PriMenuID=13&CatID=30&SubcatID=56 (accessed February 14, 2014); IRENA, Final Preparatory Conference, http://www.irena.org/menu/index.aspx?mnu=Subcat&PriMenuID=13&CatID=30&SubcatID=55 (accessed February 14, 2014).

national, European and international level.[160] Yet, NGOs and industry associations did not show major interest in the process of IRENA's creation. The political attention of the NGO community was absorbed by climate change issues, while the industry did not expect a direct economic benefit of the German initiative.[161]

In a policy paper written for the First Preparatory Conference, the German government outlined a possible design and the envisaged scope of the proposed IRENA (The Government of the Federal Republic of Germany 2008). This proposal largely built on Scheer's initial ideas. Yet, it differed in two respects: first of all, it advanced a positive 'win-win' framing on renewable energy without taking explicit opposition against other energy sources or other international organizations dealing with energy, such as IEA or IAEA. Second, it proposed an agency that would build entirely on the principle of voluntariness, taking a 'soft' approach to global governance without engaging in regulation, standard-setting or the negotiation of political declarations. The government considered these two aspects to be key for gaining broad political support for its initiative (Röhrkasten, Westphal 2013:9).[162] In addition, it stated that it wanted IRENA to "be established by a broad group of countries, including both large and small, industrialized and developing countries" (The Government of the Federal Republic of Germany 2008:6). In the policy paper, it characterized the objective and activities of the proposed agency as follows:

> "IRENA's main objective will be to foster and promote the large-scale adoption of renewable energy worldwide. This overall objective can be broken down into a number of concrete targets: improved regulatory frameworks for renewable energy through enhanced policy advice; improvements in the transfer of renewable energy technology; progress on skills and know-how for renewable energy; a scientifically sound information basis through applied policy research; and better financing of renewable energy" (The Government of the Federal Republic of Germany 2008:6).

The statement clarified that IRENA should promote renewable energy all over the world, implying that it should not concentrate on a single group of countries. It presented policy advice on regulatory frameworks as a central task for IRENA.

---

[160] Initiative for an IRENA, Explanatory Note. Work Programme. Final Preparatory Conference, Madrid, 23 and 24 October, 2008, p.5

[161] This aspect was raised by the following interviewees: senior staff member "Energy for Sustainable Development", Deutsche Gesellschaft für Internationale Zusammenarbeit (GIZ), January 2013; executive staff (#1), Division International and European Affairs of Renewable Energy, BMU, December 2012; staff member, Division International and European Affairs of Renewable Energy, BMU, November 2012; executive staff, German NGO Forum on Environment and Development, October 2013; Renewable Energy Director, Greenpeace International, October 2013; senior staff member "Energy for Sustainable Development", Deutsche Gesellschaft für Internationale Zusammenarbeit (GIZ), January 2013; board member, German Renewable Energy Federation, December 2012; staff member, international cooperation, German Solar Industry Association, November 2013.

[162] Interviews with executive staff, International Cooperation, BMU, December 2012; former IRENA special envoy (#1), November 2012; former staff member, MdB Hermann Scheer and SPD Parliamentary Group, German Bundestag, November 2012.

Furthermore, it stated that IRENA should engage in technology transfer and applied policy research and improve skills, know-how and renewable energy financing. Yet, the policy paper did not specify how IRENA should improve renewable energy financing, i.e. if IRENA should provide financing for renewable energy itself or only give policy advice on these matters. According to the German government, IRENA should furthermore act as a "facilitator and catalyst" for different programs dealing with renewable energy, national governments and the private sector (The Government of the Federal Republic of Germany 2008:6). The administration assured that IRENA would only offer its services when requested to do so, respecting the sovereignty of its member states:

> "IRENA will not aim to draw up international regulations or treaties. It will provide its services as and when requested by member states or groups of member states. It will not involve itself in states' energy policies of its own accord or try to enforce policies. All its activities will be decided upon by members" (The Government of the Federal Republic of Germany 2008:6).

It further stated that IRENA should act as a driving force and focal point for renewable energy at international level. It should enhance renewable energy with more weight in international political processes, preventing other institutions from erecting barriers to the worldwide spread of renewable energy, guide and coordinate the international action on renewable energy while also engaging in donor coordination (The Government of the Federal Republic of Germany 2008:7f.).

After the second preparatory conference in October 2008, the German government issued an updated version of the policy paper outlining the design and scope of the proposed agency (The Government of the Federal Republic of Germany, Initiative for an IRENA 2008). The deliberations presented in this paper were in line with the policy paper prepared for the first preparatory conference. Yet, IRENA's proposed objective slightly differed from that presented in the first policy paper. The second paper specified that IRENA should not promote renewable energy *per se*, but its *sustainable* use:

> "Mandated by governments worldwide, IRENA aims at becoming the main driving force in promoting a rapid transition towards the widespread and sustainable use of renewable energy on a global scale" (The Government of the Federal Republic of Germany, Initiative for an IRENA 2008:6).

Later, the government clarified that IRENA would not discriminate between different renewable energy sources, but aim at "fostering all types of renewable energy" (The Government of the Federal Republic of Germany, Initiative for an IRENA 2008:7). In addition, the policy paper specified IRENA's field of activities as follows:

> "Acting as the global voice for renewable energies, IRENA will provide practical advice and support for both industrialised and developing countries, help them improve their regulatory frameworks and build capacity. The agency will facilitate access to all relevant information including reliable data on the potential of renewable energy, best practices, effective financial

mechanisms and state-of-the-art technological expertise" (The Government of the Federal Republic of Germany, Initiative for an IRENA 2008:6).

In line with the previous policy paper, the proposed field of activities included policy advice, capacity building, technology transfer and knowledge services. Yet, it now specified IRENA's role vis-à-vis renewable energy financing: the agency would advise governments about renewable energy financing without providing funding itself. In addition, this paper also elaborated further on its institutional design (The Government of the Federal Republic of Germany, Initiative for an IRENA 2008:18f.). It proposed that IRENA should comprise three main organs: an Assembly, a Council and a Secretariat. The Assembly, composed of all member states, should adopt the budget and work program on an annual basis. The Council would comprise a limited number of representatives from its member states. It should facilitate cooperation and consultation among member states and substantiate the work program. The implementation of the work program should be the task of the Secretariat, headed by a Director-General. The paper also addressed IRENA's relation vis-à-vis the UN. The German government argued against the establishment of IRENA within the UN context, as it did not consider this option to be politically viable. Nevertheless, the government stated that it could be possible to integrate IRENA into the UN framework later on:

> "For the time being, the idea to make IRENA a new United Nations or United Nations-affiliated organisation does not appear as a realistic option. To ensure that the Agency can begin operating swiftly, it needs to be set up independently. However, in the long term the integration of IRENA into the United Nations should be considered" (The Government of the Federal Republic of Germany, Initiative for an IRENA 2008:14).

At the end of the preparatory process, agreement was reached on the Statute, financing matters, criteria and procedures for selecting the Interim Director-General and the Interim Headquarters, and the design of IRENA's initial phase. The statute establishes IRENA's main objectives as follows:

> "The Agency shall promote the widespread and increased adoption and the sustainable use of all forms of renewable energy, taking into account:
>
> a.) national and domestic priorities and benefits derived from a combined approach of renewable energy and energy efficiency measures, and
>
> b.) the contribution of renewable energy to environmental preservation, through limiting pressure on natural resources and reducing deforestation, particularly tropical deforestation, desertification and biodiversity loss; to climate protection; to economic growth and social cohesion including poverty alleviation and sustainable development; to access to and security of energy supply; to regional development and to inter-generational responsibility" (Conference on the Establishment of the International Renewable Energy Agency 2009:Art. II).

IRENA's organization structure – encompassing an Assembly, Council and Secretariat – resembles UN structures. In contrast to UN proceedings, its statute contains an important rule to diminish the veto power of single states. In the Assembly, consensus is considered as having been achieved if no more than two members object (Conference on the Establishment of the International Renewable Energy Agency 2009: Art IX,F).[163]

During the preparatory process, only a few issues sparked controversial debates. Heated debates focused on questions relating to the types of energy IRENA should promote. Some governments wanted to include nuclear power into IRENA's mandate. This demand was strongly opposed by other governments.[164] Debates also focused on the question as to what kind of renewable energy the agency should promote. Some governments raised sustainability concerns with regard to biofuels and large hydropower. Within the German administration, these concerns were particularly shared by the BMZ.[165] In an internal paper, the German government even considered defining renewable energy for the purpose of IRENA as follows:

> "Renewable energies are those energy supplies that are based on solar radiation, solar heat, wind power, tidal and wave power, sustainable biomass, geothermal energy, small hydropower and large hydropower if the latter is operated in accordance with the recommendations of the Dams and Development Project of the UN."[166]

Here, only biomass is provided with the restriction that it should be sustainable. In addition, the paper differentiated between small and large hydropower, clarifying that the latter should only be promoted if it complied with the recommendations of the UN Dams and Development Project. Several governments strongly argued against discriminating among different renewable energy sources. It was finally agreed that IRENA should promote all forms of renewable energy, if these were produced and used in a sustainable way. In line with Scheer's initial proposals, some governments wanted IRENA to engage in standard-setting. Regulative activities were not included into IRENA's mandate as many governments strongly opposed this due to sovereignty concerns.[167] With regard to the institutional set-up the following were debated: budgetary issues, the composition of IRENA's council and the agency's scope (coalition of the willing vs. broad membership). Finally, it was underlined that IRENA should be open to as many applicants as possible.[168]

---

[163] Interview with executive staff, International Cooperation, BMU, December 2012.
[164] Interviews with executive staff, International Cooperation, BMU, December 2012 and former IRENA special envoy (#3), AA, November 2012.
[165] This information is based on the following interviews: executive staff, International Cooperation, BMU, December 2012; executive staff, Division International Energy Policy, AA, December 2012; former IRENA special envoy (#1), AA, November 2012; former IRENA special envoy (#2), AA, November 2012; former IRENA special envoy (#3), AA, November 2012.
[166] Internal document of the German government, Present working-definition of "Renewable Energy" for the purpose of IRENA, undated.
[167] Interviews with executive staff, International Cooperation, BMU, December 2012 and energy policy advisor, Gesellschaft für Internationale Zusammenarbeit (GIZ), August 2012.
[168] Interview with executive staff, International Cooperation, BMU, December 2012.

In Germany, the use of large hydropower had been contested for a long time. Development and environment NGOs had been criticizing the negative social and environmental implications of large hydropower plants in developing countries.[169] In the German government, particularly the BMZ shared this criticism. The promotion of biofuels became the center of heated debates in 2007/08 when rising international food prices were ascribed to expanding biofuels deployment. The food versus fuel debate received broad media coverage in Germany and several development and environmental NGOs pronounced openly against biofuels. They not only criticized the possible competition with food production but increasingly questioned the environmental impacts of biofuels as well. Their opposition against biofuels was joined by diverse groups: churches, research institutions, and representatives of the oil industry, food industry and car industry (Beneking 2011, Brand-Schock 2010:313-317, Kohlhepp 2010:231, Selbmann, Kaup 2010:30).[170] The rising opposition against biofuels made the German government withdraw from ambitious blending quotas for biofuels in 2008. It officially justified this step by stating that a large number of cars in Germany had engines that were incompatible with high blending requirements (Beneking 2011:96 100, Kohlhepp 2010:231). Within the government, the BMZ became a strong critic of biofuels, while the BMELV remained the most supportive ministry (Beneking 2011:96-100). In 2009, the government stated in its Coalition Agreement that it would push for international sustainability certificates for bioenergy (CDU, CDU & FDP 2009:35f.). At the global level, the German government today tries to use GBEP to establish ambitious sustainability requirements for biofuels. It actively engages in GBEP, but at a low political profile.[171] Thus, GBEP does not receive much political attention within the German administration.

In 2009, the German government organized IRENA's Founding Conference and the First Session of the Preparatory Commission in Bonn (Röhrkasten, Westphal 2013:9f.). The Danish and Spanish governments served as co-chairs. This time, all UN member states were invited to participate. The founding of IRENA met broader international support. Representatives from more than 120 countries

---

[169] Interviews with executive staff, German NGO Forum on Environment and Development, October 2013 and with Renewable Energy Director, Greenpeace International, October 2013.

[170] This information builds on the following interviews: team, German Biofuels Association (VDB), November 2013; executive staff, German NGO Forum on Environment & Development, October 2013; Renewable Energy Director, Greenpeace International, October 2013.

[171] The ministries in charge of GBEP, BMELV and BMU, have engaged government agencies and research institutes to accompany GBEP's work. The BMU has commissioned the Federal Environment Agency (UBA), the Institute for Energy and Environmental Research (IFEU) and the Institute for Applied Ecology (Öko-Institut), the BMELV the Fachagentur Nachwachsende Rohstoffe (FNR) and the Deutsches Biomasseforschungszentrum (DBFZ). This information builds on interviews with staff member, Division Energy Issues/Bioenergy, BMELV, February 2013 and staff member, UBA, February 2013.

joined the conference and 75 governments signed IRENA's statutes.[172] The German government was represented by delegations from the ministries in charge: BMU, BMZ and AA. BMU and BMZ were represented at ministerial level, the AA by its Minister of State. The BMWi and BMELV sent one desk officer. The German delegation furthermore comprised nine Members of Parliament: five SPD parliamentarians, including Hermann Scheer, and one parliamentarian from each Bundestag fraction.[173] Other German institutions participated as guests. Among these were the three implementing agencies of the German development cooperation: Deutsche Gesellschaft für Technische Zusammenarbeit (GTZ, technical cooperation), International Weiterbildung und Entwicklung (InWEnt, personnel cooperation) and Kreditanstalt für Wiederaufbau (KfW, financial cooperation). The civil society organizations Forum Umwelt und Entwicklung (the German NGO Forum on environment and development), Germanwatch and Eurosolar joined as well. The German renewable energy industry was represented by the German Renewable Energy Federation (Bundesverband Erneuerbare Energien, BEE), the German Bioenergy Industry Association (Bundesverband Bioenergie, BBE) and the German Biofuels Association (Verband der deutschen Biokraftstoffindustrie, VDB).[174]

In his welcome speech, AA Minister of State, Gernot Erler (SPD, 2005–2009), emphasized that transforming the energy system was a task that concerned all countries of the world:

> "It has long been obvious that our economic system, based on fossil fuels, cannot be sustained indefinitely, and that no single country, regardless of its size, can meet this challenge alone. (…) We've recognized that we must completely change our methods of generating that energy. This is a process which concerns us all. Only if everyone contributes can we establish sustainable new energy systems before the old ones break down" (Erler 2009).

He added that promoting renewable energy not only reduced the risk of economic and energy crises but also provided crucial business opportunities:

> "We saw last summer what happens when we fail to assume that responsibility. The record price of oil created economic turmoil across the globe and alarmed people everywhere. High oil prices don't stop at borders, they affect everyone who relies on oil. The subsequent fall in the price of oil is, according to the analysts, solely due to the global economic and financial crisis. When the economy recovers, the price of energy is bound to reach new heights. It's clear, therefore, that if we don't want to spend the next few decades permanently lurching back and forth from energy crisis to economic crisis, we must reduce our dependence on fossil fuels. (…)

---

[172] Conference on the Establishment of the International Renewable Energy Agency, Conference Report, January 26, 2009 (IRENA/FC/CR)
[173] Conference on the Establishment of the International Renewable Energy Agency, List of Delegations, January 26, 2009 (IRENA/FC/proc.1)
[174] Conference on the Establishment of the International Renewable Energy Agency, List of Guests, January 26, 2009 (IRENA/FC/proc.2)

But I don't wish to only speak here about the risks of business as usual but also about the opportunities a switch will present. Just as the use of coal was instrumental for the first industrial revolution and the use of oil for the second, the breakthrough of renewable energy will kickstart a third industrial revolution. We have the unique opportunity to jointly promote and shape this new global industrial revolution. It's in all our countries' interests to participate in it and to let other countries do likewise. IRENA will help us address this major future task in a determined way" (Erler 2009).

Environment Minister Siegmar Gabriel outlined the planned fields of activity of the newly created international organization, highlighting in particular the provision of policy advice to governments:

"IRENA will give concrete advice to both industrialised and developing countries to aid their introduction of political and legal frameworks. The goal is to create the right incentives and securities for investment. This needs to be steered by governments, since a distorted market is not capable of initiating the transformation of energy systems. IRENA will act as a catalyst to facilitate technology and knowledge transfer, and to support capacity building. Using positive examples, the Agency will clear away deep-rooted concerns. IRENA will be an international platform for renewable energies. It will focus more resources on renewable energies than any other organisation to date. But others need not worry - there is more than enough work for everyone. IRENA will cooperate with other organisations and institutions to exploit synergies" (Gabriel 2009:103f.).

In her closing address, Development Minister Heidemarie Wieczorek-Zeul interpreted IRENA's creation as a signal that renewable energy would have a major contribution to future energy supply:

"We have thus been able to send a message, signalling the new, cooperative global policy that we look forward to and that we need more urgently than ever to resolve the current crises (food, fuel, finance). Together we can change the "fossilised" approach to energy policy. Renewable energies will soon move out of their niche existence to take their place at the heart of international energy policy. We can make sure that happens" (Wieczorek-Zeul 2009:108).

The German government aspired to locate IRENA's headquarters in Bonn (Röhrkasten, Westphal 2013:10). This goal was particularly important for the AA staff engaged in IRENA's founding process. At IRENA's Founding Conference, the German government purposely left the question of headquarters location open and did not announce its application for hosting IRENA. It only did so after the founding process was successfully completed. The government did not want to raise suspicion that it had only pushed for IRENA's creation because it wanted to host another international organization. Besides, it wished to create an additional incentive for governments to join IRENA. At the founding conference, Austria and the UAE formally submitted their candidacies to host IRENA's headquarters. Gaining the agency's headquarters became a clear foreign policy priority for the OPEC member UAE which considered the hosting of an international organization to be a question of national prestige. To gain support for its candidacy, the UAE visited several developing countries, particularly in Africa and the Arab world, and promised generous funds for IRENA. Next to its domestic commitment for renewable energy, its representatives underlined that no international organization was located in the Arab world. In contrast to the UAE, the German foreign policy did not consider IRENA and its location as a priority issue because it aimed at becoming a permanent member of the UN Security Council. Neither the Foreign Minister nor the Chancellor actively engaged in the German campaign. The German government hoped that its leading role during IRENA's founding process would give sufficient credentials for its own application.[175] Yet, this was not the case. When the signatory states decided on the location of IRENA's headquarters at the Second Session of the Preparatory Commission in Sharm el Sheikh, Egypt, in June 2009, the UAE clearly won the race. As consolation for the German government, it was decided that Bonn would host the IRENA Innovation and Technology Center. The third applicant, Austria, would get an IRENA Liaison office. However, this office has not been established as Austria has not yet ratified IRENA's statutes.[176] The decision on IRENA's headquarters was a serious blow for the German government. As it had aspired to locate IRENA's headquarters in Bonn, it had refrained from naming a German candidate for IRENA's interim director-general. Now, it ended without German headquarters or a German interim director-general. Yet, as a leading BMU official highlights,[177] the German application to host IRENA's headquarters also fulfilled important domestic functions. With the application, it was easier to gain political support for IRENA's creation within Germany and mobilize financial means for IRENA's founding process.

Ever since, the German administration has been actively involved in IRENA's institution-building and activities. Until the official founding of IRENA in April 2011, the German government led the Administrative Committee which oversaw issues such as IRENA's financing, personnel and its work program. The German government has been IRENA's second largest financier, after the UAE.[178] Both governments together provide by far the most financial support for IRENA. Their voluntary contributions alone amount to almost 40 percent of IRENA's budget (IRENA 2012 [16.12.2012]:6). Until December 2013, the BMU held the main responsibility for engaging with IRENA.[179] It closely cooperated with the BMZ.

---

[175] Interviews with executive staff, Division International Energy Policy, AA, December 2012; former IRENA special envoy (#1), AA, November 2012; former IRENA special envoy (#2), November 2012; energy policy advisor, Gesellschaft für Internationale Zusammenarbeit (GIZ), August 2012.
[176] IRENA, IRENA membership, https://www.irena.org/menu/index.aspx?mnu=cat&PriMenuID=46&CatID=67 (accessed January 12, 2014).
[177] Interview with executive staff, International Cooperation, BMU, December 2012.
[178] Interviews with executive staff, International Cooperation, BMU, December 2012 and energy policy advisor, Gesellschaft für Internationale Zusammenarbeit (GIZ), August 2012.
[179] In December 2013, the newly elected 'grand coalition' of CDU/CSU and SPD decided to re-transfer the competencies for renewable energy from the BMU to the BMWi (since then called Federal Ministry for Economic Affairs and Energy).

4.1.3 Launch of the Renewables Club

The newest initiative of the German government in global renewable energy governance has been the Renewables Club, founded in June 2013. Initiated by the German Environment Minister, Peter Altmaier (CDU, 2012–2013), this Club was regarded as a high-level political alliance between countries that the government considered as leading in the transformation of energy systems. It is a political initiative without institutional structure or secretariat (The Government of the Federal Republic of Germany 2013:1). Within the German government, the BMU was in charge of this initiative (The Government of the Federal Republic of Germany 2013:6). The Club comprises the governments of China, Denmark, France, Germany, India, Morocco, South Africa, Tonga, UAE and UK. The BMU underlined that the Renewables Club united countries with very diverse energy mixes and renewable energy potentials. The Renewables Club was not thought as an exclusive body but open for other countries to join. The main uniting criterion was a strong national commitment for the expansion of renewable energy:

> "The Renewables Club is a political initiative of pioneering countries that are united by an important goal: a worldwide transformation of the energy system" (BMU 2013a, cites Altmaier).

Altmaier emphasized that leading by example was an important goal of club members. He added that "we in Germany do not stand alone with our Energiewende, but are a part of a strong group of leaders" (BMU 2013a). The BMU (2013a) underlined that the Renewables Club had role model functions and served as a driver of ideas. In their communiqué, club members stated that they would continue to act as agenda-setters for an increased use of renewable energy, implementing renewables-friendly policies, "sending a strong political message of support for renewable energy's business case", raising awareness, and supporting IRENA and other entities engaged with promoting renewable energy.[180] To enhance close cooperation between the Club and IRENA, IRENA's Director-General Adnan Amin was also called to join the club (The Government of the Federal Republic of Germany 2013:3f.).

Altmaier had announced his intentions to launch such a club in an article for the *Financial Times Deutschland* three months after taking office (Altmaier 2012). In this article, he underlined the importance of transforming energy systems by promoting renewables. He stated that an energy transition offered the opportunity to align economic growth with environmental protection – not only in industrialized countries, but all over the world. He presented Germany as a pioneering country in transforming its energy system and underlined that there was broad international support for renewable energy. According to Altmaier, 118 countries had national

---

[180] Renewables Club – Communiqué, http://www.bmu.de/en/climate-energiewende-download/artikel/club-der-energiewende-staaten-kommunique/?tx_ttnews%5BbackPid%5D=289 (accessed July 15, 2013).

targets to promote renewables. He particularly emphasized that major oil-producing countries, and emerging countries too, promoted renewables. He added that Japan, too, though about transforming its energy system as a lesson from the Fukushima nuclear accident. He furthermore pointed out that global investments in renewable energy had a positive trend despite the financial crisis. He emphasized that the worldwide promotion of renewable energy provided the German renewables industry with export opportunities, and stated that emerging and developing countries needed help in order to expand their electricity supply based on renewable resources. He claimed that an international vanguard was needed, taking the lead for a global energy transition and demonstrating the benefits brought about by transforming energy systems.

The background of Altmaier's initiative was a stormy phase of German energy policy at the domestic level, with two u-turns within a very short period of time.[181] The federal elections in 2009 led to a change in the composition of the German government. This time, CDU/CSU entered into a coalition with the Free Democratic Party (Freie Demokratische Partei, FDP). In their coalition agreement, CDU/CSU and FDP decided to extend the life span of German nuclear power plants (CDU, CDU & FDP 2009:34-38). Yet, they maintained the ban on the construction of new power plants and the goal of continuously expanding renewable energy, stating that "renewable energies should make up the main part of our energy supply" (CDU, CDU & FDP 2009:34). In autumn 2010, the government implemented the decision to expand the life span of nuclear power plants. This decision was widely seen as a u-turn of the energy transition, even though its main pillars (phasing out nuclear energy and expanding the use of renewable energy) were maintained in force. The decision met widespread opposition within German society. After the Fukushima nuclear accident in March 2011, the German government immediately reacted with a volte-face in domestic energy policy-making, announcing a moratorium on nuclear energy for a period of three months and retreating from the decision to expand the life span of nuclear power plants. According to Schreurs (2013:85), "in no other country was the political reaction to the Japanese nuclear crisis as swift as in Germany". When Chancellor Merkel justified this policy change, she pointed to "a consensus across German society on phasing out (…) nuclear power" after the Fukushima accident (Merkel 2012b). In June 2011, the German government decided to decommission eight nuclear power plants and phase out nuclear energy until 2022. In addition, it set ambitious targets for increasing the share of renewable energy. With this decision, it returned to Germany's pre-2010 energy policy, yet with an important change. The decision to terminate the use of nuclear energy now built on a multi-party consensus, also borne by those parties

---

[181] For a comprehensive overview of the German nuclear phase out and the politics behind see Schreurs 2013.

that had formerly opposed this decision (see also Röhrkasten, Westphal 2012a:332f., Schreurs 2013). Whereas the phasing out of nuclear energy has met broad public support, it is still highly controversial in the CDU, CSU and FDP parties and the German industry. When Altmaier was appointed as Environment Minister, he repeatedly reinforced that the German government would maintain its decision on transforming the energy system. The founding of the Renewables Club can be interpreted as an attempt to mobilize domestic support for the German energy transition by highlighting its pioneering character and by demonstrating that Germany was not the only country aiming at transforming its energy system.[182]

4.2 Ideas on Global Renewable Energy Governance

4.2.1 Global Challenges: Predominance of Climate Protection

Climate protection is the global challenge that dominates the German deliberations on global renewable energy governance. References to climate change are never missing when underlining the need to promote the worldwide spread of renewable energy. The primacy of environmental protection becomes obvious in Chancellor Schröder's 2002 government declaration in which he underlines his commitment for *ecological* energy use at the WSSD (Schröder 2002a:15). At the Renewables 2004 Conference, he emphasizes the need to combat climate change by spreading the use of renewable energy:

> "Climate change continues to be, by far, the greatest environmental threat we face. The nightmare scenario in which deserts would expand and large parts of the world would be flooded can be avoided only if we radically reduce greenhouse gas emissions. In light of this fact, increasing the use of renewable energies will be a means of providing environmental security and protecting the lives of millions of people" (Schröder 2004).

Foreign Minister Guido Westerwelle (FDP, 2009–2013) reinforces in his speech at the third IRENA Council meeting that, for the German government, renewable energy is closely related to mitigating climate change – as he states, "a matter of high political priority in Germany" (Westerwelle 2012). In the policy papers promoting the foundation of IRENA, the German administration underlines the goal of limiting global warming to two degrees Celsius (The Government of the Federal Republic of Germany 2008:3, The Government of the Federal Republic of Germany, Initiative for an IRENA 2008:9). At the 2011 Eurosolar Conference, Development Minister Heidemarie Wieczorek-Zeul (2009 [2001]:35) points out that climate change is a global danger that affects developing countries in particular. Environ-

---

[182] Interview with staff member, Division International and European Affairs of Renewable Energy, BMU, November 2012.

ment Minister Jürger Trittin reinforces in his speech at the Renewables 2004 Conference that global warming poses a threat to "development progress, especially in the countries of the South" (Trittin 2004a). Underlining the interdependency between poverty reduction and climate protection, he adds:

> "Neither of these two global challenges can be dealt with individually. Climate protection without economic development will not be able to eliminate poverty. Development without climate protection will destroy the foundations of development itself" (Trittin 2004a).

Heidemarie Wieczorek-Zeul (2009 [2001]:35) points out that successful poverty reduction results in increasing global energy consumption. In order to avoid substantial increases in global GHG emissions, renewable energy needs to replace coal and oil. Apart from climate protection, the German government highlights further environmental benefits of renewable energy, such as improving air quality and combating deforestation, desertification and the loss of biodiversity (BMZ 2007:10, The Government of the Federal Republic of Germany 2008:4, The Government of the Federal Republic of Germany, Initiative for an IRENA 2008:9).

Yet, climate change is not the only global challenge that the German government underlines with regard to renewable energy promotion. In the policy papers promoting IRENA, cushioning energy price increases receives substantial attention (Initiative for an IRENA 2008:2, The Government of the Federal Republic of Germany 2008:3f., The Government of the Federal Republic of Germany, Initiative for an IRENA 2008:10).[183] It is also mentioned in the speeches of Angela Merkel at a symposium of the German Advisory Council on Climate Change (Wissenschaftliche Beirat der Bundesregierung Globale Umweltveränderungen, WBGU) and of Gerhard Schröder at the Renewables 2004 Conference (Merkel 2012b, Schröder 2004). In his speech at that conference, he justifies the need to promote renewable energy foremost with the adverse effects of high energy prices:

> "High energy prices, such as those we are currently seeing for oil, destroy opportunities for economic development throughout the world. They pose a threat to economic growth in developed economies. They are a hindrance to global efforts to combat poverty and hunger" (Schröder 2004).

In a policy paper promoting IRENA, the government states that conventional energy supply is becoming more and more expensive:

> "Satisfying the growing demand for energy with conventional energy sources is becoming not only increasingly difficult but also increasingly costly. Oil prices almost doubled over the last year. For the first time the price per barrel reached USD 147 in 2008. We are shifting from cheap oil to peak oil. Prices for other conventional energy sources are also rising fast. Although oil prices are difficult to predict, clear indications show that prices will continue to grow in the

---

[183] Please note that, according to interviewed BMU staff, the German government was the *de facto* author of the policy paper that was published under the name of "Initiative for an IRENA".

future" (The Government of the Federal Republic of Germany, Initiative for an IRENA 2008:10).

As with regard to climate change, the government emphasizes the adverse effects on the global economy as a whole and the economic burden put on developing countries in particular. It states that rising energy prices negatively affect consumers and businesses (The Government of the Federal Republic of Germany, Initiative for an IRENA 2008:10). It adds that developing countries, which are dependent on fossil fuel imports, are especially vulnerable to oil price increases as these deteriorate their payment balance (Schröder 2004, The Government of the Federal Republic of Germany 2008:3f., The Government of the Federal Republic of Germany, Initiative for an IRENA 2008:10). Due to its declining production costs, renewable energy offers a solution to rising energy prices:

> "Renewable energy brings energy prices under control. With declining production costs, renewable energy guarantees stable energy prices. Indeed, many renewable energy options – particularly small-scale applications – are already competitive" (The Government of the Federal Republic of Germany 2008:4).

The German government also points to the challenge of satisfying globally rising energy demand. This the global challenge it mentions in the first place in the policy papers promoting the creation of an IRENA (Initiative for an IRENA 2008:2, The Government of the Federal Republic of Germany 2008:3f., The Government of the Federal Republic of Germany, Initiative for an IRENA 2008:8). It states that population growth and economic development will lead to a substantial increase in global energy demand. Yet, increasing energy demand is faced with absolutely limited deposits of fossil fuels:

> "The world's population is forecast to grow by 2.5 billion by 2050, reaching a total of some 9.2 billion. In addition, many economies are currently experiencing rapid expansion and industrialisation. As population grows and industry expands, so does the demand for energy. If governments around the world maintain their current policies, the world's energy needs may increase by 50 % or more by 2030. In the past, these needs have been satisfied largely by finite energy sources. These will be exhausted in the future" (The Government of the Federal Republic of Germany 2008:3).

According to the German administration, renewable energy offers the potential to diversify global energy supply and thus reduce worldwide dependency on fossil fuels. In contrast to fossil fuels, renewable energy sources are not faced with absolute limits and are therefore able to cover even substantial increases in global energy demand:

> "Renewable energy (...) will never run dry. Even substantial increases in demand can be met by the enormous energy potential of wind, solar and other renewable energy sources" (The Government of the Federal Republic of Germany 2008:4).

In her speech at the international symposium of the WBGU, Chancellor Angela Merkel (2012b) emphasizes that growing energy scarcity is associated with two major global risks: it leads to price increases and to a global competition over resources. Apart from increasing energy supply on a global scale, the government underlines that renewable energy is universally available and therefore enables countries to meet their energy demand with domestic energy sources, thus reducing dependency on imports:

> "Renewable energy is home-grown, universally available and not reliant on an electricity grid or oil/gas pipeline infrastructure. It reduces dependency on rapidly diminishing fossil fuel resources. Renewable energy is thus an appropriate option for diversifying supply and increasing domestic supply" (The Government of the Federal Republic of Germany 2008:4).

In this context, it is important to note that the German government advances a predominantly economic understanding of energy security. This means that energy security, according to the German administration, is mainly about how to handle an increasing global energy scarcity. This view is widely shared among German decision-makers and experts (see also Röhrkasten, Westphal 2012a).

The government furthermore underlines that the promotion of renewable energy helps to combat poverty by enhancing access to energy. In his speeches at the WSSD 2002 and the Renewables 2004 Conference, Chancellor Schröder emphasizes that two billion people lack access to energy (Schröder 2002c, 2004). At the Renewables 2004 Conference, Chancellor Schröder illustrates the negative impacts of lacking access to energy:

> "Two billion people, about a third of the world's population, have no access to normal energy supplies. They don't have even the basics, such as the energy they would need to drive water pumps in order to have clean drinking water. They don't have the electricity they would need to have access to information, communication, and education. They also don't have the energy they would need to be able to process raw materials and, in doing so, advance their economic development. This makes it clear that those who want to combat poverty and promote development will need to invest in decentralized, renewable energy sources" (Schröder 2004).

In the policy papers promoting an international renewable energy agency, the German government also emphasizes the importance of enhancing access to energy in order to combat poverty. This is illustrated by the following statement:

> "Access to energy services is an important precondition for meeting basic needs and for developing a modern economy. A smoothly functioning energy supply system is important for a country's economic stability. Energy is thus an indispensable element to overcome poverty and to achieve the United Nations' Millennium Development Goals" (The Government of the Federal Republic of Germany, Initiative for an IRENA 2008:10).

The government repeatedly points out that renewable energy is especially suited to provide energy to the poorest in the world as it can be locally employed and offers stand-alone solutions in rural areas without needing to construct expensive new grids. The decentralized character of renewable energy thus facilitates access to energy (BMWi, Dena 2010:5f., BMZ 2008:5, The Government of the Federal Republic of Germany 2008:3f., The Government of the Federal Republic of Germany,

Initiative for an IRENA 2008:8,10, Schröder 2004, Wieczorek-Zeul 2004c). Poverty reduction also receives substantial attention in the Renewables 2004 speeches by Development Minister Heidemarie Wieczorek-Zeul (2004a, 2004b,2004c) and Environment Minister Jürgen Trittin (2004a, 2004b, 2004c), as well as in BMZ statements (BMZ 2008).[184]

The government adds that the promotion of renewable energy goes along with economic opportunities for developing and developed countries alike. In his speech at the IRENA founding conference, the AA Minister of State Erler (2009) speaks of renewable energy as a booster of a third industrial revolution. The government particularly highlights the potential of job creation by renewable energy deployment (Altmaier 2012, BMWi, Dena 2010:5, The Government of the Federal Republic of Germany, Initiative for an IRENA 2008:10f., Schröder 2002b, Trittin 2004a, Wieczorek-Zeul 2004a, Wieczorek-Zeul 2009 [2001]:34f.).[185] It also emphasizes the growth potential of renewable energy, stating that it is becoming economically competitive and attracting high investments (The Government of the Federal Republic of Germany, Initiative for an IRENA 2008:10). With regard to economic development in developing countries, it states that improving access to energy helps to stimulate income-generating activities and macroeconomic growth (The Government of the Federal Republic of Germany, Initiative for an IRENA 2008:10f.). Besides, the government underlines further economic advantages of renewable energy, which appear as especially relevant for developing countries:

> "Since renewable energy installations are often less complex than conventional power facilities, they can be manufactured in many countries and so generate local jobs. Moreover, they are relatively simple to operate and can be managed by trained members of the local workforce" (The Government of the Federal Republic of Germany 2008:4).

The potential of renewable energy to local value creation is also underlined. As renewable energy is universally available, it can be produced domestically. Countries therefore do not need to import energy and thus save foreign currency (Altmaier 2012, BMWi, Dena 2010:5, Trittin 2004a, Wieczorek-Zeul 2004a). The Economics Ministry reinforces the economic advantages of renewable energy, underlining that the renewables sector has worldwide high growth rates (BMWi, Dena 2010:5). The BMELV underlines the economic opportunities especially for rural areas.[186]

Sometimes the German government draws a connection between renewables promotion and security and peace (BMWi, Dena 2010:5f., Erler 2009, Merkel

---

[184] See also BMZ, Renewable energies: from the gas stove to the hydropower plant, http://www.bmz.de/en/what_we_do/issues/energie/renewable_energies/index.html (accessed July 16, 2013).
[185] See also BMELV, Renewable resources, http://www.bmelv.de/EN/Agriculture-RuralAreas/RenewableResources/renewable-resources_node.html (accessed January 14, 2013).
[186] BMELV, Renewable resources, http://www.bmelv.de/EN/Agriculture-RuralAreas/Renewable Resources/renewable-resources_node.html (accessed January 14, 2013).

2012b, Schröder 2004, Wieczorek-Zeul 2004c). In his speech at the Renewables 2004 Conference, Chancellor Gerhard Schröder (2004) emphasizes that reducing the worldwide dependence on oil also diminishes the vulnerability to terrorist attacks, as these often target oil infrastructure. Development Minister Wieczorek-Zeul goes into a similar direction, stating that "In the past there have been wars for the access to oil. There will never be a war on the access to the Sun" (Wieczorek-Zeul 2004c). AA Minister of State Erler (2009) underlines that renewable energy contributes to global peace as it helps to reduce global rivalries over fossil fuels. Chancellor Angela Merkel emphasizes that growing energy scarcity may lead to global competition over resources that is "capable of triggering conflicts across our civilization which will cost us dear" (Merkel 2012b). The BMZ website[187] also mentions the contribution of renewables promotion to security and peace. The BMWi points to the risks of political conflict and military confrontation. It states that fossil fuel reserves are predominantly located in politically unstable regions, as for example the Middle East and the Caspian Region (BMWi, Dena 2010:6).

4.2.2 Renewable Energy Options: Sustainability and Electricity Markets

According to the German government, transboundary cooperation should promote all renewable energy sources – if these are produced in a sustainable manner. In its speeches and documents on global renewable energy governance, sustainability appears as the core criterion for deciding which forms of renewable energy to promote (see, for example, BMZ 2008, Gabriel 2009, Merkel 2012b, Schröder 2002c, Schröder 2004, Westerwelle 2012, Wieczorek-Zeul 2004c, Wieczorek-Zeul 2009 [2001]).[188] Consequently, the German government envisaged as IRENA's main task the promotion of widespread and *sustainable* use of renewable energy (see, for example, Initiative for an IRENA 2008:2, The Government of the Federal Republic of Germany, Initiative for an IRENA 2008:6). Representatives repeatedly emphasize that the ecological, economic and social dimensions of sustainability have to be taken into account equally (see, for example, Gabriel 2009, Merkel 2012a, Merkel 2012b). Applying the principle of sustainability to the energy system, the German government states that it aspires to a worldwide energy supply that is secure/reliable, affordable/cheap and clean (The Government of the Federal Republic of Germany 2008:7, The Government of the Federal Republic of Germany, Initiative for an IRENA 2008:6). Yet, a closer look at the documents on global

---

[187] http://www.bmz.de/en/what_we_do/issues/energie/renewable_energies/index.html (accessed January 21, 2013).
[188] BMELV, Renewable resources, http://www.bmelv.de/EN/Agriculture-RuralAreas/Renewable Resources/renewable-resources_node.html (accessed January 14, 2013).

renewable energy governance reveals that the ecological dimension of sustainability receives most attention: sustainability is foremost associated with protecting the climate and reducing pressure on the environment in general (see, for example, BMZ 2013a, Gabriel 2009, Schröder 2002c, The Government of the Federal Republic of Germany 2008:4, The Government of the Federal Republic of Germany, Initiative for an IRENA 2008:8f.).[189] The affordability of renewable energy also receives attention. In her speech at the 2011 Eurosolar conference, Heidemarie Wieczorek-Zeul underlines the necessity of renewable energy being affordable, especially in developing countries:

> "We must also accept that a minimum income is initially required for people to be able to actually afford "clean" energy. What is true for the industrialised countries is doubly true for the developing countries: sustainable energy for the future must be affordable. I see the task for development policy as one of working with the partner countries to ensure that the huge gap is bridged between, on the one hand, the high prices of anything available on the market connected with renewable energies and efficient equipment for end-users and, on the other, the low purchasing power of poor people" (Wieczorek-Zeul 2009 [2001]:35).

Trittin (2004c) and Schröder (2004) also point to the need for cost reductions in order to make renewable energy affordable to developing countries. In a policy paper promoting IRENA, the government states that many renewable energy technologies, particularly small-scale options, are already competitive (The Government of the Federal Republic of Germany 2008:4,9). BMZ (2008:7) adds that "renewable energies are in many instances a lower-cost alternative to conventional forms of energy."

Although the German administration mostly speaks about renewable energy in general without differentiating between different renewable sectors and sources,[190] a closer look at the speeches and documents reveals that the electricity sector is dominating the German view on global renewable energy governance. When presenting the market development of renewable energy in one of the policy papers on IRENA, the government for example either speaks of primary energy supply in general or of the power sector in particular. Developments in the realm of heating/cooling and transport are not mentioned (The Government of the Federal Republic of Germany 2008:5). At the Renewables 2004 conference, Trittin underlines that many countries have set themselves ambitious targets for increasing the share of renewable energy. Almost all the examples he gives cover the electricity sector (Trittin 2004b, Trittin 2004c). Within the electricity sector, wind and solar energy receive most attention. A BMU press statement on an IRENA meeting in 2013, Environment Minister Altmaier singles out solar and wind energy when illus-

---

[189] See also BMELV, Renewable resources, http://www.bmelv.de/EN/Agriculture-RuralAreas/RenewableResources/renewable-resources_node.html (accessed January 14, 2013).
[190] See, for example, Initiative for an IRENA 2008, The Government of the Federal Republic of Germany 2008, The Government of the Federal Republic of Germany, Initiative for an IRENA 2008.

trating the worldwide expansion of renewable energy: "From oil to solar energy, from coal to wind energy – more and more countries are switching to renewable energies" (BMU 2013b).

The German government advances a win-win framing primarily with regard to solar and wind energy, and to a lesser degree, to small hydropower and geothermal energy. When it mentions the disadvantages of these renewable energy options, it quickly points at ways these can be tackled. When the aministration underlines the growth potential and the achieved and predicted costs reductions of renewable energy, it gives wind and solar energy as examples (BMZ 2008:28-30, The Government of the Federal Republic of Germany 2008:5, The Government of the Federal Republic of Germany, Initiative for an IRENA 2008:11, Trittin 2004b, Trittin 2004c). This is also illustrated by the following statement on the AA website: "even considerable increases in demand could be met if we harness the enormous potential of the wind, the sun and other renewable sources."[191] BMZ (2008:29f.) presents wind energy as the renewable energy option with the highest growth rate. Yet, the German government also admits that the overall contribution to the global energy supply of solar, wind, tidal and geothermal energy still remains very small (BMZ 2008:28-30,35, The Government of the Federal Republic of Germany 2008:5). When presenting positive examples for the promotion of renewable energy in different parts of the world, Trittin (2004b, 2004c) points at solar and wind energy, small hydropower and geothermal energy. The Government of the Federal Republic of Germany (2008:4) presents hot water from solar collectors, electricity from small hydropower and wind power as examples for competitive renewable energy options. The BMZ (2008:29f.) emphasizes that with very good wind conditions and state-of-the-art technology, wind energy is already competitive vis-à-vis conventional forms of energy. It adds that wind energy also makes economic sense in many developing countries. Even though the transport, installation and maintenance of wind energy might be more expensive in those countries due to less developed infrastructures, the wind conditions are often very good. BMWi & Dena (2010:11-14) reinforce that wind energy is a clean source of energy with competitive prices, has a broad scope of application and can be easily mixed with other renewable energy options. It also points at the potential of creating jobs and value added in poorer regions. The BMZ (2008:28-30) furthermore highlights the advantages of solar and wind energy for enhancing access to energy in remote areas. It states that in remote areas without connection to the national grid, solar energy can already be the most efficient option – due to the robust character of solar cells and their low maintenance requirements. BMWi, Dena (201015-20) mention further advantages

---

[191] AA, Promoting renewable energies worldwide: International Renewable Energy Agency founded, http://www.auswaertiges-amt.de/EN/Aussenpolitik/GlobaleFragen/Klima/Aktuelles/090123-Irena-Gruendung_node.html (accessed January 14, 2013, last updated: January 27, 2009).

of solar cells: emission-free electricity generation of low noise, broad scope of application, long period of application and environmental friendliness. Yet, Wieczorek-Zeul (2009 [2001]:37f.) and BMZ (2008:28f.) acknowledge that the costs of photovoltaic cells are still relatively high. In order to achieve cost reductions, Wieczorek-Zeul (2009 [2001]:37f.) calls for improved technological developments and increasing demand. The BMZ (2008:28-30) presents a further disadvantage of solar and wind energy: its variability depending on weather, time of day and season. Yet, it states that this can be compensated by combining the use of solar and wind energy with other sources and by improving storage technologies. With regard to geothermal power, the BMZ (2008:35) acknowledges that the high initial investment costs and the associated financial risks pose an important barrier to the expansion of geothermal power. Yet, it also points out that once the potential of geothermal energy has been identified, it is a low-risk and low-cost alternative for energy provision. On its website it states that "apart from the initial investment, though, alongside hydropower geothermal energy is one of the least expensive and most profitable sources of power".[192] It furthermore underlines the inexhaustible potential and permanent availability of geothermal power (BMZ 2008:35). BMWi & Dena (2010:31f.) reinforce the inexhaustible potential and permanent availability. In addition to low costs, it also underlines the environmental friendliness of geothermal power

The German government raises sustainability concerns and thus trade-offs with regard to large hydropower and biofuels. In the official statements and pronunciations on global renewable energy governance, these concerns are raised by the BMZ. Yet, such concerns are communicated by representatives of other ministries as well.[193] The BMZ (2007:10) singles out hydropower and biofuels when stating that the use of certain forms of renewable energy can also have negative impacts on the environment if sustainability is not taken into account. In its deliberations on hydropower, the BMZ mentions two major advantages: its low production costs and uninterrupted energy supply (except for droughts). It also points at possible secondary benefits with regard to irrigation, water supply, flood protection and river navigability. BMWi & Dena (2010:29f.) reinforce that hydropower is a reliable and low-cost renewable energy option that contributes to grid stability. It reduces dependence on energy imports and can provide the basis for economic development in remote areas. However, the BMZ underlines that the sustainability of new hydropower plants needs careful examination, as large dams go hand in hand with major interventions into natural environments and the lives of the local

---

[192] http://www.bmz.de/en/what_we_do/issues/energie/renewable_energies/geothermal_energy/index.html (accessed January 21, 2013).
[193] Interviews with executive staff, Division International Energy Policy, External Energy Policy, BMWi, November 2012 and various staff members of the Division International and European Affairs of Renewable Energy, BMU.

population. If a local population needs to be resettled, compensation and further income-generating activities should be provided for (BMZ 2008:31f., BMZ 2007:23)[194]. BMWi & Dena (2010:29f.) also differentiate between large and small hydropower. However, it addresses possible disadvantages of large hydropower only indirectly, stating that small hydropower does not pose any disadvantages to the environment or the society and that the modernization of existing plants can lead to the harmonization of large hydropower plants with the environment. In BMZ publications, biofuels receive much more attention than hydropower. In his foreword to the BMZ strategy paper on biofuels, state secretary Beerfeltz (FDP, 2009-2013) chooses drastic words when referring to the trade-offs involved in promoting biofuels:

> "The rising worldwide demand for non-food agricultural products and renewable energies presents a major opportunity for rural regions in developing countries. But good intentions are not enough to guarantee good outcomes. Well-meant but superficial "eco-romanticism" in Germany can cause hunger, forced displacement and even death in developing countries" (BMZ 2011:3).

The BMZ (2007:22f., 2011:13) underlines the potential of biofuels to guaranteeing energy supply and providing export opportunities while saving expenses on energy imports. The BMZ (2011:4) adds that the production of biofuels can improve living conditions by generating income, providing access to energy and rehabilitating degraded land. It highlights that the growing demand for agricultural feedstock brings with it several problems. It increases the pressure on natural resources and contributes to rising food prices – which poses a serious threat to food security in developing countries:

> "It is now undisputed that the additional demand for farmed feedstock stimulated by targeted support policies in the USA and EU, coupled with other factors (harvest losses, financial speculation, climate change), have already contributed to raising the prices of agricultural commodities" (BMZ 2011:10).[195]

The BMZ (2013b) furthermore points to the ambiguous effects on emissions reduction and competition with natural resources. It also mentions additional social and ecological risks, such as the displacement of small farmers, poor working conditions on plantations and destruction of environment and biological diversity (BMZ 2011:4, 2008:33f.). According to the BMZ, this especially applies when biofuels are produced for the purpose of export:

> "There is the danger that plantations geared mainly to export will place small farmers at a disadvantage and be detrimental to the ecosystem" (BMZ 2007:23).

---

[194] BMZ, Renewable energies: hydropower, http://www.bmz.de/en/what_we_do/issues/energie/renewable_energies/hydropower/index.html (accessed January 21, 2013)
[195] To substantiate its argument, the BMZ cites a study commissioned by the European Commission (Fonseca et al. 2010).

However, in a later publication the BMZ also acknowledges that producing biofuels for export can make economic sense for developing countries, as export markets can provide stable demand and relatively high prices. In addition to an increased economic output from rural regions, developing countries can also profit from rising tax revenues (BMZ 2011:13). BMWi & Dena (2010:28) present a more favorable view on biofuels. They point out that biofuels are less dangerous than fossil fuels as biofuel accidents do not pose environmental threats. They underline that biofuels help to replace expensive fossil fuels imports and that they are almost neutral in greenhouse gas emissions. They further state that sustainability standards have been developed to ensure public approval for biofuels.

Traditional biomass is the only renewable energy option that is presented as being unworthy of support. In a policy paper promoting IRENA, the German government states that traditional biomass is used in an inefficient and nonsustainable way, producing negative implications for human health and the environment (The Government of the Federal Republic of Germany, Initiative for an IRENA 2008:8). These aspects are also reinforced by the BMZ. It adds that traditional biomass remains the most prevalent source of renewable energy, and for many people the only source of energy. As a consequence, it calls for improvements in the efficiency and environmental friendliness of biomass use (BMZ 2008:33f., 2007:23).[196]

4.2.3 Barriers: Markets and Policies Favoring Conventional Energy

The German government points out that current energy market structures are still geared towards conventional forms of energy, confronting renewable energy with several disadvantages. In the policy papers promoting IRENA (The Government of the Federal Republic of Germany 2008:5-7, The Government of the Federal Republic of Germany, Initiative for an IRENA 2008:6-14), it highlights the fact that fossil and nuclear energy rely on fully developed technologies, established industries, strong market structures and powerful companies. It adds that the requirements and structures of current energy markets build on conventional forms of energy and hinder the spread of renewable energy, as renewable energy needs different technology and service structures and requires energy costs to be calculated differently. The government adds that the competitiveness of renewables is furthermore undermined as energy prices do not include the external costs of energy production and consumption. At the Renewables 2004 Conference, the Development Minister Wieczorek-Zeul (2004a) highlights that that "distorted prices are one

---

[196] BMZ, Biomass, http://www.bmz.de/en/what_we_do/issues/energie/renewable_energies/biomass/index.html (accessed July 16, 2013).

of the most effective barriers to an expansion in the market for renewables." In her speech at the Eurosolar conference in 2001, she particularly refers to the external costs that energy consumption poses on the environment:

> "When prices come to include the environmental costs of energy consumption we will have made a decisive step forward. If the true ecological costs were included in the price of environmentally harmful energy consumption, then the consumer would be able to make an informed choice about the best form of energy in economic and environmental terms" (Wieczorek-Zeul 2009 (2001):37).

She reinforces that it is no longer technical but economic issues that dominate concerns on renewable energy (Wieczorek-Zeul 2009 [2001]:36). BMWi & Dena (2010:5) reinforce that market prices for fossil fuels and nuclear energy only represent a fraction of the real costs to society. If external costs for environmental damages or political conflicts were included, renewable energy would easily compete with conventional forms of energy and often even represent the lower-cost alternative.

According to the German government, political structures and policy-making also favor conventional forms of energy. Financial subsidies for fossil fuels serve as a common example for the political discrimination of renewable energy vis-à-vis fossil fuels (BMZ 2008:27, Gabriel 2009, The Government of the Federal Republic of Germany 2008:7, The Government of the Federal Republic of Germany, Initiative for an IRENA 2008:12, Wieczorek-Zeul 2009 (2001):36f.). Wieczorek-Zeul (2009 [2001]:36) adds that renewable energy is also faced with direct discrimination, "for example when special taxes or duties are levied on solar installations because they are classed as luxury goods or because they are quite simply seen as a profitable source of revenue."

The government points out that lacking political will and public awareness prevents policy-makers from taking the necessary steps to change the political framework of energy markets in a way that renewable energy can actually compete with conventional forms of energy. In his speech at the Renewables 2004 Conference, Schröder (2004) identifies lack of political will and hesitance to address renewable energy as major barriers to its expansion, pointing to changing mindsets as an important precondition. In his closing address at the Renewables 2004 Conference, Trittin goes in a similar direction. He states that as a result of the conference "renewables can no longer be ignored" (Trittin 2004b). Schröder (2002b) and Altmaier (2012) point at prejudices that hinder the spread of renewable energy: sustainability and environmental protection are assumed as obstructing economic growth and jobs creation; yet, they each underline that the promotion of renewable energy shows that the opposite is true. The German government claims that political will is missing both within states and at international level. With regard to the international level, it highlights that there is a need to change mindsets and enshrine the promotion of renewable energy in international processes on trade, investment, environment and energy (The Government of the Federal Republic of Germany 2008:7, The Government of the Federal Republic of Germany, Initiative for an IRENA 2008:14). At the Renewables 2004 Conference, Schröder states that renewable energy does not receive sufficient attention in international policy-making:

> "In the United Nations framework, for example, renewable energies continue to play a subordinate role. There are lots of organizations that include them in their agendas but I still don't see a source of impetus that would steadily move things forward towards achieving this objective on a global scale" (Schröder 2004).

Yet, Schröder (2004) testifies to "a growing willingness in the international community to assume responsibility for a sustainable energy future". In her speech at the IRENA founding conference, Heidemarie Wieczorek-Zeul also claims that international energy policy does not pay sufficient attention to renewable energy:

> "Together we can change the "fossilised" approach to energy policy. Renewable energies will soon move out of their niche existence to take their place at the heart of international energy policy. We can make sure that happens" (Wieczorek-Zeul 2009:108).

When promoting the founding of IRENA, the German government points at inadequate national policies as being a major barrier to the spread of renewable energy. The reason behind this is not only the absence of political will, but also the lack of capabilities:

> "To date, only a minority of states have shown themselves willing or able to introduce efficient renewable energy policies (including appropriate legislation and institutional frameworks), develop effective industries, assess their national potential and promote research, development, education and training" (The Government of the Federal Republic of Germany 2008:7).

In the policy papers promoting IRENA, the government highlights that information and technological know-how is often missing (The Government of the Federal Republic of Germany 2008:5, The Government of the Federal Republic of Germany, Initiative for an IRENA 2008:13). Policy-makers often lack information on efficient renewable energy policies, industry requirements and on research, development and training. Misinformation also prevails. Thus, governments are not able to develop appropriate strategies to structurally reform their energy systems. An additional barrier is the lack of technological know-how. A high share of renewable energy puts new demands on the design and management of energy systems and grids. In particular, developing countries often do not have the expertise to produce renewable energy technologies nor the means to buy, maintain and repair the necessary equipment. The lack of finance in developing countries is an aspect also mentioned by Wieczorek-Zeul (2004b) at the Renewables 2004 Conference. At the same conference, Chancellor Schröder underlines the need for cost reductions in order to enhance the affordability of renewable energy for developing countries:

"But it is only when we are able to bring about a significant reduction in the cost of renewable energies that we will have improved opportunities to promote their use in poorer countries" (Schröder 2004).

Further obstacles mentioned by the German government include import restrictions, as well as technical barriers and insecure financing of renewable energy projects (The Government of the Federal Republic of Germany, Initiative for an IRENA 2008:6).

4.2.4 Tasks: Improving Domestic Regulatory Frameworks

According to the German government, the most important step for worldwide promotion of renewable energy is to improve regulatory frameworks within countries. It especially emphasizes the importance of time-bound targets on increasing the renewable energy share within the national energy supply (Altmaier 2012, BMU, BMZ & AA 2008, Schröder 2004, The Government of the Federal Republic of Germany 2008, Trittin 2004b, Trittin 2004c). At the Eurosolar conference in 2001, Development Minister Wieczorek-Zeul presents the German feed-in legislation as a positive example for other countries:

> "Here in Germany we have drawn the appropriate conclusions and, with the Renewable Energy Sources Act, have created a regulatory framework that is helping to increase sales of electricity from renewable sources, thereby contributing to a lowering of the generation costs. I consider this route in particular to be of interest for the developing countries as well. It opens up opportunities for renewable energies because it makes it possible to charge different prices depending on the technology used. It does not place any burden on the public purse since the additional costs are borne by the electricity consumers" (Wieczorek-Zeul 2009 [2001]:39).

She furthermore emphasizes that the legislation has clear advantages over quota models and other policy instruments:

> "This unbureaucratic price-regulating mechanism will, in the medium to long term, help all the different regenerative forms of energy compete with conventional sources of energy. I see here a crucial advantage over other promotional models, such as the quota systems which the USA for example is pursuing. This is also our starting point for bringing our influence to bear on the World Bank, an institution that in other instances has shown a tendency to follow the American example" (Wieczorek-Zeul 2009 [2001]:39).

The positive effects of feed-in legislation are also underlined by further representatives of the German government (BMWi, Dena 2010:4, Schröder 2002b, Schröder 2002c, Schröder 2004, Trittin 2004a, Trittin 2004c).[197]

In the government's view, an important task for transboundary policy-making is to enhance the commitment of governments to promote renewable ener-

---

[197] AA, energy security, http://www.auswaertiges-amt.de/EN/Aussenpolitik/GlobaleFragen/Energie/Energiesicherheit_node.html (accessed January 10, 2014).

gy within their countries. These commitments can take different forms. At the WSSD 2002, it argued for the establishment of internationally agreed targets on increasing the worldwide share of renewable energy. As no international consensus on such targets could be reached, it later on pushed for voluntary commitments as a second best option. At the Renewables 2004 Conference, Chancellor Gerhard Schröder (2004) underlines that it was the failure of the attempt to establish internationally agreed renewable energy targets that inspired him to organize an international conference on renewable energy promotion in Germany. In his speech, he makes the case for voluntary commitments as a conference outcome. The importance of voluntary commitments and actions is also underlined by Wieczorek-Zeul (2004a) in her closing speech at the Renewables 2004 Conference. She adds that these commitments should be followed by a voluntary monitoring process. At the same conference, Trittin (2004b) emphasizes that multilateralism and voluntary commitments are not opposites, but belong together. Out of the commitments achieved, Trittin particularly highlights time-bound targets that countries set themselves for increasing the share of electricity based on renewable sources within their countries (Trittin 2004b, 2004c).

In addition, transboundary policy-making shall support countries to improve their policy frameworks for renewable energy promotion (BMU, BMZ 2008, BMU, BMZ & AA 2008, Schröder 2004, The Government of the Federal Republic of Germany 2008:7f.). In a policy paper promoting IRENA, the German government states that transboundary cooperation should help countries to achieve their national targets on renewable energy promotion (The Government of the Federal Republic of Germany 2008:7). In a joint press statement on IRENA's creation, BMU, BMZ and AA elaborate further on the areas on which global renewable energy governance should focus in order to support governments. They particularly refer to policy advice, capacity-building, financing and technology transfer:

> "Worldwide many countries have set themselves ambitious targets for increasing the share of renewable energies in national energy consumption. A large majority of these countries would like detailed consultation and advice on this process. IRENA will aid its member states in adapting their political framework conditions, capacity building and improving financing and technology transfer for renewable energies" (BMU, BMZ & AA 2008).

In the policy papers promoting IRENA, the German government adds two more areas to the tableau: knowledge and research, and networking. At the Renewables 2004 Conference, Wieczorek-Zeul (2004a) goes in a similar direction, proposing the envisaged multistakeholder network REN21 focus on policy frameworks, financing options, research and capacity development.

With regard to knowledge and research, the German government emphasizes that a comprehensive understanding of existing activities promoting renewable energy and available resources is needed in order to be able to advise governments and inform the public on "the benefits and potential offered by renewable energy"

(The Government of the Federal Republic of Germany, Initiative for an IRENA 2008:15). Transboundary policy-making should promote best practices and lessons learned, disseminate reliable data on renewable energy potential, and enhance information on effective financial mechanisms and technological expertise. According to the government, transboundary cooperation should strengthen research on the following issues: uses and potentials of renewable energy, policy instruments, economic incentives, investment, technology, grids, conservation, storage and efficiency (Initiative for an IRENA 2008:2-4, The Government of the Federal Republic of Germany 2008:6-8, The Government of the Federal Republic of Germany, Initiative for an IRENA 2008:15,17). At the Renewables 2004 Conference, Wieczorek-Zeul (2004b, 2004c) reinforces the importance of learning from successful examples and experience.

Another area of international cooperation the German government highlights is policy advice (Initiative for an IRENA 2008:2-4, The Government of the Federal Republic of Germany 2008:6-8, The Government of the Federal Republic of Germany, Initiative for an IRENA 2008:15f.). According to the German government, many countries lack knowledge as to their renewable energy potential and available policy options. International cooperation shall help to fill this gap. It presents the following areas as relevant for policy advice: selection and adaptation of energy sources, technology and system configurations, business models, organizational and regulatory frameworks, and financing. In his speech at IRENA's founding conference, Environment Minister Gabriel highlights that both developing and industrialized countries need policy advice on political and legal frameworks in order "to create the right incentives and securities for investment" (Gabriel 2009:103f.). The importance of developing policy recommendations is also highlighted by Trittin (2004b) and the BMZ.[198]

With regard to technology transfer and financing, the German government emphasizes that "enormous efforts are required to facilitate technology transfers and investments in developing countries" (The Government of the Federal Republic of Germany, Initiative for an IRENA 2008:16). In the policy papers promoting IRENA, the government repeatedly underlines the importance of transferring technologies (Initiative for an IRENA 2008:2-4, The Government of the Federal Republic of Germany 2008:6-8, The Government of the Federal Republic of Germany, Initiative for an IRENA 2008:16). Schröder (2004), in his speech at the Renewables 2004 Conference, also argues for North-South transfers of finance and technology. In this context, he underlines the potential of the Kyoto-Protocol for incentivizing industrialized countries to invest in the promotion of renewable energy in

---

[198] BMZ, International energy policy: Follow-up process to the International Conference for Renewable Energies –Renewables 2004, http://www.bmz.de/en/what_we_do/issues/energie/international_energy_policy/renewables/index.html (accessed January 23, 2013).

developing countries. Publications by different ministries reinforce the importance of the Clean Development Mechanism, which was established by the Kyoto-Protocol, for transferring finance and technology (BMU, BMZ 2012b:22, BMWi, Dena 2010:10, BMZ 2008:13). The need to transfer finance, technology and know-how to developing countries is also emphasized by Wieczorek-Zeul (2009 [2001]:39-41) at the Eurosolar conference in 2001. At the Renewables 2004 Conference, Trittin (2004a, 2004c) points to the importance of development finance and development cooperation in general.

With regard to capacity building, the government points to the need to form human capital for expanding renewable energy sectors. Thus, transboundary cooperation should improve training on renewable energy, encompassing a wide range of qualifications and educational institutions at different levels (Initiative for an IRENA 2008:2-4, The Government of the Federal Republic of Germany 2008:6-8, The Government of the Federal Republic of Germany, Initiative for an IRENA 2008:17).

The German government adds that transboundary cooperation should also focus on networking activities to facilitate synergies between different organizations and institutions dealing with renewable energy and ensure that international political processes on issues such as trade, investment, environment and energy take account of renewable energy (Initiative for an IRENA 2008:2-4, The Government of the Federal Republic of Germany 2008:6-8, The Government of the Federal Republic of Germany, Initiative for an IRENA 2008:17). In her speech at the 2001 Eurosolar conference, Wieczorek-Zeul underlines the importance of effective networking:

> "What matters most is that these institutions and initiatives are effectively networked, that a common vision is strengthened and that political stimuli are developed for coordination" (Wieczorek-Zeul 2009 [2001]:41).

With regard to bioenergy and biofuels, the German government points at a further task for transboundary policy-making: establishing minimum sustainability requirements. Sustainability requirements shall ensure that the potential of bioenergy can be explored without having negative side effects. This is also established in the coalition agreement of 2009:

> "We want to bring forward initiatives for an internationally valid sustainability certificate for biomass that would cover fuel and electricity production as well as its use for both foodstuff and feedstuff" (CDU, CDU & FDP 2009:35f.).

In his foreword to a BMZ publication on biofuels, BMZ State Secretary Beerfeltz particularly refers to possible conflicts between biofuel production and food supply that should be tackled:

> "Let me therefore make one point quite clear: where conflict arises, and adequate food supply is always our priority before seeking additional resources for a sustainable energy supply. Yet at

the same time, I am convinced that any such goal conflict can be resolved productively; in fact, we need not let it arise in the first place.

Our core message is clear: we want to make resolute use of the opportunities arising for developing countries from the growing worldwide demand for biofuels. At the same time, we want to eliminate possible risks and side-effects. To this end, transparent standards on human rights and social and environmental responsibility must become the touchstone of all action taken. If this can be accomplished – and I am convinced that it can – the cultivation of energy crops becomes a development-policy gain for all involved. In that event, biofuel production and food security will not be at odds – quite the opposite: with the additional income created, and through general stimulation of rural regions, we will then make a significant contribution to both to climate protection and to the fight against hunger!" (BMZ 2011:3).

To prevent goal conflicts, he calls for standards on human rights and social and environmental responsibility. The BMZ (2011:4) elaborates further on this issue, claiming that international institutions dealing with biofuels should base their action on the following principles: 1) primacy of local food security and the supply of drinking water, 2) greenhouse gas reductions and conservation of biodiversity and other ecosystem services, 3) compliance with minimum social standards, 4) involvement of local communities, 5) respect of existing land and water rights, and 6) participation of the population in local value creation. The BMZ (2011:16f.) adds that the risks involved in biofuel production concern agricultural production in general; however, since biofuels are the object of targeted policy incentives, biofuels markets are particularly suited for implementing sustainability requirements. This can have positive side-effects on the production of other agricultural goods as well.

4.2.5 Global Governors: Providing Information and Advice on Renewable Energy

In its deliberations on global governors who deal with renewable energy, the German government particularly points at IRENA, IEA and REN21, underlining their role in enhancing information and advice on renewable energy. It also refers to different UN organizations, the G8 and global partnerships that deal with renewable energy, such as REEEP and GBEP.

IRENA

In the German government's deliberations, IRENA is given a central role. The AA and BMU websites mention IRENA in the first place when referring to global governors on renewable energy.[199] And on the BMZ and BMWi websites, IRENA is

---

[199] AA, Energy, http://www.auswaertiges-amt.de/EN/Aussenpolitik/GlobaleFragen/Energie/Uebersicht_node.html (accessed July 18, 2013); BMU, renewable energy – EU/international, http://www.erneuerbare-energien.de/en/topics/eu-international/ (accessed July 18, 2013).

mentioned in second place.[200] The German administration particularly emphasizes three characteristics of IRENA: its institutional structure as an intergovernmental organization, its global approach (focusing on industrialized and developing countries alike) and its exclusive focus on renewable energy.[201] In a joint publication on the German contribution to SE4All, BMU and BMZ underline IRENA's support to its member states and its function as a global voice for renewable energy:

> "With analyses, advisory and networking activities, IRENA supports its roughly 100 Member States in creating an enabling environment that is conducive to the expansion of renewable energy. The Agency is the voice of renewable energy in international debates" (BMU, BMZ 2012b:23).

In a press statement on the 2013 meeting of the IRENA Assembly, Environment Minister Altmaier highlights the achievements of IRENA, underlining in particular IRENA's function as an international hub for renewable energy:

> "In recent months there has been an impressive increase internationally in the political commitment to the expansion of renewable energies. Many countries have launched a transformation of their energy systems. This is also thanks to the work of IRENA, which is becoming an increasingly important hub internationally, as we had hoped when we set it up" (BMU 2013b, cites Altmaier).

On their websites, BMU, BMZ and AA specify that IRENA advises its member countries on policy frameworks, capacity-building, funding, technology transfer and renewable energy knowledge.[202] The outlined activities widely match with the scope the German government envisaged when initiating IRENA's creation (The Government of the Federal Republic of Germany, Initiative for an IRENA 2008:6). The BMU additionally highlights that IRENA facilitates access to technological exper-

---

[200] BMZ, Energy policy – development-policy involvement in international initiatives, http://www.bmz.de/en/what_we_do/issues/energie/international_energy_policy/index.html (accesed July 18, 2013); BMWi, Internationale Energiepolitik, http://www.bmwi.de/DE/Themen/Energie/Energiepolitik/internationale-energiepolitik.html (accessed July 18, 2013). The BMZ mentions REN21 in first place, the BMWi the IEA.
[201] BMU, The International Renewable Energy Agency (IRENA), http://www.erneuerbare-energien.de/en/topics/eu-international/irena/ (accessed July 18, 2013); BMZ, International energy policy – International Renewable Energy Agency (IRENA), http://www.bmz.de/en/what_we_do/issues/energie/international_energy_policy/irena/index.html (accessed July 18, 2013); BMZ, International energy policy – International Renewable Energy Agency (IRENA), http://www.bmz.de/en/what_we_do/issues/energie/international_energy_policy/irena/index.html (accessed July 18, 2013); AA, IRENA: Promoting renewable energy worldwide, http://www.auswaertiges-amt.de/EN/Aussenpolitik/GlobaleFragen/Energie/IRENA-Gruendung_node.html (accessed July 18, 2013).
[202] BMU, The International Renewable Energy Agency (IRENA), http://www.erneuerbare-energien.de/en/topics/eu-international/irena/ (accessed July 18, 2013); BMZ, International energy policy – International Renewable Energy Agency (IRENA), http://www.bmz.de/en/what_we_do/issues/energie/international_energy_policy/irena/index.html (accessed July 18, 2013) ; AA, IRENA: Promoting renewable energy worldwide, http://www.auswaertiges-amt.de/EN/Aussenpolitik/GlobaleFragen/Energie/IRENA-Gruendung_node.html (accessed July 18, 2013)

tise, economic and resource data and renewable energy scenarios. BMU staff add that IRENA should improve its capacities in the modeling of renewable energy markets – so that it can develop alternative scenarios to those provided by the IEA.[203] In his Financial Times Deutschland article, Environment Minister Altmaier (2012) reinforces that IRENA serves as a network hub for international cooperation, as a driving force for innovation and as an important voice within the global public. Again, this interpretation widely matches with the ideas advanced by the German government during IRENA's founding process. Here, the German government emphasized that IRENA should influence international policy-making, assuming coordinative functions and creating a "momentum for renewable energy on an international level" (The Government of the Federal Republic of Germany 2008:7).

Altmaier (2012) interprets IRENA's widespread membership as an indicator for worldwide advancement on the promotion of renewable energy. On its website, the BMZ presents a similar interpretation:

> "The rapid growth in membership demonstrates the importance that governments around the world attach to renewable energies. They have recognized that climate change and the growing scarcity of fossil resources necessitate a fundamental restructuring of energy systems."[204]

This interpretation is in line with the character the German administration had envisaged for IRENA: being a governmental organization that incorporates interested and like-minded countries (The Government of the Federal Republic of Germany, Initiative for an IRENA 2008:4). However, in the expert interviews several representatives of the government recognize that IRENA does not only involve like-minded countries and that member states held diverse motivations to join IRENA. As a consequence, among the IRENA members are governments that merely expected access to financial means or planned to weaken IRENA's role from within (see also chapter 3.1.3).[205] Altmaier furthermore states that IRENA's existence provides important policy signals for a global energy transition:

> "IRENA (…) is a signal that the start into a new energy era has already begun all over the world. It is a symbol that governments cooperate and are willing to learn from each other. And

---

[203] Interviews with executive staff, International Cooperation, BMU, December 2012; executive staff (#1), Division International and European Affairs of Renewable Energy, BMU, December 2012; staff member, Division International and European Affairs of Renewable Energy, BMU, November and December 2012.
[204] BMZ, International energy policy – International Renewable Energy Agency (IRENA), http://www.bmz.de/en/what_we_do/issues/energie/international_energy_policy/irena/index.html (accessed July 18, 2013).
[205] This aspect was highlighted by BMU and AA staff in the interviews.

it is an exemplary instrument to incarnate a major political and economic vision" (Altmaier 2012, translation S.R.).[206]

In the interviews, representatives of the BMWi assess IRENA in a way that clearly differs from the German government's public deliberations. They question IRENA's value added, particularly vis-à-vis the IEA. As the BMWi was not in charge of IRENA during the analyzed period of time, it did not pronounce publicly on the issue. Therefore, its view did not enter official government statements and the German administration's pronouncements on the agency only reflect the perspective of BMU, BMZ and AA.

IEA

The German government furthermore stresses the importance of the IEA. It particularly attributes much space to the IEA in the policy papers promoting the founding of an IRENA (Initiative for an IRENA 2008:4f., The Government of the Federal Republic of Germany 2008:9). On their websites, the AA and the BMWi refer to the IEA as one of the main global governors on (renewable) energy.[207] In one of the policy papers promoting IRENA, the German government summarizes IEA's scope of activities as follows:

> "The IEA has designed arrangements for emergency preparedness, analyses and monitors developments on the international oil and gas market, undertakes policy analysis and cooperation, collects and processes data (World Energy Outlook), fosters energy technology, and among others focuses on energy efficiency and environmental issues. It also produces extensive energy statistics that also cover some non-IEA countries. The World Energy Outlook is the IEA's main publication, quoted as an important reference worldwide. IEA's regular country reports reviews the respective energy policies" (Initiative for an IRENA 2008:5)

In the policy papers on IRENA's creation, the German administration praises IEA's role in guiding energy research and development priorities. It also accredits the IEA's comprehensive know-how and long-standing experience in the field of renewable energy, highlighting in particular the contribution of the IEA Working Party on Renewable Energy Technologies (Initiative for an IRENA 2008:4f., The Government of the Federal Republic of Germany 2008:9).

---

[206] Original quotation in German: „Irena ist damit ein Signal dafür, dass der Aufbruch in ein neues Energiezeitalter weltweit begonnen hat. Es ist ein Symbol dafür, dass die Regierungen dabei zusammenarbeiten und bereit sind, voneinander zu lernen. Und es ist ein beispielhaftes Instrument, um einer großen politischen und wirtschaftlichen Vision Gestalt zu geben".
[207] In the case of the BMWi, the IEA is mentioned in the first place. See BMWi, Internationale Energiepolitik, http://www.bmwi.de/DE/Themen/Energie/Energiepolitik/internationale-energiepolitik.html (accessed July 18, 2013); AA, Energiesicherheit, http://www.auswaertiges-amt.de/DE/Aussenpolitik/GlobaleFragen/Energie/Energiesicherheit_node.html (accessed July 19, 2013).

Yet, the government points to two major shortcomings of the IEA which made IRENA's creation necessary: the IEA's membership being restricted to OECD countries and its insufficient support for renewable energy. The German government underlines that IEA's main mandate is to secure energy supply and advise on energy issues in general. According to the German government, this does not allow for a "visionary leadership on renewable energy" (The Government of the Federal Republic of Germany 2008:9). It claims that – due to IEA's "traditional focus on conventional energy" – conventional energy stands at the heart of its activities, while only a small portion of its core budget is attributed to renewable energy (Initiative for an IRENA 2008:5). The government criticizes the fact that IEA's activities do not pay sufficient attention to the economic, political and social aspects of renewable energy. It states that IEA country reviews do not mirror the full potential and regulatory needs of renewable energy and that its policy advice is biased towards large-scale energy solutions. Further, it does not offer enough expertise "on how to adapt energy markets towards more decentralized energy sources such as renewable energy" (The Government of the Federal Republic of Germany 2008:9). In the expert interviews, representatives of BMU, AA and BMZ reinforce that IEA still favors conventional energy over renewable sources. According to their assessment, this is particularly apparent in the modeling of scenarios for IEA's key publication, the World Economic Outlook. Representatives of the BMWi do not share this critique.

REN21

The German government also underlines the contribution of the policy network REN21. The websites of the AA, BMU and BMZ mention REN21 among the first global governors on renewable energy.[208] On its website, the BMU underlines that REN21 is a follow-up network of the Renewables 2004 Conference. It emphasizes that it pushes for the worldwide expansion of renewable energy, promoting the exchange of ideas and information:

> "Providing a forum for international leadership on renewable energy, REN21 connects the wide variety of dedicated stakeholders that came together at the Bonn conference. Its goal is to allow the rapid expansion of renewable energies in developing and industrial countries by bolstering policy development and decision-making on international, national, and sub-national levels.

---

[208] AA, Energiesicherheit, http://www.auswaertiges-amt.de/DE/Aussenpolitik/GlobaleFragen/Energie/Energiesicherheit_node.html (accessed July 19, 2013); BMU, renewable energy – EU/international, http://www.erneuerbare-energien.de/en/topics/eu-international/ (accessed July 18, 2013); BMZ, Energy policy – development-policy involvement in international initiatives, http://www.bmz.de/en/what_we_do/issues/energie/international_energy_policy/index.html (accesed July 18, 2013).

REN21 helps create an environment in which ideas and information are shared and cooperation and action are encouraged to promote renewable energy worldwide."[209]

The BMZ also highlights the character of REN21 as a multi-stakeholder forum – comprising governments, international institutions, NGOs and the private sector – that facilitates the sharing of information and ideas (BMZ 2008:23f.).[210] With regard to the network's activities, the BMZ singles out the publication of the Global Status Report which provides a comprehensive overview of worldwide renewable energy trends (BMZ 2008:23f.).[211] When elaborating on REN21 in one of its policy papers promoting IRENA, the German government also refers to the Global Status Reports. In addition, it underlines REN21's role in organizing the International Renewable Energy Conferences. It also points at the network's shortcomings: REN21 has no legal status, no mandate to advise governments and only very limited financial and human resources (Initiative for an IRENA 2008:7f., The Government of the Federal Republic of Germany 2008:10). In the expert interviews, BMU staff underline that REN21 also has some competitive advantages over IRENA. They point out that REN21 unites like-minded policy actors that are highly committed to the worldwide promotion of renewable energy and that a multi-stakeholder network offers more flexibility than an intergovernmental organization.[212] A BMU staff member points at a further advantage for the German government: as the German government provides most of REN21's funds, it has a strong influence on the direction of REN21 activities.

UN Organizations

When elaborating on global governors of renewable energy, the German government also refers to UN organizations. In a policy paper promoting IRENA, the government points out that the organizations within the UN framework do not pay sufficient attention to renewable energy (The Government of the Federal Republic of Germany 2008:10). At the Renewables 2004 Conference, Chancellor Schröder (2004) underlines that the UN is the framework where global challenges should be addressed. However, he adds that the UN does not realize its potential and gives

---

[209] BMU, REN21 - A Renewable Energy Policy Network for the 21st Century, http://www.erneuerbare-energien.de/en/uebrige-seiten-ohne-verlinkung/ee-ren21-a-renewable-energy-policy-network-for-the-21st-century/ (accessed July 18, 2013).

[210] See also the BMZ website on REN21: BMZ, International energy policy – Renewable Energy Policy Network for the 21st Century (REN21) http://www.bmz.de/en/what_we_do/issues/energie/ international_energy_policy/ren21/index.html (accessed July 18, 2013).

[211] See also the BMZ website on REN21: BMZ, International energy policy – Renewable Energy Policy Network for the 21st Century (REN21) http://www.bmz.de/en/what_we_do/issues/energie/ international_energy_policy/ren21/index.html (accessed July 18, 2013).

[212] Interviews with executive staff, International Cooperation, BMU, December 2012; executive staff (#1), Division International and European Affairs of Renewable Energy, BMU, December 2012; staff member, Division International and European Affairs of Renewable Energy, BMU, November 2012.

renewable energy only a subordinate stance. According to Schröder, different UN organizations put renewable energy on the agenda, but without providing the "impetus that would steadily move things forward." In the policy papers promoting an IRENA, the government mentions UNEP, UNDP, UNIDO, UNDESA and UNIFEM as UN organizations dealing with renewable energy (The Government of the Federal Republic of Germany 2008:10, Initiative for an IRENA 2008:5-7). Initiative for an IRENA (2008:5-7) covers more in detail the activities of UNEP, UNDP and UNIDO. It praises "UNEP's advanced expertise on financing mechanisms" (Initiative for an IRENA 2008:6), as UNEP's renewable energy activities place special emphasis on promoting investment opportunities. Despite this, it also points out that UNEP devotes only a very small portion of its budget to energy. The paper underlines UNDP's experience in providing rural energy services. It adds that UNDP activities mainly focus on the project level and that for the UNDP, promoting renewables is only one option next to supporting conventional energy sources and improving energy efficiency. The policy paper stresses UNIDO's expertise on industrial applications of renewable energy. Yet, it underlines that UNIDO dedicates only a small fraction of its activities to renewable energy. Wieczorek-Zeul (2009 [2001]:35) and BMU & BMZ (2012b:22) refer to another UN body they consider relevant for the worldwide promotion of renewable energy: the UNFCCC. Wieczorek-Zeul (2009 [2001]:37) states that the UNFCCC provides effective instruments to internalize the climate-related external costs of energy production and thus improve the competitiveness of renewable energy vis-à-vis fossil fuels. BMU & BMZ (2012b:22) add that UNFCCC's flexible mechanisms – Clean Development Mechanism (CDM) and Joint Implementation – help to mobilize private funds for renewable energy projects. BMU and BMZ underline the importance of the UN initiative "Sustainable Energy for All" which was launched by the UN Secretary General Ban Ki-Moon in 2011 to promote renewable energy, access to energy and energy efficiency on a global scale. In a joint publication, BMU and BMZ illustrate the German contribution in the achievement of the initiative's goals (BMU, BMZ 2012b). They emphasize that Germany promotes the implementation of these goals both at international and domestic level. With regard to the domestic level, they point to the transformation of the German energy system. The publication clearly expresses the support of the German government to the initiative:

> "The German government welcomes the initiative and its goals and is committed to promoting its implementation within Germany as well as at the international level. Sustainable Energy for All is completely in keeping with the German government's diverse strategies and activities" (BMU, BMZ 2012b:8).

In his foreword, Environment Minister Altmaier emphasizes the initiative's contribution to environmental protection, particularly climate change mitigation (BMU, BMZ 2012b:4). Development Minister Niebel states that the initiative is "an excellent instrument to address energy scarcity in a sustainable way", combining climate protection with poverty alleviation (BMU, BMZ 2012b:5).

The German administration also refers to the contribution of the World Bank to global renewable energy governance. At the Renewables 2004 Conference, Schröder (2004), Wieczorek-Zeul (2004b, 2004c) and Trittin (2004a) present the World Bank as an important global governor on renewable energy. When underlining the need to improve the financing of renewable energy in developing countries, Schröder (2004) points to the activities of the World Bank and states that regional development banks, too, should engage in renewable energy financing. When speaking about international financing cooperation, Wieczorek-Zeul (2004a) refers to the World Bank and regional development banks as being in "an excellent position to provide leadership, given their activities and their leverage power." In a policy paper promoting IRENA (Initiative for an IRENA 2008:7), the German government mentions the World Bank Group as one of the largest multilateral lenders within the energy sector. It states that although the importance of renewable energy is increasing in World Bank activities, in 2007 renewable energy accounted for less than 30 percent of World Bank energy lending. It adds that of the World Bank's renewable energy funding, more than 60 percent is attributed to large hydro-power. The BMZ claims that the World Bank should set itself more ambitious goals on the promotion of renewable energy and increase significantly the financing of "new" forms of renewable energy (BMZ 2008:19f.). With regard to biofuels, the BMZ states that the various World Bank organs apply different standards when assessing biofuels projects. It points out that human rights, environmental and social principles should serve as "fundamental preconditions" of World Bank projects in this area (BMZ 2011:6). In her speech at the Eurosolar conference in 2011, the Development Minister Wieczorek-Zeul (2009 [2001]:39) characterizes the World Bank as "an institution that (...) has shown a tendency to follow the American example" and emphasizes that Germany should influence it to promote feed-in tariffs.

G8

Both BMZ and BMWi furthermore mention the G8 as a global governor on renewable energy.[213] The BMZ (2008:17f., 2007:16)[214] states that energy and climate are among the major concerns of the G8. It highlights that the G8 dialogue on climate change, clean energy and sustainable development with the emerging countries Brazil, China, India, Mexico and South Africa also includes the promotion of re-

---

[213] The BMWi only mentions the G8 without elaborating further on its activities. See BMWi, Internationale Energiepolitik, http://www.bmwi.de/DE/Themen/Energie/Energiepolitik/internationale-energiepolitik.html (accessed July 18, 2013)
[214] BMZ, International energy policy – Energy policy of the Group of Eight, http://www.bmz.de/en/what_we_do/issues/energie/international_energy_policy/g8/index.html (accessed July 19, 2013).

newable energy. The BMZ furthermore underlines that the G8 members committed themselves to increasing the use of renewable energy and protecting the climate. Additionally, it states that the G8 recognized the importance of enhancing access to energy in developing countries and called upon the World Bank to increase support for clean forms of energy. At her speech at the Eurosolar conference in 2001, Wieczorek-Zeul (2009 [2001]:41) presents the activities of the G8 working group on renewable energy. She highlights that the mixed membership of the working group, comprising both government and private sector representatives, illustrates the "strategic importance of the alliance with the business world." In addition, she underlines that the working group "will emphasise in particular the role and the obligations of the major industrialised countries with regard to intensifying their efforts and making better use of the possibilities of joint institutions", even though the main focus of the working group is on how to increase the share of renewable energy in developing countries.

REEEP, GBEP and further Partnerships

In a policy paper promoting the creation of IRENA, the German government underlines that more than twenty partnerships and networks promote the worldwide spread of renewable energy (The Government of the Federal Republic of Germany 2008:10). Next to REN21, it mentions REEEP and GBEP as examples. According to the government, these partnerships have contributed to "driving the global discourse forward" through capacity-building and agenda-setting (The Government of the Federal Republic of Germany 2008:10). It also praises their flexibility. However, it points at one major disadvantage of these partnerships: the lack of a mandate to advise governments. Initiative for an IRENA (2008:7f.) covers REN21 and REEEP in more detail, emphasizing that REEEP's main focus is the private sector. It appraises REEEP's extensive network of partners and its internet platform REEGLE as a valuable information source on renewable energy. In this policy paper, GBEP is not mentioned.

In a press statement issued at GBEP's launch, the BMU demands that GBEP should focus on the development of international standards and certification systems for environmentally friendly biomass production (BMU 2007). It cites Agriculture Minister Horst Seehofer (CSU, 2005 – 2008) who states that environmental damages caused by bioenergy must be avoided. Environment Minister Gabriel adds that "it is totally unacceptable that extensive areas of tropical rain forests are deforested to produce cheap palm oil" (translation S.R.).[215] In publications on biofuels, the BMZ (2011:5, 2013b) states that it supports the development and implementation of sustainability indicators and the sustainable and development-

---

[215] Original text of this quotation: „Es ist völlig unakzeptabel, dass der tropische Regenwald großflächig gerodet wird, um billiges Palmöl zu erzeugen".

friendly production and use of biofuels in the context of GBEP. In a press statement for a GBEP meeting in Germany in May 2013 (BMZ 2013a), BMZ State Secretary Hans-Jürgen Beerfeltz reinforces the importance of environmental protection when promoting the sustainability of bioenergy. The BMZ furthermore highlights that it puts issues such as rural development and the involvement of small farmers on GBEP's agenda. In one of its biofuel publications, the BMZ also refers to other transboundary initiatives on biofuel sustainability, namely the Roundtable for Sustainable Palm Oil, the Roundtable for Sustainable Soy and the Roundtable for Sustainable Biofuels (BMZ 2011:12). In the expert interviews, a representative of the BMWi furthermore points to CEM as a global governor on renewable energy.[216] He emphasizes that CEM's focus on the ministerial level and its involvement of the private sector provides important opportunities for the worldwide promotion of renewable energy.

4.2.6 Salient Features: the Responsibility of Industrialized Countries to Lead

According to the German government, transboundary policy-making should promote renewable energy all over the world, i.e. both in developing and in industrialized countries (Altmaier 2012, BMU 2013a, BMU 2008, BMU, BMZ 2008, Initiative for an IRENA 2008:11, Schröder 2004, The Government of the Federal Republic of Germany 2008:11, Wieczorek-Zeul 2004a). As a consequence, the Germany envisaged IRENA, REN21 and the Renewables Club to focus on both groups of countries. At the Renewables 2004 Conference, Chancellor Schröder states that industrialized, emerging and developing countries all together bear the responsibility to achieve a sustainable energy policy that preserves natural resources. He underlines that high oil prices pose a threat to economic development all over the world (Schröder 2004). Wieczorek-Zeul adds that a sustainable energy future must be created for the whole world, "for people in the North who are rich in energy, and for people in the South who are energy-poor" (Wieczorek-Zeul 2004a). Altmaier (2012) underlines that the worldwide promotion of renewable energy can lead to significant cost reductions of renewable energy technologies.

The German administration commonly differentiates between the role of developing and industrialized countries in global renewable energy governance. It states that industrialized countries take the lead in the worldwide promotion of renewable energy. In his speech prior to the WSSD in 2002, Chancellor Schröder underlines that industrialized countries must take the first steps in implementing

---

[216] Interview with executive staff, Division International Energy Policy, External Energy Policy, BMWi, November 2012.

renewables-friendly policies. Successful policies can later on serve as positive examples for developing countries:

> "We in the industrial countries can show the developing countries how successful economic development can be combined with the protection of natural resources in a manner appropriate to respective local conditions. In my view it makes no sense to demand of the developing countries, the poorest of the poor, something that in and of itself is sensible, i.e. protecting their resources, if we in the world's most wealthy countries don't set an example and make our own contribution in advance" (Schröder 2002b).

Jürgen Trittin (2004b, 2004c) and Heidemarie Wieczorek-Zeul (2009 [2001]:38) refer to industrialized countries as the locations in which renewable energy technologies are being developed. After having developed appropriate technologies, industrialized countries should help to bring these to the developing world. The BMZ (2007:13) claims that it was industrialized countries that initiated the use of renewable energy sources. Schröder (2004) states that industrialized countries have a duty to lead due to their high share in global energy demand. Merkel (2012a) attests that industrialized countries carry "a burden of history" of high emissions; therefore, they bear a special responsibility to take action in protecting the climate. Yet, industrialized countries alone cannot save the climate but require other emitters to cooperate. BMZ (2008:12) adapts this argument directly to the worldwide promotion of renewable energy: industrialized countries bear most responsibility but alone cannot achieve the desired results. As a consequence, transboundary policy-making on renewable energy increasingly focuses on developing countries as well:

> "The agreements reached were initially directed in particular at industrialized countries, which began launching modern, renewable energy sources. (...) The debate on sustainable energy is now increasingly focusing on developing countries, because they also use energy inefficiently and their share of overall consumption is rising rapidly" (BMZ 2007:13).

The historic responsibility of industrialized countries as the main causes of human-induced climate change is also emphasized by the BMZ (2008:4,11) and BMU & BMZ (2012b:6). However, the responsibility of emerging countries is also underlined. At the Renewables 2004 Conference, Schröder states that

> "emerging countries like China, India, Brazil, and Mexico also have a responsibility to see to it that their fast-growing economies make an effort to protect natural resources" (Schröder 2004).

BMZ (2008:3) adds that satisfying increasing energy demand has become an important issue for emerging countries. When elaborating on the promotion of renewable energy, Merkel (2012b) presents emerging countries as economic competitors. She claims that a country like Germany should be cautious not to rest "too complacently on its laurels" but to actively ensure that it maintains its prosperity by staying innovative, whereas "there are emerging economic powers doing exactly that extremely well." Merkel (2012a) also refers to a problem faced by highly industrialized countries such as Germany when reforming their energy systems: the population is skeptical about new infrastructure projects and does not understand the need to build new electricity grids when all citizens already enjoy access to energy.

The government emphasizes that developing countries are confronted with special needs. It states that high and fluctuating oil prices pose a special economic burden on developing countries, especially African countries, as these need to spend a high portion of their export earnings on oil imports (BMZ 2008:4, BMZ 2007:9, Schröder 2004). At the Eurosolar conference, Heidemarie Wieczorek-Zeul underlines that for developing countries, it is more difficult to cope with the import-dependency on fossil fuels than it is for industrialized countries:

> "For many developing countries, in particular, their dependency on imports of fossil fuels for energy is a serious obstacle to development. Since these countries have few other alternatives to turn to, this dependency is a far greater burden for them than it is for the industrialised nations. The developing countries in particular therefore stand to benefit in the long term if they move away from using fossil fuels for energy towards renewable energies" (Wieczorek-Zeul 2009 [2001]:34).

The BMZ (2008:12) adds that developing countries are most affected by climate change, although they have done least to cause it. Wieczorek-Zeul (2009 [2001]:35) emphasizes that industrialized countries have committed themselves to greenhouse gas reductions while recognizing that developing countries have a right to development and thus to increased greenhouse gas emissions. The BMZ (2008:7) states that many developing and emerging countries reveal the symptoms of poor energy supply, such as large areas without connections to the national grid, and prevailing use of traditional biomass. Besides, the German government asserts that developing countries often lack access to renewable energy, having difficulties to buy, maintain and repair the necessary equipment (The Government of the Federal Republic of Germany, Initiative for an IRENA 2008:13). As a consequence, the affordability of renewable energy is a bigger issue in developing countries than in industrialized countries (Wieczorek-Zeul 2009 [2001]:35). In a policy paper promoting IRENA, the German government furthermore underlines that renewable energy could help developing countries to "leapfrog directly into a clean energy scenario" (The Government of the Federal Republic of Germany, Initiative for an IRENA 2008:11).

The German government mostly attributes to developing countries a reactive role in global renewable energy governance. This is particularly salient in the speeches held by German government representatives. Schröder (2004), Altmaier (2012) and Wieczorek-Zeul (2004a, 2009 [2001]) speak of developing countries as being on the receiving end of financial resources or development aid in general. Trittin (2004b, 2004c) and Wieczorek-Zeul (2009 [2001]:38) characterize such countries as beneficiaries of the cost reductions achieved in countries like Germany and the USA. Schröder (2002b) adds that developing countries need the help of industrialized countries as "it is perfectly clear that the developing countries will not be able to implement an energy policy of this kind on their own strength." In a policy paper promoting IRENA, the German administration underlines that it is not enough to grant developing countries market access; rather, they should be enabled to produce renewable energy technologies autonomously. Thus, it states that "enormous efforts are required to facilitate technology transfer and investments in developing countries" (The Government of the Federal Republic of Germany, Initiative for an IRENA 2008:16). Schröder (2002b) adds that industrialized countries must increase renewables-related development finance. He cites the Kyoto-Protocol as a positive example of North-South transfers of financial resources, stating that "there are major opportunities here for tangible progress on the road to sustainable energy supplies both for industrial and developing countries" (Schröder 2004).

4.3 The German Government's Action and Ideas at a Glance

The German initiatives on global renewable energy governance were fueled by disappointments with UN processes. The government several times attempted to establish time-bound targets for increasing the global share of renewable energy within the UN framework. As no advancements could be achieved, it opted for initiatives outside the formal UN context. The main focus of the government's action was institution-building. It launched the multi-stakeholder network REN21, initiated IRENA's creation and finally founded the Renewables Club. While these initiatives oscillated between 'coalitions of the willing' and broad participation and representativeness, they all focused on one major goal: supporting governments around the globe to increase the deployment of renewable energy sources. The initiatives were grounded in domestic policy efforts to promote renewable energy, which primarily focus on the electricity sector. The main driving force behind the domestic policies has been a strong anti-nuclear movement. Concerns about climate change also came in.

The BMU has been the main driver behind the German initiatives on global renewable energy governance. It has been joined by the BMZ. The AA only came in when Germany pushed for IRENA's creation. It is interesting to note that not even IRENA's creation, which so far has been the government's major initiative to shape transboundary cooperation on renewable energy, has enjoyed much high-level political attention within the German government. Neither the Foreign Minster nor the Chancellor was actively engaged. The German government's action has been strongly influenced by Hermann Scheer, the former SPD Parliamentarian, who had been lobbying for an IRENA since the 1980s. NGOs and industry associations showed no major engagement. The deliberations on global renewable energy governance by different government actors rarely reveal expressed conflict. There is one notable exception. IEA's and IRENA's role in global renewable energy governance is one of the very few topics that actually meet highly controversial assessments within the German government. While the ministries that initiated IRENA's creation regard the IEA as being biased towards conventional energy and present IRENA as a much-needed counterweight, the BMWi praises the IEA and questions IRENA's value added.

According to the government, transboundary policy-making on renewable energy is intrinsically linked to global environmental protection. In the German deliberations on global renewable energy governance, references to climate protection are never missing. Sustainability, particularly in its ecological sense, is presented as the core criterion to guide the choice between different renewable energy options. Discussions strongly focus on the electricity sector. The administration presents solar and wind energy as a win-win option for transboundary cooperation, while it raises sustainability concerns with regard to large hydropower and biofuels. The German administration states that market structures and policies that discriminate against renewable energy pose a major barrier to the worldwide spread of renewables. Transboundary policy-making should primarily aim at improving domestic regulatory frameworks. To do so, it should commit governments to promote renewables and support governments in realizing their commitments. The admnistration presents IRENA, IEA and REN21 as the most relevant global governors. All of these provide information and policy advice on renewable energy. It emphasizes that transboundary policy-making should focus on the worldwide promotion of renewables, thus targeting both industrialized and developing countries. It underlines that industrialized countries should take the lead in the worldwide promotion of renewable energy, while it attributes developing countries with a rather reactive role.

5 Brazilian Ideas on Global Renewable Energy Governance

This chapter presents the Brazilian government's action and ideas on global renewable energy governance. Brazil has one of the worldwide highest shares of energy generated from renewable sources. Next to the high rate of hydropower utilized in its electricity supply, this is due to the widespread use of bioenergy. The Brazilian government has been a pioneer in the worldwide promotion of biofuels. It had already begun to substitute oil for ethanol in the 1970s. Its initiatives in the realm of global renewable energy governance have been concentrated on biofuels. From 2006 to 2010, biofuels were a priority issue of Brazilian foreign policy and the government engaged in highly visible ethanol diplomacy.

The first section of this chapter traces the government's action in the realm of global renewable energy governance. After presenting the initial steps, which were taken in the context of the WSSD 2002 and the Renewables 2004 Conference, it illustrates the three main phases of Brazilian ethanol diplomacy: the initial phase in times of expanding ethanol production and usage, the intensification of this diplomacy in light of growing international criticism of biofuel sustainability and the continuation of ethanol diplomacy at a lower political profile since 2010/11. As outlined in the theoretical-analytical framework, this section traces how and why the Brazilian government engaged in transboundary policy-making on renewables, specifying which policy actors were involved and what policy goals they pursued. It furthermore illustrates the contextual factors and political processes at the domestic and global level that influenced the administration's action.

The second section presents the governemment's ideas on global renewable energy governance. As in the case study on Germany, this analysis is structured along the following elements: global challenges addressed by renewable energy promotion, specification of and differentiation between renewable energy options, barriers to the worldwide promotion of renewables, tasks for transboundary policy-making, relevant global governors and salient features in transboundary policy-making.

5.1 Tracing the Government's Action on Global Renewable Energy Governance

5.1.1 Initial Steps: WSSD 2002 and the Renewables 2004 Conference

At the WSSD in Johannesburg 2002, the Brazilian government took the first steps in shaping transboundary policy-making on renewable energy. As the country had hosted the preceding summit, the Earth Summit in Rio de Janeiro 1992, the Brazilian administration was strongly commited to advance UN negotiations on sustainable development. In the run-up to the WSSD 2002, renewable energy was one of its fields of action. During the international preparation process for the WSSD, the international Preparatory Committee had proposed to set a worldwide target for the use of renewable energy. According to this proposal, all countries should achieve a renewable energy share of at least five percent of their total energy use by 2010 (Goldemberg 2002a:1). At the WSSD, President Fernando Henrique Cardoso (Brazilian Social Democracy Party, Partido da Social Democracia Brasileira, PSDB, 1995–2002) presented a proposal, the Brazilian Energy Initiative, which suggested to increase this target to ten percent. To achieve the target, countries should be allowed to trade renewable energy certificates. In addition, an intergovernmental body of Energy Ministers should make recommendations on how to implement the target and track the implementation process (Goldemberg 2002a).

According to the Support Report to the Brazilian Energy Initiative, the reasoning behind the Brazilian proposal was the "Brazilian experience - global landmark on renewable energy sources implementation - and the significant advantages of the accelerated introduction of new renewables in the world profile of energy consumption" (Goldemberg 2002b:17). The proposal particularly referred to Brazilian sugarcane ethanol promotion as an example of success. It stated that sugarcane ethanol had replaced one half of the gasoline that would have otherwise been used in Brazil (Goldemberg 2002a). The Support Report added that the country's administration regarded its proposal as a contribution to global climate protection, in particular to the implementation of the Kyoto Protocol, and the global fight against poverty in the context of the MDGs (Goldemberg 2002b:12).

The Brazilian Energy Initiative was launched by the Interministerial Commission for the Preparation of Brazil's Participation at the WSSD. As with UNCED processes in general, the Ministry of Foreign Relations (Ministério de Relações Exteriores, MRE, commonly referred to as Itamaraty because its original headquarter was the Itamaraty Palace in Rio de Janeiro) took the lead, cooperating closely with the Ministry of the Environment (Ministério de Meio Ambiente, MMA) and the Ministry of Science, Technology and Innovation (Ministério de Ciência, Tecnologia e Inovação, MCTI). The Commission was coordinated by Everton

Vargas, Head of Itamaraty's Environment Department (Corrêa do Lago 2009:169f.). The Brazilian Energy Initiative was developed under the leadership of the São Paulo State Secretary of the Environment and renowned energy scientist, Professor José Goldemberg. In the months prior to the WSSD, the Brazilian government proposed that Latin American and the Caribbean states take joint action in order that Brazilian and Latin American priorities would gain more visibility within the UN process. Together, they endorsed the Latin American and Caribbean Initiative on Sustainable Development (Iniciativa Latino-Americana e Caribenha sobre Deselvolvimento Sustantável, ILAC) which also included a regional ten percent target for the use of renewable energy (Corrêa do Lago 2009:158f., MRE 2003a, Vargas 2004:1).

At the WSSD, the Brazilian delegation did not succeed with its proposal to establish a worldwide ten percent target for renewable energy use. While Latin American countries and the EU called for the establishment of a global goal for renewable energy deployment, the USA and major oil-producing countries opposed it. Yet, these countries finally accepted that the Johannesburg Plan of Implementation referred to changes needed in the energy field, such as eliminating environmentally harmful energy subsidies and increasing the worldwide share of renewable energy. According to Ambassador Corrêa do Lago, Director General of Itamaraty's Department of Environmental and Special Affairs and former Head of the Energy Department, this reference to renewable energy "may be one of the most significant advancements in relation to Rio, where oil-producing countries had managed to block references to greater incentives for renewable energies" (Corrêa do Lago 2009:162). He adds that the Brazilian leadership role in the effort to establish a renewable energy target contributed to the fact that "the Brazilian Delegation was generally recognized as one of the most active at the Johannesburg Summit" (Corrêa do Lago 2009:164).

To reaffirm its leading role in international negotiations as well as its regional leadership in this area, the Brazilian administration decided to head the preparation process for the Renewables 2004 Conference[217] within Latin America (Corrêa do Lago 2009:158f., MRE 2003a). In October 2003, the Itamaraty organized, together with the Ministry of Mines and Energy (Ministério de Minas e Energia, MME) and the MMA, the Latin America and Caribbean Regional Conference on Renewable Energy. The conference consisted of two segments: a technical segment comprising Latin American and Caribbean specialists, representatives of multilateral organizations and development banks in addition to civil society representatives; and a high-level segment, uniting the region's Energy and Environment Ministers. The Minister endorsed the declaration 'Brasilia Platform on Renewable Energies', committing

---

[217] The Renewables 2004 Conference was an international multi-stakeholder conference organized by the German Government. For further information on this conference, see chapters 3.1.3 and 4.1.1.

themselves to the promotion of renewable energy and reaffirming the regional ten percent target (MRE 2003b). From the Brazilian side, further ministries that participated at the elaboration of the declaration were the Ministry of Development, Industry and Foreign Trade (Ministério do Desenvolvimento, Indústria e Comércio Exterior, MDIC), the MCTI and the Ministry of Agriculture, Livestock and Food Supply (Ministério da Agricultura, Pecuária e Abastecimento, MAPA) (MRE 2003a).

At the Renewables 2004 Conference, Dilma Rousseff – by then Minister of Mines and Energy (Workers' Party, Partido dos Trabalhadores (PT), 2003–2005) – presented the Latin American and Caribbean position on the worldwide promotion of renewable energy. She underlined the fact that renewable energy contributes to sustainable development and poverty alleviation by creating jobs, generating income and mitigating climate change. Rousseff also emphasized the comparative advantages of Latin American countries in the realm of hydropower and biofuels, underlining their leading role in the promotion of renewable energy. In addition, she reinforced the principle of sovereignty with regard to the use of natural resources: Latin American countries would not accept conditionalities on the use of their hydropower potential (Vargas 2004:1f.). According to Vargas (2004:2), Latin American and Caribbean governments furthermore supported African countries when these underlined the importance of hydropower.

5.1.2 Ethanol Diplomacy Begins: Time of Expansion

In 2006, biofuels became a priority issue for the Brazilian government's transboundary action. President Luiz Inácio Lula da Silva (PT, 2003–2010) began active ethanol diplomacy. The former Head of Itamaraty's Energy Department, Antônio Simões, underlines Lula da Silva's commitment for biofuels with the following words:

> "In all President Lula's foreign trips and in all contact with foreign visitors to Brazil, biofuels always play a leading role. The determination of the President in disseminating the worldwide use of biofuels is without a doubt one of the distinguishing marks of his administration" (Simões 2007a:11, translation S.R.).[218]

The Head of Itamaraty's Division for New and Renewable Energy emphasizes that biofuels topped Lula da Silva's presidential foreign policy agenda, next to getting a permanent seat in the UN Security Council and agricultural cooperation with Afri-

---

[218] Original text of this quotation: "Em todas as viagens do Presidente Lula ao exterior e em todos os contatos com visitantes estrangeiros ao Brasil, os biocombustíveis sempre têm papel de destaque. A determinação do Presidente em difundir o emprego de biocombustíveis no mundo é, sem dúvida, uma das marcas de sua administração".

can countries.[219] In the government program for his first term (2003-2006), Lula da Silva had already announced his support for ethanol due to its "national and environmental value" (Lula da Silva 2002:69, translation S.R.).[220] He was closely linked to the Brazilian sugarcane industry, a traditional and politically influential industry which is particularly strong in the State of São Paulo – Lula da Silva's home state.[221] Sugarcane is the main feedstock of ethanol production in Brazil and the state of São Paulo has been the traditional hotspot of sugarcane production. It also has the most technically advanced production systems. In Brazilian politics, states play a very important role. In this huge and culturally diverse country with pronounced socio-economic inequalities, states also serve as an important point of reference in political identify formation.[222]

In the government program for his second term (2007–2010), Lula da Silva specified two aims for the Brazilian transboundary action on biofuels: 1) Enhancing biofuel-related technology exports to Latin American and African countries and 2) promoting Brazilian ethanol exports and establishing ethanol as a commodity[223] (Lula da Silva 2006:19). The main goal of ethanol diplomacy was to turn Brazil into a major exporter of the product [224] In addition, the government regarded the promotion of biofuels as a suitable tool to combine climate protection and environmental friendliness in general with socioeconomic opportunities for developing countries, as the tropical regions offer particularly favorable natural conditions for the production of biofuels. Therefore, biofuels were also seen as a promising area for South-South cooperation, strengthening the links between Brazil and other developing countries, especially on the African continent.[225] In addition, the administration regarded biofuels as one of the few areas in which it could interact with the USA and EU on equal ground and thus challenge OECD-leadership in

---

[219] Interview in March 2013. Several other interviewees emphasize that Lula da Silva was a strong supporter of biofuels. Interviews with former president, Sugarcane Industry Association (UNICA), April 2013; secretary and team, Secretariat of Planning and Energy Development, MME, March 2013; team, Bioenergy Projects, Getulio Vargas Foundation (FGV), February 2013; Biodiesel Federal Program Coordinator and staff member, Analysis and Following up of Government Policies, Casa Civil, March 2013; policy advisor, Brazilian Office, Oxfam, April 2013.
[220] Original text of this quotation: "valor nacional e ambiental".
[221] Interviews with Biodiesel Federal Program Coordinator and staff member, Analysis and Following up of Government Policies, Casa Civil, March 2013; Professor, Centre for Graduate Studies in Agricultural Development (CPDA), Federal Rural University of Rio de Janeiro (UFRRJ), March 2013; Professor, Economics Institute, Federal University of Rio de Janeiro (UFRJ), March 2013; Professor (#2), Energy Institute, Universidade de São Paulo (USP), April 2013.
[222] On Brazilian Federalism see, for example, Costa 2004.
[223] A commodity is traded as a homogenous good without qualitative differentiations across markets.
[224] Interview with Head of the New and Renewable Energy Resource Division, Itamaraty, March 2013; Professor (#1), Energy Institute, USP, April 2013; Vice-President, Brazilian Center for International Relations (CEBRI), April 2013.
[225] Interviews with former president, UNICA, April 2013, and team, Bioenergy Projects, FGV, February 2013.

transboundary policy-making (Kloss 2012:58). This ethanol diplomacy built on strong domestic support both within the media and the political sphere (Giersdorf 2011:153, Viola 2013:7).[226] In an expert interview, an academic in a Brazilian university emphasizes that in Brazil, ethanol is an element of national pride.[227] The ethanol industry is one of the most industrialized agricultural sectors in the country, and the possibility of expanding Brazilian exports and exercising an international leadership role met widespread support.

In his ethanol diplomacy, Lula da Silva cooperated closely with the Itamaraty which is responsible for Brazilian foreign policy by constitution. As such, it is also in charge of global energy governance. This distinguishes the Brazilian government from many others which often have divided responsibilities for transboundary energy relations (MRE 2010b:3f.). In 2006, the Itamaraty created an Energy Department. Gas disputes with Bolivia were the main reason behind this step.[228] However, this institutional reform was also instrumental in strengthening Itamaraty's action on biofuels. Itamaraty's Energy Department is split into two divisions. The Non-Renewables Division deals with oil, gas, coal, mining, energy efficiency and clean energy technologies such as carbon capture and storage.[229] The Renewables Division is in charge of biofuels and non-conventional energy sources such as wind, solar, hydro and geothermal energy.[230] Since its creation, it has dedicated most of its time to bioenergy.[231] Next to the two divisions, the Energy Department has a General Coordination on Nuclear Energy.[232] The Itamaraty highlights that an important task of its Energy Department is to serve as a broker between transboundary policy-making on renewable energy and national policy-making and ensure coherence and continuity in the Brazilian positions (MRE 2010b:3f.). Within the Brazilian administration, the Itamaraty closely coordinates its action on renewables with the MME.[233] With regard to biofuels, MAPA is a central cooperation partner, as it rep-

---

[226] Interviews with Head of Energy Department, Itamaraty, April 2013; Professor, CPDA, UFRRJ, March 2013; Professor, Luiz Queiroz College of Agriculture, USP, April 2013; Professor (#1), Energy Institute, USP, April 2013.
[227] Interview with professor, CPDA, UFRRJ, March 2013.
[228] Interview with Head of the New and Renewable Energy Resource Division, Itamaraty, March 2013.
[229] MRE, Divisão de Recursos Energéticos Não-Renováveis – DREN, http://www.itamaraty.gov.br/o-ministerio/conheca-o-ministerio/subsecretaria-geral-politica-i/dren-divisao-de-recursos-energeticos-nao-renovaveis (accessed June 20, 2013).
[230] MRE, Divisão de Recursos Energéticos Novos e Renováveis – DRN, http://www.itamaraty.gov.br/o-ministerio/conheca-o-ministerio/tecnologicos/drn-divisao-de-recursos-energeticos-novos-e-renovaveis (accessed June 20, 2013)
[231] Interview with Head of the New and Renewable Energy Resource Division, Itamaraty, March 2013.
[232] MRE, Divisão de Recursos Energéticos Novos e Renováveis – DRN, http://www.itamaraty.gov.br/o-ministerio/conheca-o-ministerio/tecnologicos/drn-divisao-de-recursos-energeticos-novos-e-renovaveis (accessed June 20, 2013).
[233] The following remarks on Government entities involved in Brazilian ethanol diplomacy rely on MRE 2010b:3.

resents the agroindustry. Outside of the government, the São Paulo-based Sugarcane Industry Association (União da Indústria de Cana-de-Açúcar, UNICA) has a very influential role within Brazilian ethanol diplomacy.[234] UNICA opened two international offices, one in Washington and another in Brussels. The Head of Itamaraty's New and Renewables Energy Division points out that UNICA advises the Itamaraty and that it is very receptive to Itamaraty's demands.[235] In addition, the Institute for International Trade Negotiations (Instituto de Estudos do Comércio e Negociações Internacionais, ICONE) has also been an active player within Brazilian ethanol diplomacy, providing policy advice.[236] ICONE was created by agribusiness in 2003 to provide the Brazilian government and private sector with applied research on trade and agriculture.[237] The Itamaraty also consults further ministries. The MCTI deals with the scientific and technological aspects of biofuels. It has an important role to play, as ethanol is an area of advanced research in Brazil.[238] The MDIC represents the automotive industry and is responsible for agroindustrial exports and export specifications.[239] The MMA is in charge of climate policy and environmental licensing of energy projects.[240] Lastly, the Ministry for Rural Development (Ministério do Desenvolvimento Agrário, MDA) acts in relation to social issues and biofuel production, especially in relation to the biodiesel program.[241] Within Brazilian agricultural policy, MDA serves as the counterpart to MAPA which is responsible for the development of agribusiness, i.e. large-scale and economically competitive agricultural production. As such, MAPA is responsible for

---

[234] Interviews with Head of Energy Department, Itamaraty, April 2013; Head of the New and Renewable Energy Resource Division, Itamaraty, March 2013; Biodiesel Federal Program Coordinator and staff member, Analysis and Following up of Government Policies, Casa Civil, March 2013; Director of the Department of Renewable Fuels, MME, March 2013; Director of Sugarcane and Agroenergy, MAPA, March 2013; General Coordinator for Sugar and Ethanol, MAPA, March 2013; Manager, Climate Change Department, MMA, March 2013; Foreign Trade Analyst, Secretariat for Innovation, MDIC, March 2013; Director, Centro de Estudos de Integração e Desenvolvimento (CINDES), March 2013; Consultant, FASE Brasil, March 2013; Professor, Energy Planning Program, Federal University of Rio de Janeiro (UFRJ), February 2013; Researcher on applied economics, FGV, March 2013; professor (#1), Institute for International Relations, Universidade de Brasília (UnB).
[235] Interview in April 2013.
[236] Interviews with Head of Energy Department, Itamaraty, April 2013; Professor, CPDA, UFRRJ, March 2013; Professor, Faculty of Philosophy, Languages and Literature, and Human Sciences (FFLCH), USP, April 2013.
[237] ICONE, the institute, http://www.iconebrasil.org.br/the-institute (accessed February 14, 2014).
[238] Interview with professor, CPDA, UFRRJ, March 2013.
[239] Interviews with Biodiesel Federal Program Coordinator and staff member, Analysis and Following up of Government Policies, Casa Civil, March 2013; Foreign Trade Analyst, Secretariat for Innovation, MDIC, March 2013; Consultant, FASE Brasil, March 2013.
[240] Interviews with diplomat, Itamaraty, March 2013 and Manager, Climate Change Department, MMA, March 2013.
[241] Interviews with Director of the Department of Renewable Fuels, MME, March 2013; Biodiesel Federal Program Coordinator and staff member, Analysis and Following up of Government Policies, Casa Civil, March 2013; diplomat, Itamaraty, March 2013.

the ethanol sector. In addition, government-related research institutions such as the Brazilian Bioethanol Science and Technology Laboratory (Laboratório Nacional de Ciência e Tecnologia do Bioetanol, CTBE), founded by the MCTI, and the Brazilian Agricultural Research Corporation (Empresa Brasileira de Pesquisa Agropecuária, Embrapa), linked to MAPA, play a role. The Itamaraty furthermore underlines the contribution of the Working Group of Bioenergy within the Social and Economic Development Council (Conselho Nacional de Desenvolvimento Econômico e Social, CDES) which fosters dialogue between the government, labor unions, businesses and civil society (MRE 2010b:3).[242]

Brazil's ethanol diplomacy builds on its long-standing experience with the domestic promotion of ethanol. The government began to promote ethanol during the 1970s. As a response to the oil price shock in 1973, the Brasília took two measures to reduce its dependency on oil imports: it invested in domestic oil production and, in 1975, started the program Proálcool. This program had two main objectives: to introduce anhydrous ethanol as a blend to gasoline, and incentivize the production of cars that could run on ethanol only (hydrated ethanol). While energy security concerns were important, they were not the only driver behind this program; in light of low world market prices for sugar, Proálcool was also thought of as a support program for the sugarcane industry. Until 1988, several fiscal and financial incentives supported the production and consumption of ethanol. The process of economic deregulation, which started after the end of the Brazilian military government (1964–1985) and the adoption of the democratic constitution in 1988, also affected energy markets. Hydrated ethanol practically vanished from the market as consumers lost confidence in cars running on pure ethanol after the occurrence of supply shortages of hydrated ethanol. In 1993, the government established minimum blending requirements for anhydrous ethanol ranging between 20 and 25 percent. Until 2003, ethanol demand was practically restricted to this obligatory blending.[243]

The market introduction of flex fuel cars in 2003 gave the Brazilian ethanol market a decisive boost (Bastos Lima 2012:350, Kohlhepp 2010:228, Schutte, Barros 2010:35). Flex fuel cars enable consumers to choose between any mixture of gasoline and hydrated ethanol. As such, hydrated ethanol directly competes with gasoline, while the demand of anhydrous ethanol is determined by obligatory blend-

---

[242] For a comprehensive overview of the participants, scope and activities of the working group see CDES, GT Bioenergia: Etanol, Bioeletricidade e Biodiesel, http://www.cdes.gov.br/grupo/141/gt-bioenergia-etanol-bioeletricidade-e-biodiesel.html (accessed June 24, 2013).
[243] See Bastos Lima 2012:348f., Goldemberg 2006, Johnson 2010:15-21, Kloss 2012:140, Kohlhepp 2010:226-228, Scheibe 2008:40-44, Schutte, Barros 2010:34, Simões 2007b:17-20.

ing requirements.[244] The government incentivized ethanol consumption with tax reductions for ethanol and flex fuel cars (Kloss 2012:70). The broad coverage of filling stations providing pure ethanol – a heritage of the Proálcool Program – was instrumental in the market penetration of flex fuel cars (Kloss 2012:140). After 2003, domestic ethanol demand increased significantly – also due to rising oil prices (Herrera, Wilkinson 2010:751). The expanding ethanol production was accompanied by high levels of private investments and technological modernization in the sugarcane industry (Simões 2007a:20). The market introduction of flex fuel cars also had a further effect, forming a politically influential coalition between agribusiness and the automotive industry.[245]

In 2004, the Lula da Silva government launched the Biodiesel Program. Before that, Brazilian biofuel production was restricted to ethanol only (Bastos Lima 2012:352). The Biodiesel Program was thought of as a tool to foster regional development and social inclusion. It aims to promote socio-economic development in economically deprived rural areas by including small and family farmers. In addition, the use of biodiesel was supposed to improve air quality in cities. Rising fossil fuel prices also contributed to the launch of the program. It established mandatory biodiesel blending of five percent. Out of the total biodiesel volume, 80 percent must be supplied by producers that contract with small farmers for their feedstock. The Casa Civil, the Office of the Head of State and Government, has overall oversight over the program whereas the MME coordinates its implementation. The MDA, which was created in 1999 to promote land reform and family agriculture, is responsible for the social strategy of the program (Giersdorf 2011:37-52, Herrera, Wilkinson 2010:757-762, Pousa, Santos & Suarez 2007, Simões 2007b:21-23).

Between 2003 and 2005, interest in biofuels also grew outside Brazil (Bastos Lima 2012:349, Kloss 2012:183). This provided a special window of opportunity for the Brazilian government as the international pioneer in ethanol production. In 2003, the European Commission established goals on biofuel blending (Directive 2003/30/EC). Two years later, in 2005, the US Environmental Protection Agency introduced the Renewables Fuels Standard to promote biofuels. And in the same year, the G8 decided to launch a Global Bioenergy Partnership (GBEP) to promote the deployment of biomass and biofuels, particularly in developing countries. The introduction of policy incentives for biofuels in other countries opened new export opportunities for Brazilian ethanol producers. The Brazilian experience of ethanol promotion received much international attention as no other country had gone so far.[246] Many countries and international organizations took Brazil as a point of ref-

---

[244] As ethanol has lower energy content than gasoline, there is the rule of thumb in Brazil that it is advantageous for the consumer to choose ethanol instead of gasoline if the ethanol price is at least 30 percent lower than the gasoline price.
[245] Interview with professor, CPDA, UFRRJ, March 2013.
[246] Interview with former president, UNICA, April 2013.

erence for biofuel promotion, particularly countries from the developing world which approached Brazil for cooperation, information and experience exchange (Kloss 2012:183). The attention was not limited to the political sphere as foreign companies also started to invest in the Brazilian ethanol sector.[247]

In 2006, the government launched the International Biofuels Forum (IBF). It wanted the IBF to create the technical base for transforming biofuels into an international commodity and thus focus on the following issues: adopting blending goals for ethanol, increasing its sustainable production, developing internationally accepted standards and technical specifications and introducing the quotation of ethanol into the main stock markets. Brasília invited the USA, EU, South Africa, China and India to join as it wished to gather the foremost producers and consumers of biofuels as well as major emerging countries. For the government, the participation of emerging countries was very important. According to Kloss (2012:188), Brazil sought to ensure that the perspective of developing countries was taken into account when promoting biofuels as a new sector in the worldwide economy. At the first preparatory meeting, participating governments agreed that cooperation should respect each country's particularities and aim at integrating biofuels into policies to promote sustainable economic growth. The Brazilian administration envisaged that IBF would distinguish itself from other initiatives through a pragmatic and results-based character. Whereas the Brazilian government initially wanted to focus on ethanol only, the US and the EU demanded the inclusion of biodiesel as well. The first IBF meeting was held in Brussels in 2007 within the context of Lula da Silva's visit to the European Commission (Feres 2010:57). In the IBF declaration, member states agreed that the IBF should pursue two major aims: assessing, preparing and disseminating recommendations to expand biofuel production and consumption, and promoting the harmonization of biofuel standards and norms. These objectives differ only slightly from the Brazilian proposal which had also foreseen to work on stock markets. IBF activities finally evolved around two major areas: the harmonization of technical standards and the preparation of an international biofuels conference in Brazil (Kloss 2012:25, 125-131).

Only Brazil, the US and the EU effectively engaged in IBF's work on harmonizing standards which resulted in the publication of a White Paper on Internationally Compatible Biofuels Standards in 2007.[248] China, South Africa and India were not interested in joining the taskforce although Brazil tried to convince them to do so. The EU commission, the governments from Brazil and the USA and the standardization institutions from the three countries/regions named several biofuel

---

[247] Interviews with former President, UNICA, April 2013 and professor, Luiz de Quiroz College of Agriculture, USP, April 2013.
[248] White Paper on Internationally Compatible Biofuels Standards, http://www.globalbioenergy.org/bioenergyinfo/global-initiatives/detail/en/c/2756/ (accessed May 15, 2013).

experts to join the taskforce on harmonizing standards. Whereas the USA and the EU only named representatives from the private sector and some academics, the Brazilian government directly engaged in the taskforce which was split into one group dealing with bioethanol and another with biodiesel. The Itamaraty served as co-leader of both groups. From the Brazilian side, the Brazilian Petroleum, Natural Gas and Biofuels Agency (Agência Nacional do Petróleo, Gás Natural e Biocombustíveis, ANP), the National Institute of Metrology (Instituto Nacional de Metrologia, Qualidade e Tecnologia, INMETRO), the National Association of Automobile Manufacturers (Associação Nacional dos Fabricantes de Veículos Automotores, ANFAVEA) and the state-owned energy corporation Petrobras participated in both groups. A further participant of the biodiesel group was the São Paulo State Government Institute of Technological Research. UNICA and the Chemical Science Institute of the State University of Campinas joined the group on bioethanol. According to Kloss (2012:134), the taskforce's work allowed for a comprehensive mapping of the differences between biofuel standards in Brazil, the USA and the EU and a deeper understanding of the technical difficulties of possible harmonization. In addition, it strengthened the links between the metrology institutions in the three countries/regions (Kloss 2012:132-134, 140-146).

In 2007, the Brazilian government announced its interest in establishing ethanol as an environmental good under the World Trade Organization (WTO). In its first term (2003–2007), the Lula da Silva government had tried to advance multilateral trade negotiations on agriculture in general, but without success.[249] In the Doha Declaration of 2001, WTO member states had agreed to negotiate the reduction of trade barriers on environmental goods and services – without including a definition of such goods and services. According to Kloss (2012:159f.), the discussions on environmental goods and services in the WTO Committee on Trade and Environment in Special Sessions (CTESS) mainly focused on technology-intensive products that are produced by companies in industrialized countries. The Brazilian government, represented by the Itamaraty, wanted to include the environmentally friendly products that Brazil produced with comparative advantages and exported to other countries. It argued that the list of environmental goods and services had to comply with the purpose of the Doha Development Round. Therefore, it should encompass products that developing countries produce with comparative advantages, such as biofuels. To support its argument, the Itamaraty underlined the benefits of biofuels in the realm of the environment (such as the substitution of pollutants and greenhouse gas emission reductions), development (such as economic benefits for rural areas and the possibility of integrated food and energy production) and trade (such as diversifying the foreign currency income of developing countries). The Brazilian proposal was supported by Colombia but opposed by the USA and EU.

---

[249] Interview with professor, FFLCH, USP, April 2013.

Whereas the USA argued that agricultural goods *per se* could not be counted as environmental goods, the EU explicitly opposed the inclusion of ethanol (Kloss 2012:159-164).

5.1.3 Intensification of Ethanol Diplomacy in the Light of Growing International Criticism

In 2007/2008 the international debate on biofuels underwent important changes. The sustainability of biofuels was increasingly questioned. Food prices rose rapidly, and many analysts regarded the production of biofuels as an important driver behind this development. An intensive discussion about food versus fuel started. The discussion initially concentrated on the relation between increasing international maize prices and the growing production of maize-based ethanol in the USA, yet it also touched the biofuel production of other feedstocks and in other countries. According to Kloss (2012:79-82), a decisive turning point in the worldwide debate on biofuels was the Report of the Special Rapporteur on the Right to Food, Jean Ziegler (2007), at the UN General Assembly 2007. In this report, Ziegler openly criticized the production of biofuels, demanding a five year moratorium in order to develop new technologies and regulative frameworks that could avoid the negative impacts of biofuel production. According to Kloss (2012:87), even OPEC began to mobilize against biofuels, arguing in the OPEC World Oil Outlook 2008 that biofuel production was not only responsible for rising food prices, but also for oil price increases.[250] The environmental impacts were also increasingly questioned. In particular, it was feared that biofuel production could lead to deforestation of tropical rainforests (Kloss 2012:133).

As a response to international criticism on biofuels, the Brazilian government began to campaign actively for biofuels, defending in particular the sustainability of Brazilian sugarcane-based ethanol production. Lula da Silva addressed the issue in his speeches at the meetings of the UN General Assembly 2007–2009 and several other UN meetings (see Lula da Silva 2009a, 2009b, 2008b, 2008c, 2008d, 2008e, 2007b). The Itamaraty created the General Subsecretary of Energy, Environment, Science and Technology to respond to the growing importance of biofuels within Brazilian diplomacy (Kloss 2012:184). The Brazilian administration questioned the causality between biofuel production and rising food prices. It stated that such a problem assessment was not based on sound empirical evidence but rather on precipitous judgments. It further argued that it was not possible to assess the impacts of biofuel production in a generic and global way, and claimed that the

---

[250] While the OPEC World Oil Outlook 2008 indeed links rising food prices to increasing biofuels production, it does not establish a direct relation between oil price increases and biofuels production. See OPEC 2008.

criticism did not apply to the Brazilian case of sugarcane-based ethanol production. The following statement by Lula da Silva at the FAO conference on world food security in 2008 illustrates the main points of the Brazilian argument:

> "In short, sugar-cane ethanol in Brazil is not a threat to the Amazon, it does not take land out of food production, nor does it take food off the tables of Brazilians or other peoples in the world" (Lula da Silva 2008c).

In addition, the administration repeatedly underlined the benefits of biofuel production for sustainable development, emphasizing that sustainable development not only comprised environmental protection, but also socio-economic development (Kloss 2012:83f., MRE 2010b:4).

In addition to its global action, the Brazilian government decided to take domestic policy measures to reduce its vulnerability with regard to the sustainability concerns raised at the global level. At the 2007 meeting of the UN General Assembly, Lula da Silva (2007b) announced that Brazil would implement an agroecological zoning to identify farmland that best suited biofuel production. In 2009, a presidential decree established agroecological zoning for sugarcane. Agroecological zoning is coordinated by MAPA and involves MME and the Ministry of Finance. Technical assistance is provided by the Public Enterprise of Agricultural Research, Embrapa.[251] ICONE also had a crucial role in developing the zoning.[252] According to a representative from MAPA,[253] the government established agroecological zoning as a preventative action in a strategic sector. It wanted to demonstrate that Brazilian ethanol production was compatible with food production and environmental protection. The agroecological zoning identifies areas within Brazilian territory that comply with the following criteria: they have the climate and soil potential for sugarcane production with mechanical harvesting, they have previously been used for livestock production and are not faced with environmental restrictions. 7.5 percent of Brazilian territory complies with these criteria. Agroecological zoning serves as guidance for public and private funding, the installation of new ethanol plants and the environmental licensing procedure.[254] In addition, the administration signed an agreement with the representatives of businesses and workers to ensure decent working conditions on sugarcane plantations. These domestic measures were pre-

---

[251] Presentation by Marlon Arraes J. Leal, Coordinator -Renewable Fuels Department, MME, at IEA – Biofuels Roadmap Workshop Discussion in Paris, September16th 2010, http://www.iea.org/media/bioenergyandbiofuels/06_arraes.pdf (accessed February 14, 2014).
[252] Interview with Professor, CPDA, UFRRJ, March 2013.
[253] Presentation by Luis Job at the GBEP Bioenergy Week in Brasília, March 2013, http://www.globalbioenergy.org/fileadmin/user_upload/gbep/docs/2013_events/GBEP_Bioenergy_Week_Brasilia_18-23_March_2013/2.2_JOB.pdf (accessed February 14, 2014).
[254] Presentation by Luis Job at the GBEP Bioenergy Week in Brasília, March 2013, http://www.globalbioenergy.org/fileadmin/user_upload/gbep/docs/2013_events/GBEP_Bioenergy_Week_Brasilia_18-23_March_2013/2.2_JOB.pdf (accessed February 14, 2014).

sented by Lula da Silva at the 2009 meeting of the UN General Assembly and the International Biofuels Conference in São Paulo 2008 (Lula da Silva 2009b, 2008a).

In 2008, the Brazilian government organized the International Biofuels Conference "Biofuels as a driving force of sustainable development" in São Paulo – according to Kloss (2012:136f.), the first intergovernmental conference on biofuels that was held in a developing country. Among the conference participants were 93 government delegations and 23 representatives of intergovernmental organizations. Next to intergovernmental sessions, the conference held multistakeholder discussions comprising academics, businesspeople, civil society representatives and parliamentarians. The government was represented at a high level: President Lula da Silva, the São Paulo State Governor José Serra, the Foreign Minister Celso Amorim, the Agriculture Minister Reinhold Stephanes, the Energy Minister Edison Lobão, the Science and Technology Minister Sérgio Rezende, the Environment Minister Carlos Minc and the Minister for Rural Development Guilherme Cassel participated. Dilma Rousseff joined in her function as the President's Chief of Staff (Amorim 2008, Kloss 2012:136-139, Lula da Silva 2008a).

By holding this conference, Brasília wished to influence the international debate on biofuels. Initially, the conference was thought to concentrate on technical and market issues, such as innovation and standardization. As biofuels were increasingly criticized, the government decided to change the agenda and stimulate a broader discussion about biofuels. It perceived the international debate as one-sided, neglecting the positive impacts of biofuel production on environment, economics, and rural development and failing to address trade issues as major barriers to the worldwide deployment of biofuels. It sought to transmit the Brazilian perspective as a developing country's viewpoint on the sustainability of biofuels (Amorim 2008, Kloss 2012:137, Lula da Silva 2008a, MRE 2010b:2). As a consequence, the conference was designed to discuss a broad range of biofuel-related issues on the international agenda, covering the following topics: energy security, climate change, sustainable development, innovation, science and technology as well as international markets. Due to the ongoing international food crisis and related criticism on biofuels, the conference also placed the link between biofuels and food production on the agenda. Prior to the conference, IBF members published thematic Green Papers which served as an input to it. These papers mapped the existing literature on the following issues: quantification of greenhouse gas emissions reduction, impact on food security, energy balances, impact on forests, water and biodiversity and social questions, especially labor rights (Kloss 2012:133f.).

The government regarded the conference as a success. Kloss (2012:138f.) underlines that it produced important results in the realm of energy security, environmental protection, sustainable development and on the link between biofuels and food production. He presents these results as follows: in the realm of energy security, the conference concluded that biofuels were a real alternative to fossil fuels and that a significant number of countries had the potential to produce them. In the realm of environmental protection, the conference pointed at climate change as a major global challenge that required profound transformations of energy systems. It recognized the advantages of biofuel deployment as a climate mitigating instrument in the transport sector, particularly its capacity to provide an easily expandable low-carbon option at competitive prices. With regard to sustainable development, the conference concluded that policy decisions required sound data and information. Sustainability criteria should be inclusive, transparent, based on scientific evidence and multilaterally accepted. They should not impose unjustified trade barriers, making the production of biofuels in developing countries impractical. In addition, the ministerial segment of the conference endorsed a statement that there was no direct link between biofuel production and food price increases. Kloss (2012:138f.) highlights the importance of this statement. He argues that that food security concerns seem to be the major vulnerability of biofuels as these concerns attract much international public opinion.

Sustainability issues also entered discussions at the last IBF meeting in Brussels in 2009. To the discontent of the Brazilian delegation, the sessions on sustainability covered environmental issues only. The Brazilian government argued that discussions on sustainability ought to include the social and economic pillars as well, and emphasized the socio-economic benefits of biofuel production, such as employment generation in rural areas, improved air quality and related health benefits in cities. It furthermore stressed that issues such as biofuel subsidies and trade barriers should not be omitted. The Brazilian delegation announced that it wished to continue IBF's work on standard harmonization, but the USA and EU signaled their lack of interest (Kloss 2012:151-153). In 2010, the Brazilian administration wished to hold another IBF meeting to discuss the sustainability criteria that were being developed in the USA and EU. It also sought to update the White Paper on harmonizing standards by including further countries. However, the meeting was not realized (Kloss 2012:139). Although Brazil tried several times to resume the activities of the IBF, it did not succeed in doing so. The IBF ceased to exist because the other member states were not interested in continuing its activities (Kloss 2012:189).

In 2008, Brazil became co-chair of GBEP. The G8 countries had officially launched GBEP at the XIV session of the UN CSD in 2006 (see chapter 3.1.3). Brazil had participated at the GBEP meetings from the beginning, first as an observer and from 2007 as a participant. Since 2008, the Head of Itamaraty's Energy Department has served as the vice-president of the partnership. Within GBEP, the Brazilian government sought to discuss issues that are relevant for the creation of an international biofuels market, such as trade barriers and technology transfer. It wished to ensure that GBEP discussions take the Brazilian experiences with biofuels into account.

Due to growing concerns about food security and the environmental implications of biofuel production, GBEP created a Task Force on Sustainability in 2008. In order to avoid duplication of work, IBF members decided to leave the discussion of sustainability issues to GBEP (Kloss 2012:136). However, Kloss (2012:172, 180) states that at the beginning, the Brazilian delegation viewed GBEP sustainability discussions with distrust. It considered GBEP as a forum dominated by industrialized countries and wanted to avoid sustainability debates being conducted in a way that rewards biofuel production in industrialized countries over developing countries. In particular, it feared that sustainability concerns would be used to shield the biofuel sectors of industrialized countries against more economically competitive production in developing countries. The Brazilian government therefore wished to impede that sustainability concerns in industrialized countries could obstruct the development of biofuel industries in developing countries. It furthermore feared that sustainability concerns could even impede the creation of an international biofuels market. Despite these concerns, it decided to actively engage in GBEP discussions on sustainability. It recognized that sustainability issues had a strong impact on the acceptance of biofuels. In addition, it sought to use the sustainability debate to reinforce the benefits of biofuels. Whereas the USA and EU initially wanted to concentrate sustainability discussions on environmental aspects, Brazil ensured that the GBEP taskforce considered all three pillars of sustainability. Within the taskforce's work, the Brazilian delegation was instructed to maintain a positive attitude vis-à-vis biofuels. Here, it took a counterpoint vis-à-vis European countries, particularly the Netherlands and Germany, which emphasized the possible negative impacts of biofuel production, particularly in developing countries. When European delegations wanted to include indirect land use changes (iLUC) into the sustainability assessment, the Brazilian government argued strongly against it, referring above all to the scientific uncertainties involved. Within the discussions on indirect land use change, the Brazilian government cooperated closely with the Brazilian agribusiness think tank ICONE (Kloss 2012:172-182).

In 2008, the Brazilian Technical Standards Association (Associação Brasileira de Normas Técnicas, ABNT) proposed, together with the German Institute for Standardization (Deutsches Institut für Normung, DIN), to create a body at the International Standardization Organization (ISO) dealing with sustainability issues of bioenergy. The ISO is a network of governmental and private standardization organs. Brazil is represented by the ABNT which is a private standardization entity. In 2007, ISO created a subcommittee on liquid biofuels (ISO/TC28/SC7). The ABNT took over the coordination of the subcommittee, with Sergio Fontes of

Petrobras serving as the chair.[255] In 2009, the ISO project committee on sustainability criteria for bioenergy (ISO/PC 248) was formed, co-chaired by DIN and ABNT.[256] The ABNT did not consult the Brazilian government before taking this initiative. To express its discomfort with this initiative, the Casa Civil organized a meeting with the ABNT, the Itamaraty, MME and MAPA. In the view of the Brazilian government, the sustainability criteria for bioenergy were too politically sensitive to be discussed in a private technical forum. As the WTO does not consider ISO standards as unjustified trade barriers, the administration feared that an ISO standard on sustainability could diminish the possibilities of challenging the sustainability requirements of other countries at the WTO. Since the Brazilian government feared that a formal withdrawal from the initiative could have negative reputational implications, it agreed with the ABNT to maintain the initiative, but slow it down from within (Kloss 2012:25,154-158).

The ISO project committee on sustainability held its first meeting in Rio de Janeiro in 2010 (Kloss 2012:156-158). Brazil was represented by a wide range of actors. From the government side, the Casa Civil, Itamaraty, MAPA and MDIC participated. Other public actors included the governmental standardization body Inmentro, ANP, Brazilian Development Bank (Banco Nacional do Desenvolvimento, BNDES), Embrapa and Petrobras. The private sector was represented by the ABNT, the Brazilian Association of Vegetable Oil Industries (Associação Brasileira das Indústrias de Óleos Vegetais, Abiove), UNICA and the sugar and ethanol trader Copersucar S.A. Academics from ICONE, Unicamp and CTBE also joined the meeting. Whereas European representatives wanted to develop sustainability standards for the whole production chain and include indirect effects on biodiversity and greenhouse gas emissions, Brazil argued, together with the USA, against having too rigid standards. According to Kloss (2012:157f.), the meeting showed that the sustainability debate within the ISO was heavily influenced by European perspectives based on EU norms, favoring a precautionary approach and focusing exclusively on the environmental pillar of sustainability. He furthermore claims that in order to rebalance the discussions within ISO, it would be necessary to increase the participation of developing countries which at the meeting was restricted to Brazil and Colombia.

---

[255] ISO, ISO/TC 28/SC 7 Liquid Biofuels, http://www.iso.org/iso/home/standards_development/list_of_iso_technical_committees/iso_technical_committee.htm?commid=551957 (accessed February 14, 2014).
[256] ISO, ISO/PC 248 Sustainability criteria for bioenergy, http://www.iso.org/iso/home/standards_development/list_of_iso_technical_committees/iso_technical_committee.htm?commid=598379 (accessed February 14, 2014).

5.1.4 Continuation of Ethanol Diplomacy at a Lower Political Profile

In 2010/11 Brazilian ethanol diplomacy lost visibility as biofuels ceased to be a priority issue for the government's transboundary action. In 2010, Lula da Silva ceased putting biofuels on the agenda of international meetings. At the 2010 meeting of the UN General Assembly, for instance, the Brazilian government did not mention biofuels (Amorim 2010)[257] – a stark contrast to prior years. Dilma Rousseff (PT), Brazilian President since the beginning of 2011, has not revived Brazilian ethanol diplomacy.

The economic difficulties of the ethanol sector seem to be the major reason behind this policy change.[258] As visible in Figure 7, domestic ethanol production has declined significantly in 2010/11. In 2011, Brazil for the first time had to import significant amounts of ethanol, especially from the USA. However, this does not mean that Brazilian ethanol exports were completely stalled. Brazil even remained a net ethanol exporter. Since 2010, Brazilian ethanol is recognized as an 'advanced biofuel' by the US Environmental Protection Agency. This was an important victory for the country's ethanol diplomacy as the government had lobbied intensively to achieve this recognition. As an 'advanced biofuel', Brazilian ethanol receives a premium price on the US market. As a consequence, exporting ethanol was still an attractive option for the country's producers. To fill the domestic output gap, Brazil imported corn-based ethanol from the USA which received a lower price on the Brazilian market than the Brazilian sugarcane ethanol in the USA.[259] With the domestic output gap of ethanol production, the government could no longer sell its ethanol as a reliable import source to other countries.[260] In addition, opening new export opportunities for the Brazilian ethanol producer became less of a priority and as such, biofuels lost currency within the Brazilian diplomacy.[261]

---

[257] At this meeting of the UN General Assembly, Brazil was represented by its Foreign Minister Celso Amorim. Lula da Silva did not participate due to ongoing Presidential elections.
[258] Interviews with Biodiesel Federal Program Coordinator and staff member, Analysis and Following up of Government Policies, Casa Civil, March 2013; Head of Energy Department, Itamaraty, April 2013; Secretary and team, Secretariat of Planning and Energy Development, MME, March 2013; Manager, Climate Change Department, MMA, March 2013; team, Bioenergy Projects, FGV, February 2013; researcher on applied economics, FGV, March 2013.
[259] Interview with team, Bioenergy Projects, FGV, February 2013.
[260] Interview with researcher on applied economics, FGV, March 2013.
[261] Interview with Biodiesel Federal Program Coordinator and staff member, Analysis and Following up of Government Policies, Casa Civil, March 2013; Head of Energy Department, Itamaraty, April 2013; Secretary and team, Secretariat of Planning and Energy Development, MME, March 2013; Manager, Climate Change Department, MMA, March 2013.

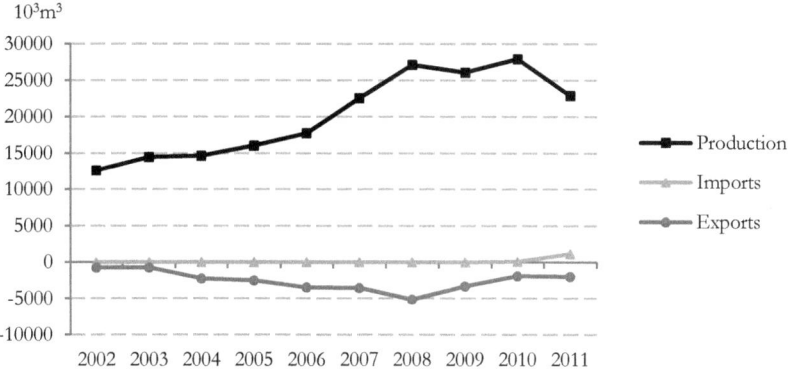

Figure 7: Brazilian Ethanol Production and Trade Balance, 2002–2011[262]

The output gap of Brazilian ethanol production in 2010/11 is commonly explained with a bundle of factors on both the supply and the demand side. The freezing of the domestic gasoline price is commonly seen as the main reason for economic difficulties in the ethanol sector.[263] In 2006/07 the government had decided to stabilize the domestic gasoline price, de-coupling it from rising international oil prices. The main reason behind this step was the fight against inflation.[264] In addition, low gasoline prices are instrumental in gaining the political support of the middle class as they subsidize people with cars.[265] As anhydrous ethanol directly competes with gasoline, a price cap on gasoline indirectly limits the selling price for anhydrous ethanol and therefore negatively affects the ethanol sector. In addition, the international financial crisis affected the Brazilian ethanol sector, as foreign investments dried up. Before the global financial crisis, the participation of foreign capital had increased significantly in the Brazilian ethanol sector while many small companies disappeared from the market. With the outbreak of the global financial crisis, the sector entered stagnation. The lack of capital impeded necessary rein-

---

[262] Source: illustration S.R. with data from Empresa de Pesquisa Energética (2012:67).
[263] Interviews with Biodiesel Federal Program Coordinator and staff member, Analysis and Following up of Government Policies, Casa Civil, March 2013; team, Bioenergy Projects, FGV, February 2013; former director, UNICA, April 2013; Vice-president, CEBRI, April 2013, Professor, COPPE, UFRJ, February 2013; staff member, Solidaridad, April 2013; professor (#1), Energy Institute, USP, April 2013; professor, Luiz Queiroz College of Agriculture, USP, April 2013; professor, CPDA, UFRRJ, March 2013.
[264] Interviews with Secretary and team, Secretariat of Planning and Energy Development, MME, March 2013; professor (#1), Energy Institute, USP, April 2013; professor, FFLCH, USP, April 2013; professor, Economics Institute, UFRJ, March 2013; staff member, Solidaridad, April 2013.
[265] Interviews with former president, UNICA, April 2013, and professor (#1), Energy Institute, USP, April 2013.

vestments into sugarcane plantations and mills. The reduced re-investment capacity negatively affected the productivity of the sugarcane sector (Kloss 2012:10).[266] On top of that, a series of bad harvests due to unfavorable climatic conditions since 2009 and rising land costs also hit the sector.[267] Yet, there are also factors on the demand side that contributed to the output gap. Economic growth, a rising middle class with the purchasing power to buy cars, and fiscal incentives for car purchases led to success in the sale of flex fuel cars which also increased domestic ethanol demand (Kloss 2012:10).[268] At the same time, high international sugar prices made it more attractive for sugarcane producers to sell sugar instead of ethanol.[269]

The oil discoveries in Brazil's pre-salt layer since 2006/07 may also have contributed to decreasing visibility of Brazilian ethanol diplomacy. With the pre-salt discoveries, fossil fuels gained much political attention within Brazil (Schutte, Barros 2010:33, Viola 2013:7). After the detections, Lula da Silva presented Brazil as a country that was heading into a bright future, freeing itself from energy limitations and establishing itself as a new energy superpower.[270] It is interesting to note that government representatives maintain that the fossil fuel discoveries did not lower the government's interest in biofuels, and that fossil fuels and biofuels are seen as two important and parallel business opportunities for the country.[271] This position was also put forward by Lula da Silva in his speeches at the International Biofuels Conference 2008 and the 2009 meeting of the UN General Assembly: He pointed to the newly discovered oil wealth but emphasized that Brazil would not change its conviction on the importance of biofuels (Lula da Silva 2008a), instead seeking to consolidate its "role as a world power in green energy" (Lula da Silva 2009b). Yet,

---

[266] Interviews with Biodiesel Federal Program Coordinator and staff member, Analysis and Following up of Government Policies, Casa Civil, March 2013; Director of the Department of Renewable Fuels, MME, March 2013; team, Bioenergy Projects, FGV, February 2013; former president, UNICA, April 2013; Vice-president, CEBRI, April 2013; staff member, Solidaridad, April 2013; professor, COPPE, UFRJ, February 2013; professor, CPDA, UFRRJ, March 2013; professor, Luiz Queiroz College of Agriculture, USP, April 2014.

[267] Interviews with Secretary and team, Secretariat of Planning and Energy Development, MME, March 2013; Biodiesel Federal Program Coordinator and staff member, Analysis and Following up of Government Policies, Casa Civil, March 2013; diplomat, Itamaraty, March 2013; former president, UNICA, April 2013.

[268] Interview with Biodiesel Federal Program Coordinator and staff member, Analysis and Following up of Government Policies, Casa Civil, March 2013.

[269] Interviews with Biodiesel Federal Program Coordinator and staff member, Analysis and Following up of Government Policies, Casa Civil, March 2013 and Vice-President, CEBRI, April 2013.

[270] Interviews with Vice-President, CEBRI, April 2013 and professor (#1), Institute for International Relations, UnB, April 2013.

[271] Interviews with Head of Energy Department, Itamaraty, April 2013; Secretary and team, Secretariat of Planning and Energy Development, MME, March 2013; Biodiesel Federal Program Coordinator and staff member, Analysis and Following up of Government Policies, Casa Civil, March 2013; Director of the Department of Renewable Fuels, MME, March 2013; General Coordinator for Sugar and Ethanol, MAPA, March 2013; professor, Luiz Queiroz College of Agriculture, USP, April 2013.

several Brazilian experts that do not belong to the administration claim that the discoveries did indeed change the government's energy priorities from renewable energy to fossil fuels.[272]

Since her accession to power in 2011, Dilma Rousseff has not taken an active stance on global renewable energy governance. In contrast to Lula da Silva who travelled frequently and was very present in international forums, Rousseff focuses on domestic issues and does not engage much in foreign affairs. Instead of following a presidential approach to foreign policy, she leaves the field to the Itamaraty.[273] She is commonly characterized as a discrete person who in general has less public presence than Lula da Silva.[274] Several experts claim that she is less interested in biofuels than Lula da Silva and that she has stronger links to the fossil fuel sector.[275] A few months after her accession to power, Rousseff signed a provisional measure that lowered the minimum volume of the ethanol blending range from 20 to 18 percent. This increased government flexibility when setting the ethanol blending rate. However, she did not change the actual blending requirement which at that time was set at the cap of 25 percent.[276]

Brazilian foreign policy on biofuels continues despite its lower visibility. The government is still interested in creating an international market for biofuels and continues to regard bioenergy as an important policy option for Brazil and other countries. Its engagement in the worldwide promotion of biofuels now concentrates on GBEP, which is still co-chaired by the Itamaraty.[277] According to Kloss (2012:172f.), the government regards itself as the voice of developing countries in GBEP, trying to emphasize "the perspective of developing countries on the use of bioenergy and its benefits" (translation S.R.).[278] Within GBEP, Brazil especially focuses on capacity building. In March 2013, the government organized a one-week

---

[272] Interviews with former president, UNICA, April 2013; Vice-President, CEBRI, April 2013; professor (#1), Institute of International Relations, UnB, April 2013; staff member, Solidaridad, April 2013; Consultant, FASE Brasil, March 2013; Professor, CPDA, UFRRJ, March 2013.
[273] Interviews with Head of the New and Renewable Energy Resource Division, Itamaraty, March 2013 and diplomat, Itamaraty, March 2013.
[274] Interviews with Head of the New and Renewable Energy Resource Division, Itamaraty, March 2013; diplomat, Itamaraty, March 2013; General Coordinator for Sugar and Ethanol, MAPA, March 2013; Foreign Trade Analyst, Secretariat for Innovation, MDIC, March 2013; researcher on applied economics, FGV, March 2013.
[275] Interviews with Biodiesel Federal Program Coordinator and staff member, Analysis and Following up of Government Policies, Casa Civil, March 2013; former president, UNICA, April 2013; professor, CPDA, UFRRJ, March 2013; team, Bioenergy Projects, FGV, February 2012.
[276] US Department of Agriculture, Brazil Biofuels Annual, Annual Report 2012, http://gain.fas.usda.gov/Recent%20GAIN%20Publications/Biofuels%20Annual_Sao%20Paulo%20AT O_Brazil_8-21-2012.pdf (accessed March 06, 2014).
[277] Interviews with Head of Energy Department, Itamaraty, March 2013, and Head of the New and Renewable Energy Resource Division, Itamaraty, March 2013.
[278] Original text of this quotation: "a perspectiva dos países em desenvolvimento quanto ao uso da bioenergia e aos benefícios dela decorrentes".

GBEP Study Tour on Capacity Building in Brasília,[279] hosted by Itamaraty, MAPA and Embrapa. Financial support was provided by the US Department of State, the Organization of American States (OAS) and the biofuels company Raizen Energia S.A., a joint venture of the Brazilian Cosan S.A. and Royal Dutch Shell. The main purpose of this 'Bioenergy Week' was to transmit the Brazilian experience on bioenergy to developing countries. In addition to GBEP members, developing countries from Latin America and the Economic Community of West African States (ECOWAS) were invited. The conference discussed the three pillars of bioenergy sustainability. With regard to the social pillar, it paid special attention to family agriculture and rural development. In the realm of economic sustainability, it especially focused on funding opportunities. Regarding the environmental pillar, it concentrated on greenhouse gas emissions, land use changes and water usage. In addition, the conference provided regional overviews of bioenergy production in different parts of the world and presented state of the art feedstock production and bioenergy conversion processes. A wide range of Brazilian actors were actively involved in the conference. From the government side, the Casa Civil, MME and MDA provided substantial input next to the conference organizers Itamaraty and MAPA. The research sector was represented by the Sugarcane Technology Center (Centro de Tecnologia Canavieira, CTC), the Institute of Energy and Environment of the University of São Paulo (Instituto de Energia e Ambiente da Universidade de São Paulo, IEE/USP), the agribusiness think tank ICONE and the Getulio Vargas Foundation (Fundação Getulio Vargas, FGV). Abiove, Belém Brazil Bioenergy, UNICA and ANFAVEA represented the private sector. In addition, INMETRO, Petrobras and BNDES were present.[280] NGO representatives did not participate in the conference.

5.2 Ideas on Global Renewable Energy Governance

5.2.1 Global Challenges: Predominance of Socio-Economic Development

The Brazilian government recognizes that for most policy actors around the world, concerns about climate change and energy security are the most important motivations in promoting renewables. However, the administration repeatedly underlines

---

[279] See GBEP, Working Group on Capacity Building meetings and activities 2013, http://www.globalbioenergy.org/events1/gbep-events-2013/working-group-on-capacity-building-meetings-2013/en/ (accessed March 06, 2014).
[280] GBEP, Bioenergy Week – provisional programme, http://www.globalbioenergy.org/fileadmin/user_upload/gbep/docs/2013_events/GBEP_Bioenergy_Week_Brasilia_18-23_March_2013/GBEP_BioenergyWeek__Draft_Agenda_EN15032013.pdf (accessed June 26, 2013).

that it prioritizes another global challenge, namely the enhancement of socio-economic development in developing countries.

The Brazilian government commonly refers to the global challenge of climate change as one of the main drivers behind the worldwide promotion of renewable energy. In his speech at the 2007 meeting of the UN General Assembly, Lula da Silva points out that the increased use of biofuels is a crucial element in the global fight against climate change:

> "We will not overcome the terrible impacts of climate change until humanity changes its patterns of energy production and consumption. The world urgently needs to develop a new energy matrix, in which bio-fuels will play a vital role. Bio-fuels significantly reduce greenhouse gas emissions" (Lula da Silva 2007b).

Lula da Silva reinforces this argument in his speeches at several international meetings (see, for example, Lula da Silva 2009b, 2008a, 2008c, 2008d, 2008e, 2007a) and in his articles on biofuels in The Guardian and Financial Times (Lula da Silva 2008b, 2007c). In a rare pronouncement on the transboundary promotion of biofuels, Dilma Rousseff (2011a, 2011b) also highlights the positive environmental impacts associated with biofuels. Several representatives of the Itamaraty emphasize that environmental concerns, in particular efforts to mitigate climate change, are major reasons behind the efforts of many governments to reduce the use of fossil fuels. These concerns thus provide a window of opportunity for the worldwide dissemination of renewable energy (see, for example, Amorim 2007b:6-8, Kloss 2012:23,26, MRE 2003a, Simões 2007b:17,31, Vargas 2004). The Brazilian Energy Proposal at the WSSD 2002 highlights that the use of biofuels and other renewable energy options does not only tackle climate change but also helps to improve local air quality (Goldemberg 2002a, 2002b). Foreign Minister Celso Amorim (PT, 2003–2010) points out that the transport sector has one of the highest emission rates of greenhouse gases and these emissions have more than doubled since the 1970s (Amorim 2007b:8). Kloss (2012:56) adds that biofuels are the only economically viable option to reduce greenhouse gas emissions in the transport sector. The quality of biofuels as a low-cost alternative to climate mitigation is also reinforced by Lula da Silva. He states that in the global fight against climate change, it is especially important for poorer countries to have access to low-cost mitigation technologies (Lula da Silva 2008a, 2007a, 2007b).

The government presents energy security as another global challenge that serves as a major driver behind worldwide efforts to increase the use of renewable energy. According to Simões (2007a:17,31) and Kloss (2012:23,26), it is insecurity about the future supply of fossil fuels next to climate concerns that make many government consider how to reduce their fossil fuel dependence. In his article in the *New York Times*, the first Head of Itamaraty's Energy Department, Antônio Simões[281], points at these two challenges:

> "The reality is that if we maintain the current rate of oil consumption without major reductions in carbon emissions, we will surely be heading in the direction of unprecedented climate change and natural disasters. It is also a fact that if oil demand continues to increase, prices will skyrocket, terribly affecting poor countries. The International Energy Agency itself admits that increasing demand and irregular supply will impose additional pressure on prices, which in turn will also be affected by higher extraction costs of new reserves (deep waters, heavy and extra-heavy oil)" (Simões 2007b).

He underlines that rising oil demand will lead to significant oil price increases which will hit poor countries in particular. In his speeches at a meeting of the UN Economic and Social Council (ECOSOC) and the International Biofuels Conference in São Paulo, Lula da Silva also points at rising energy prices. He claims that countries which depend on fossil fuel imports have an urgent need to diversify their energy matrix (Lula da Silva 2008a, 2008d). Dilma Rousseff (2011b) reinforces the contribution of biofuels for the diversification of energy supply. The Brazilian Energy Proposal for the WSSD in 2002 also underlines the fact that renewable energy contributes to a secure and diversified energy supply (Goldemberg 2002a). Simões (2007a:14) elaborates further on the increasing scarcity of fossil fuels. He claims that discoveries of new fossil fuel reserves have diminished in recent years and that the rate of fossil fuel consumption exceeds the discovery of new reserves. He observes a tendency of declining production despite technological advancement and rising fossil fuel prices and refers to the relatively high production costs of new reserves. Simões emphasizes that the comparative advantages of biofuels increase in the light of declining reserves and rising prices of fossil fuels. However, he also underlines that biofuels are not a threat to fossil fuels, but "cohabit harmoniously" (translation S.R.)[282] with them (Simões 2007a:31). As with regard to climate change, Brazilian official statements repeatedly underline that biofuels provide a low-cost alternative to fossil fuels, particularly for developing countries (Amorim 2007b:8, Lula da Silva 2008a, Lula da Silva 2007a, Lula da Silva 2007b, Simões 2007b:33). Simões (2007a:33) and André Amado (2010b), Itamaraty's former General Subsecretary on Energy and High Technologies, add that the production of sugarcane ethanol provides another positive impact on the energy security of developing countries as the byproducts of sugarcane ethanol can be easily used to generate electricity at low cost. This is particularly relevant for countries that need to satisfy a growing electricity demand.

---

[281] Itamaraty, Subsecretário-Geral da América do Sul Embaixador Antonio José Ferreira Simões, http://www.itamaraty.gov.br/o-ministerio/curriculos/subsecretario-geral-da-america-do-sul (accessed December 20, 2013).
[282] Original text of this quotation: "convivem harmoniosamente".

However, the Brazilian government makes it clear that it regards the enhancement of socio-economic development in developing countries as a key global challenge that the promotion of renewable energy should address. It repeatedly points to the socio-economic benefits of biofuel production in developing countries (see, for example, Amado 2010b, Amorim 2008, Amorim 2007b:6-8, Goldemberg 2002b:3-5, Lula da Silva 2008a, Lula da Silva 2008c, Lula da Silva 2008e, Simões 2007a:23, Vargas 2004:1). In his speech at the 2007 meeting of the UN General Assembly, Lula da Silva underlines the importance of these socio-economic benefits by highlighting that biofuels are much more than a means to climate mitigation:

> "We will not overcome the terrible impacts of climate change until humanity changes its patterns of energy production and consumption. The world urgently needs to develop a new energy matrix, in which bio-fuels will play a vital role. Bio-fuels significantly reduce greenhouse gas emissions. (…) Bio-fuels can be much more than a clean-energy alternative. They can open up excellent opportunities for over a hundred poor and developing countries in Latin America, Asia and, especially, Africa. They can enhance energy autonomy, without costly investments. They can create jobs and income and promote family farming. They can help balance trade deficits, by reducing imports and generating surplus exportable crops" (Lula da Silva 2007b).

At the International Biofuels Conference, Amorim argues in a similar vein. According to him, biofuel promotion is not only about energy security and climate change, but also about sustainability, the fight against hunger and poverty, technological innovation, and trade (Amorim 2008). At the International Biofuels Conference in Brussels 2007, Lula da Silva emphasizes the need to combine climate protection with economic growth, income and employment generation. He underlines that biofuels are not only a low-cost mitigation technology; they even foster income and job creation which is particularly urgent in poor countries:

> "Biofuels are a low-cost option of proven efficiency in the transition to an economy based on low carbon emissions. In reducing these emissions, biofuels remove the grave dilemma of choosing between the adoption of high-cost technologies on the one hand and a reduction in the rhythm of world growth on the other. This choice is especially dramatic for poor countries, which don't have the resources to adopt expensive technology but, at the same time, have an urgent need to create jobs, wealth and income" (Lula da Silva 2007a).

In his Financial Times article, Lula da Silva makes it clear that, for him, supporting socio-economic development in developing countries is more important than climate protection. He states that it would be unacceptable for him that climate mitigation measures negatively impact the fight against poverty and the aspirations of poorer countries to improve the wellbeing of their societies (Lula da Silva 2008b). André Amado adds that in developing countries, poverty is the major cause for environmental destruction. Therefore, economic growth is instrumental for environmental protection (Amado 2010b). Lula da Silva underlines that in order to realize development benefits, biofuels should not be produced at high costs in the USA and EU but rather in poor and developing countries. Rich countries should build partnerships with poor countries, especially in Africa, and import part of their biofuel consumption from there (Lula da Silva 2008a, 2007a).

The importance that the Brazilian government attaches to socio-economic development also becomes clear in its deliberations on sustainable development. In the foreign policy review of the Lula da Silva government, the Itamaraty states that sustainability is a key aspect in the worldwide debate on biofuels. It claims that there must be a balanced approach towards sustainability, so that all three pillars of the concept (environmental, economic and social development) are equally taken into account (MRE 2010b:1). By demanding a balanced approach towards sustainability, the Itamaraty insinuates that international debates about sustainability are often conducted in an unbalanced way. In the view of the Brazilian government, international debates on sustainability often prioritize the environmental pillar and forget to take the economic and social dimension into account (Kloss 2012:151-153,156-158,172-182). The Itamaraty furthermore maintains that sustainability criteria should not be used as trade barriers (MRE 2010b:2).

Out of the socio-economic development benefits, the Brazilian government especially focuses on the fight against poverty and hunger (Amorim 2008, Goldemberg 2002b:3, Lula da Silva 2008a, Lula da Silva 2008c, Lula da Silva 2007b, Vargas 2004:1). For Lula da Silva (2008c, 2007b), lack of income is the main reason behind hunger on the world. He emphasizes that hunger is not a problem of overall food scarcity, but rather a problem of unequal distribution: it is lack of income (and not lack of food production) that prevents people from accessing food and that makes them suffer from hunger (Lula da Silva 2008c, 2007a). A structural cause for hunger lies in the unbalanced international trade on agriculture, as subsidies and protectionism in rich countries impede the formation of a strong agricultural sector in poor countries:

> "There will be no structural solution for world hunger as long as we are unable to direct resources into food production in poor countries, while also removing the unfair trade practices that characterize trade in agricultural goods" (Lula da Silva 2008c).

While developing countries are confronted with trade barriers when exporting agricultural products to richer countries, they depend on aid and subsidized agricultural imports from their industrialized counterparts (Lula da Silva 2008d, 2007a). Lula da Silva relates the rising food prices in 2007/2008 to rising oil prices affecting the prices of fertilizers and transport in the agricultural sector, speculations on oil and commodity prices, and crop failures due to climatic conditions (Lula da Silva 2008a, 2008d, 2008e). In his opinion, biofuels can make an important contribution to the fight against hunger and poverty. Offering an alternative to oil, they alleviate the impact of rising oil prices on agricultural production (Lula da Silva 2008d, 2008e). In his New York Times article, Simões illustrates this impact as follows:

"The increase in oil prices will have serious consequences on the price of food products. More expensive fertilizers will become less accessible to farmers in poor countries. Sharp increases in transportation costs will reduce the access to food for millions. Therefore, higher oil prices will surely mean less food consumption" (Simões 2007b).

According to Lula da Silva (2008d, 2008e), biofuels also strengthen the agricultural sector in poor countries, enabling poorer countries to produce their own food and reducing the structural imbalances caused by distortions in agricultural trade. Simões also underlines the possibility of strengthening the agricultural sector in developing countries with the help of biofuel production:

"Large extensions of unutilized arable land in the Southern Hemisphere would be employed for highly profitable biofuel-oriented crops, restructuring the agricultural sector. Millions of jobs would be generated, thus increasing income, exports and food purchasing power of the poorest" (Simões 2007b).

In addition, representatives of the Brazilian government repeatedly underline that biofuel production tackles poverty and hunger by generating income and employment in rural areas (Amado 2010b, Amorim 2008, Goldemberg 2002b:5, Lula da Silva 2008a, Lula da Silva 2008c, Lula da Silva 2008d, Simões 2007a:23). Kloss (2012:55) explains that people living in extreme poverty often depend on agricultural activities to survive; therefore, agricultural development is key to improving the living standards of poor people. Lula da Silva (2008a) underlines that biofuel production also provides an effective means to prevent poverty migration as it offers economic opportunities to poor countries. The Brazilian Energy Proposal for the WSSD 2002 also elaborates on poverty. Yet, it underlines a different aspect. It emphasizes that renewable energy helps to enhance access to energy in remote areas, facilitating small-scale and decentralized energy production as well as energy systems that are relatively easy to operate and maintain (Goldemberg 2002b:6).

The Brazilian government furthermore points to the macroeconomic benefits of biofuel production in developing countries. Biofuel production may enhance economic growth in developing countries, enabling them to take economic advantage of natural resources and offering poor countries a product with great economic potential (Amorim 2008, Kloss 2012:31, Vargas 2004:1). By replacing costly fossil fuel imports and generating new export revenues, the domestic production of biofuels may improve the trade balance of developing countries (Amorim 2007b:8, Lula da Silva 2007b, Simões 2007a:33). Kloss (2012:55-57) underlines that developing countries have mor financial resources to spend on education, health, sanitation and housing if they save oil imports; in addition, biofuels can help to diversify and add value to the exports of developing countries, which are often concentrated on a few products with low value. The Brazilian Energy Proposal for the WSSD 2002 relates these trade benefits for developing countries not only to biofuels, but to renewable energy in general (Goldemberg 2002a:2). Lula da Silva (2008a) highlights another macroeconomic benefit of biofuel production in developing countries: as biofuel production involves low-cost technologies instead of the expensive and high-tech, it also facilitates the integration of poorer countries into transboundary science and technology cooperation.

The Brazilian government underlines that biofuel production not only helps to tackle asymmetries in agricultural trade, but also contributes to a more democratic energy system globally. The following statement by Simões is representative of the Brazilian government's line of argument on this issue:[283]

> "Nowadays, world energy resources are concentrated in 20 countries. Biofuels will allow a true democratization of the international market, as over 100 countries will be producing energy for the world. There is no doubt about the fact that this is a great change, maybe as revolutionary as the one that began in the early 20th Century. After all, the transition from animal traction to petroleum was antipodal to environmental sustainability. Today, we can correct this and, at the same time, contribute to the generation of employment and wealth in the countries of the South - much to the benefit of the global community" (Simões 2007b).

The statement underlines that oil reserves are concentrated in a limited number of countries, while biofuels can be produced in many different places. As such, they reduce asymmetries between fossil fuel exporting and importing countries and may even contribute to preventing conflicts over scarce resources (Lula da Silva 2007a, Simões 2007a:23). The Brazilian Energy Proposal for the WSSD 2002 makes a similar point with regard to renewable energy in general: whereas most fossil fuels are concentrated in politically and economically unstable regions, renewable energy sources are more evenly distributed (Goldemberg 2002b:7). Lula da Silva (2008a, 2007a) furthermore emphasizes that biofuels differ from oil in the sense that they do not require expensive technologies and thus ease the access to energy, especially for poor countries. At the International Biofuels Conference in Brussels 2007, he illustrates this difference between oil and biofuels with the following words:

> "Look at the world of fossil fuel, imagining how many countries have oil, how many countries have the technology to drill for it at a depth of three of four thousand metres, or to build oil platforms that cost a billion dollars each. The majority certainly don't.
>
> Now, look at the world and see that everyone – from the poorest country on Earth, from the poorest living person on this planet – has the technology to dig a small hole, 30 centimetres deep, and plant an oil-producing plant that could provide the energy they couldn't produce in the twentieth century" (Lula da Silva 2007a).

Itamaraty officials elaborate further on the security implications of energy supply. Kloss (2012:22) states that many countries instrumentalize fossil fuels for geopolitical purposes. Simões (2007a:14) underlines the crucial role of energy for military and economic matters. According to him, governments regard energy security as a fundamental aspect of national security, as interruption in the energy supply can hit

---

[283] See also Amorim 2008, Goldemberg 2002b:7, Simões 2007a:16,23, Simões 2007b.

any modern society severely. Often, they militarize energy policies, trying to control the access to oil and the respective transport routes.

5.2.2 Renewable Energy Options: Biofuels and Competitiveness

The Brazilian deliberations on global renewable energy governance clearly concentrate on biofuels, and more specifically on sugarcane ethanol. The terms 'biofuel' and 'ethanol' are often used as synonyms. As such, the transport sector prevails in the country's deliberations. The government repeatedly underlines that it is necessary to differentiate between different forms of ethanol when assessing the benefits of its implementation. It particularly highlights the advantages of Brazilian sugarcane ethanol vis-à-vis corn ethanol produced in the USA. When presenting the benefits of sugarcane ethanol, representatives of the Brazilian government commonly refer to effects on food security, environmental impacts, energy balance and economic competitiveness (Goldemberg 2002a:3, Goldemberg 2002b:8, Kloss 2012:126, Lula da Silva 2009b, Lula da Silva 2008c, Lula da Silva 2007c, Simões 2007a:26). At the FAO meeting in 2008, Lula da Silva clearly differentiates between good ethanol based on sugarcane and bad ethanol made of corn:

> "I am not in favour of producing ethanol from corn or other food crops. I doubt that anyone would go hungry, to fill up their car's fuel tank. Meanwhile, corn ethanol can obviously only compete with sugar-cane ethanol when it is shot up with subsidies and shielded behind tariff barriers. Sugar-cane ethanol yields 8.3 times more energy than the fossil energy used to produce it. Corn ethanol, meanwhile, yields only 1.5 times the energy it consumes. That is why some people compare ethanol to cholesterol. There is good ethanol and bad ethanol. Good ethanol helps clean up the planet and is competitive. Bad ethanol comes with the fat of subsidies" (Lula da Silva 2008c).

The government presents economic competitiveness as a core criterion that should guide the choice between different renewable energy options. Lula da Silva's statement at the FAO meeting illustrates this: whereas "good ethanol" is competitive, "bad ethanol" depends on subsidies. In his Guardian article, he reinforces the importance of economic competitiveness. He states that Brazilian sugarcane is the most productive raw material for ethanol production and directly compares its efficiency to "its main European and American competitors": ethanol made of corn and sugar beet (Lula da Silva 2007c). Kloss (2012:112) emphasizes that large-scale production of biofuels is an important pre-condition for economic competitiveness. The government furthermore presents affordability as an important criterion for the assessment of renewable energy options.[284] The former Head of Itamaraty's De-

---

[284] Interview with Secretary and team, Secretariat of Planning and Energy Development, MME, March 2013.

partment for the Environment, Everton Vargas, emphasizes the importance of cost-effectiveness and affordability when promoting renewable energy:

> "Policies for the promotion of renewable should be cost-effective and, therefore, seek the lowest possible costs and fair prices that take into account the concerns of the consumers and of our income levels" (Vargas 2004:2).

Lula da Silva (2008a, 2008d, 2007a, 2007b) repeatedly underlines that biofuels are a low-cost option in promoting renewable energy. Kloss (2012:126) even argues that out of the different renewable energy sources, biofuels are the only economically feasible option in the short run. In line with this, officials of the Casa Civil state that photosynthesis is still the least expensive way of generating energy.[285]

An Itamaraty official points out that the Brazilian engagement for the worldwide promotion of biofuels does not focus on ethanol only, but also includes biodiesel. He adds that ethanol might be Brazil's solution but other countries might have other solutions, and Brazil does not want every country to follow its example.[286] The name of the GBEP conference organized by Brazil in 2013, 'Bioenergy week', also illustrates a broader focus. During this conference, the government not only talked about ethanol as a fuel, but also touched ethanol cookstoves, biodiesel and electricity generation based on biomass.[287]

The Brazilian administration rarely pronounces on other renewable energy options. In its foreign policy review of the Lula da Silva government, the Itamaraty dedicates a section on the promotion of renewable energy in general. Yet, it clarifies from the beginning that it considers biofuels as especially important. Within the section, biofuels are the only renewable energy source that is explicitly mentioned (MRE 2010b). The government statements before the start of its active ethanol diplomacy also refer to renewable energy in general (Goldemberg 2002a, Goldemberg 2002b, Vargas 2004). When specific renewable energy options are mentioned, bioenergy prevails. In the support report to the Brazilian Energy Initiative, Goldemberg (2002b:2-4) predicts that by 2010 solar energy, wind energy and bioenergy will have the potential to significantly contribute to the worldwide energy supply whereas hydropower and ocean energy will only have a modest share. He adds that in the short term, bioenergy has the largest potential to displace fossil fuels.

---

[285] Interview with Biodiesel Federal Program Coordinator and staff member, Analysis and Following up of Government Policies, Casa Civil, March 2013.
[286] Interview with diplomat, Itamaraty, March 2013.
[287] GBEP, Working Group on Capacity Building meetings and activities 2013, http://www.globalbioenergy.org/events1/gbep-events-2013/working-group-on-capacity-building-meetings-2013/en/ (accessed July 02, 2013).

5.2.3 Barriers: Prejudices and Trade Restrictions

The Brazilian government presents unjustified concerns about possible negative impacts of biofuels as major barriers to their worldwide promotion. These concerns relate above all to a possible competition between food and biofuel production, but also to environmental degradation. In his New York Times article, Simões states that there is a discrepancy between reality and myth with regard to biofuels: "Today, as we are again facing the challenges of changing our energy matrix, it is important to clearly establish what is reality and what is myth regarding biofuels" (Simões 2007b). The Itamaraty states that skepticism about the sustainability of biofuels prevails in developed countries, particularly in Europe (MRE 2010b:3). However, Kloss (2012:127) admits that concerns about food versus fuel are also expressed in countries like China and India.

The Brazilian government dedicates much attention to the food versus fuel debate. Kloss (2012:79,138f.) emphasizes that food security concerns are the major vulnerability of biofuels as these easily attract attention in international debates and may pose ethical and moral barriers to biofuel utilization. As he states, "to feed the world population and producing biofuels with the same resources are for some morally and physically irreconcilable goals" (Kloss 2012:79, translation S.R.).[288] According to Simões (2007b), "one of the most common myths is that biofuels will necessarily compete with food production." Lula da Silva and Itamaraty officials argue that the Brazilian experience shows that there is no opposition between biofuels and food production: instead of causing hunger, biofuel production help to fight hunger and poverty (see Amorim 2008, Lula da Silva 2008a, Lula da Silva 2008c, Lula da Silva 2008e, Lula da Silva 2007a, Lula da Silva 2007b, Simões 2007a:27, Simões 2007b). In his New York Times article, Simões (2007b) underlines this point by stating that "experience has proven that biofuel production generates income, increasing food production". Lula da Silva (2008d) reinforces this argument by highlighting that biofuel production in developing countries strengthens their agricultural sector and thus enables them to produce their own food instead of importing it. Simões (2007a:27) adds that in the Brazilian case, biofuel production increased proportional to augmenting food production. Kloss (2012:112) underlines that not only Brazil but the whole region of Latin America has enough available land to produce biofuels without negatively impacting food production. Yet, in his speech at the UN General Assembly 2009, Lula da Silva (2009b) constrains the argument about a positive relation between biofuels and food production to the Brazilian case of sugarcane ethanol, stating that the Brazilian sugarcane ethanol in contrast to other biofuels does not affect food security.

---

[288] Original text of this quotation: "nutrir a população mundial e produzir biocombustíveis com os mesmos recursos seriam objetivos moral e fisicamente irreconciliáveis para alguns".

The Brazilian government argues that neither environmental concerns apply to the Brazilian case of ethanol production. As an Itamaraty official points out, these concerns relate mainly to deforestation and biodiversity loss, but also to negative impacts on soil and water. He adds that the Brazilian example shows that there are good policies to prevent negative impacts on the environment.[289] In their statements, Lula da Silva and Itarmaraty officials pay special attention to concerns regarding the deforestation of the Amazonas rain forest. Lula da Silva (2009b, 2007a) underlines that the Brazilian ethanol production is distant from the Amazon. Instead of causing environmental degradation, the plantation of sugarcane helps to recover degraded lands:

> "Ethanol use does not threaten the environment. Neither does sugarcane cause damage to rainforests, for it grows poorly in Amazonian soil. Sugarcane does, however, help to recover degraded pasture lands elsewhere in the country, which can then be brought back into agricultural use" (Lula da Silva 2007c).

Itamaraty officials reinforce this argument. Simões (2007a:25) states that most of the sugarcane plantations are based in the state of São Paulo, far away from the Amazonas rainforest. He claims that "large sucar cane plantations are located at least 1,000 kilometer away from the Amazonas region" (Simões 2007b). According to an Itamaraty official, the Brazilian government always shows a map in international meeting that illustrates that the distance between the ethanol production in São Paulo and the Amazon equals the distance between Lisbon and Moscow.[290] Simões (2007a:25, 2007b) claims that low soil fertilities and excessive rains within the Amazonas impede the formation of saccharose, making this region unsuitable for sugarcane plantations. To strengthen his argument, he adds that the deforestation rate of the Amazonas rain forest was reduced by 52 percent between 2004 and 2006, while biofuel production increased significantly during this period of time (Simões 2007b) (for the production increase, see Figure 7 on page 189). Amado (2010b) also refutes the argument that indirect land use changes due to biofuel production may cause deforestation. According to Amado, these concerns, raised by actors in the USA and Europe, dominate the global debate about biofuel sustainability but do not apply to the Brazilian case.[291]

Lula da Silva (2008e) claims that "attempts to tie high food prices to the dissemination of biofuels do not stand up to an objective analysis of reality." Instead of jumping to conclusions, analyses should be conducted in a thorough and calm way (Lula da Silva 2008d).[292] Kloss (2012:81) reinforces this point, claiming that analysts started to attack biofuels without verifying the causality between biofuel

---

[289] Interview with Head of the New and Renewable Energy Resource Division, Itamaraty, March 2013.
[290] Interview with diplomat, Itamaraty, March 2013.
[291] This point is also reinforced by the Director of Sugarcane and Agroenergy, MAPA in March 2013.
[292] A former UNICA president makes a similar point, stating that in the worldwide debate on biofuels, aprioristic positions and misinformation prevail (interview in April 2013).

promotion and food price increases. In the eyes of Lula da Silva (2008c) blaming ethanol production for rising food prices is not only an over-simplification, but also an affront. Lula da Silva (2008a) admits that there are legitimate concerns in the global debate about biofuels.[293] Yet, what prevailed were false ideas, distortions and prejudices (Lula da Silva 2008a). Kloss (2012:54) also speaks of preconceived and distorted ideas that negatively influence political decision-makers when assessing possible policies to stimulate biofuel production or consumption. According to Lula da Silva (2008a), misinformation is one reason behind false ideas on biofuels. The Itamaraty also emphasizes this point, stating that misinformation is a major reason behind the skepticism about biofuels (MRE 2010b:3). Kloss (2012:54) points at a lack of knowledge about the economic, scientific and environmental aspects of ethanol production. Lula da Silva adds that actors are looking for easy answers, blaming others in their problem analyses:

> "Unfortunately, it is easier to issue warnings than to change consumption habits and eliminate waste. It is easier to blame others than to make necessary changes that harm vested interests" (Lula da Silva 2008c).

According to Lula da Silva (2008a, 2008c, 2008d, 2007a, 2007b), powerful lobby groups also play a role. An Itamaraty official points out that there are two groups of actors campaigning against biofuels: those that are well-intended and really concerned, and those who use the will and name of the former to advance their economic interests.[294] Lula da Silva particularly points at the agriculture and the oil lobbies that campaign against biofuels, as the following two statements illustrate:

> "We need to unmask those campaigns, fostered by trade protectionism and the vested interests of oil groups, that seek to demonise biofuels. These campaigns blame biofuels for the rise in food prices, and for global warming" (Lula da Silva 2008d).

> "If we are to fully understand the true causes of today's food crisis, we must therefore clear away smokescreens raised by powerful lobbies who try to blame ethanol production for the recent inflation in food prices" (Lula da Silva 2008c).

Kloss (2012:27,86f.) also points at the lobbying influence of the oil industry, which wants to avoid a further expansion of biofuels. He claims that it is also due to the influence of powerful lobbying groups that industrialized countries do not address what the Brazilian government regards as the true causes for hunger: income poverty and distortions in agricultural trade by industrialized countries. Instead, industrialized countries blame biofuels for rising food prices:

> "It is frightening, therefore, to see attempts to draw a cause-and-effect relationship between biofuels and the rise in food prices. (…) Such behaviour is neither neutral nor unbiased. It offends me to see fingers pointed at clean energy from biofuels – fingers dirty with oil and coal. I

---

[293] According to Itamaraty's Head of the New and Renewable Energy Resource Division, some of the sustainability concerns are justifiable, for example in relation with palm oil (interview in March 2013).
[294] Interview with diplomat, Itamaraty, March 2013.

am desolated to see that many of those who blame ethanol - including ethanol from sugar cane - for the high price of food are the same ones who for decades have maintained protectionist policies to the detriment of farmers in poor countries and of consumers in the entire world" (Lula da Silva 2008c).

Simões (2007b) makes a similar point, stating that "it is lack of income that fuels hunger, not the use of biofuels". Amorim (2008) adds that it is an irony in the debate about food versus fuel that nobody refers to the fact that those countries that are most affected by food crises have never been involved in biofuel production. An Itamaraty official furthermore underlines that the biggest biofuel producers in the world, the USA and Brazil, do not have any problems with food scarcity.[295] Simões blames the trade distortions by industrialized countries as a major cause for hunger, as these prevent most developing countries from building up large scale food production:

"Nowadays, the largest food producers are the developed countries that strongly subsidize their agriculture. In developing countries, with few exceptions, large scale food production does not occur: They simply cannot compete with rich countries' agricultural subsidies. It is more cost-effective to import products offered as food aid from developed countries, or sold at subsidized prices, than to produce locally" (Simões 2007b).

Lula da Silva furthermore underlines that analyses on the causes of rising food prices strangely overlook the impact of rising oil prices:

"Another essential factor in rising food prices is high oil prices. It is curious that many speak about rising food prices but are silent about the impact of oil prices on the cost of food production. It is as if one factor had nothing to do with the other. Yet any well-informed person knows this is not the case" (Lula da Silva 2008c).

The Brazilian government criticizes attempts to measure the indirect effects of biofuel production on food production or environmental degradation. In her speech at the Bioenergy Week 2013 in Brasília, the Head of Itamaraty's Energy Department, Mariangela Rebua, highlights that there are no scientific criteria to measure these effects.[296] Kloss (2012:181f.) adds that there are significant discrepancies between iLUC models, depending on the assumptions chosen. An Itamaraty official[297] reinforces that no model agrees with the other; he claims that European countries merely employ these models to defend their own interests.

The Brazilian government points to a further barrier to the worldwide promotion of biofuels: protectionist policies by industrialized countries. According to the Brazilian government, this barrier does not get enough attention in international debates (Amorim 2008, Kloss 2012:138f.). Amorim (2008, 2007a) argues that agricultural subsidies and prohibitive tariff and non-tariff trade barriers of rich countries

---

[295] Interview with Head of the New and Renewable Energy Resource Division, Itamaraty, March 2013.
[296] Participant observation at the conference.
[297] Interview with diplomat, Itamaraty, March 2013.

form the two major obstacles in the biofuels sector, making the production of biofuel in poor countries impracticable. Kloss (2012:55) reinforces that the economic viability of biofuel production in developing countries often depends on their access to the markets of industrialized countries. Vargas (2004) points to subsidies that distort international trade with renewable energy. At a UNFCCC meeting, Amorim chooses drastic words to condemn these protectionist policies:

> "Biofuels produced in developing countries present a big potential, still unexplored, to reduce greenhouse gas emissions. Yet, big energy consumers in the developed world have applied all kinds of barriers to biofuels from developing countries. At the same time, they spend billions of Euros and Dollars to subsidy their inefficient producers. Such measures distort markets, augment energy prices, disseminate poverty, threaten food security and are totally inconsistent with concern on climate change. If we want to seriously tackle climate change, these measures need to be removed immediately and unconditionally" (Amorim 2007a, translation S.R.).[298]

According to Amorim, agricultural protectionism leads to rising energy prices, poverty and hunger while subverting climate protection. Lula da Silva (2008a, 2008d, 2007a) adds that international trade policies favor oil over biofuel, as the latter is confronted with import barriers that do not apply to oil. Kloss (2012:21,57) reinforces this point, highlighting that ethanol does not enjoy the same market conditions as oil or agricultural commodities. Whereas fossil fuels are globally traded, the absence of a global biofuels market inhibits the further development of the ethanol industry. As a consolidated international market on biofuels is lacking, large scale deployment still depends on government support. Kloss (2012:30) and Simões (2007a:29) add that depending on the scope and design, sustainability criteria could also pose unjustified barriers to the biofuel trade. They thus want to prevent the establishment of discriminatory sustainability criteria. Lula da Silva even states that it is important to eliminate trade barriers "under any appearance or pretext" (Lula da Silva 2008a).[299]

The Brazilian government also names further barriers to the worldwide promotion of biofuels. Lula da Silva (2008a) emphasizes that there is a general reluctance to change which leads to the continuity of fossil fuel based energy systems. Kloss (2012:91) refers to conditions in developing countries, particularly on the African continent, that make the production of biofuels difficult, such as land own-

---

[298] Original text of this quotation: "Os biocombustíveis produzidos nos países em desenvolvimento apresentam grande potencial, ainda inexplorado, para reduzir as emissões de gases de efeito estufa. No entanto, grandes consumidores de energia no mundo desenvolvido têm colocado todo tipo de barreira aos biocombustíveis dos países em desenvolvimento. Ao mesmo tempo, gastam bilhões de euros e dólares subsidiando seus produtores ineficientes. Tais medidas distorcem os mercados, aumentam os preços da energia, disseminam a pobreza, ameaçam a segurança alimentar e são totalmente inconsistentes com as preocupações com a mudança do clima. Se quisermos tratar da mudança do clima com seriedade, essas medidas devem ser removidas imediata e incondicionalmente".
[299] Original text of this quotation: "sob qualquer roupagem ou pretexto".

ership, production scale, missing infrastructure for transport and distribution and the lack of knowledge and resources for research and development.

In contrast to the deliberations after the beginning of ethanol diplomacy, the Brazilian Energy Proposal for the WSSD in 2002 elaborates on a bundle of factors that hinder the worldwide spread of renewable energy in general (i.e. not only of biofuels):

> "In the special case of renewable energy, it [the current trend of energy usage and that rate of change] faces several tough barriers, such as: (a) uncompetitive costs; (b) need for efficient, cheap, environmentally sound energy conversion technologies; (c) lack of financing programs, in the case of most developing countries; (d) required development of dedicated fuel supply; (e) socioeconomic and organizational barriers; (f) changes in national policies and; (e) public acceptability" (Goldemberg 2002b:9).

The proposal adds that many countries lack appropriate regulatory and legal frameworks, posing bureaucratic obstacles to renewable energy. Besides, renewable energy is confronted with the barriers faced by any new product or service when entering established markets:

> "As with any new product or service, there are barriers to entering established markets. These barriers are particularly important in the energy sector, where the use of new technologies are tied to lifestyle choices (e.g., preferences for sport utility vehicles versus smaller vehicles), government regulated tariff structures (e.g., utility rates that decline with increasing consumption and below market prices set for fossil fuels), and sometimes require large investments in associated infrastructure (e.g., fueling stations for alternative-fuel vehicles). A minimum business volume is typically required to efficiently supply and service a market. The expectation that costs will decline with economies of scale is often the basis for forward pricing in manufacturing, but this is more problematic when a product has yet to be been proven in the market" (Goldemberg 2002b:2).

According to this quote, the market introduction of renewable energy depends on changes in lifestyle, new policies and infrastructure investments – and these steps must be made without knowing if renewable energy will be an economically competitive solution. Goldemberg (2002b:9) adds that old technologies often survive due to market distortions, such as the subsidies or external costs that society as a whole bears instead of the consumer.

5.2.4 Tasks: Creating an International Biofuels Market

According to the Brazilian government, the main task for transboundary policy-making is to create an international market for biofuels, establishing them as a commodity, i.e. a homogenous good that is globally traded at a uniform market price.[300] The Itamaraty points out that the number of biofuel consuming and producing countries has to increase in order to create an international biofuels market (MRE 2010b:1f., Simões 2007a:23). To exploit their full potential and to maximize the socio-economic benefits, the market should rely on South-North trade flows, with rich countries consuming and poor countries producing biofuels (Amorim 2007b:8, MRE 2010b:2). The Brazilian government stresses that countries in the tropical zone are especially suited for biofuel production, as these areas provide the most favorable natural conditions for economic competitiveness (Kloss 2012:12).[301] Kloss (2012:23) estimates that approximately 120 countries in the tropical region could produce biofuels in a sustainable way. To create an international biofuels market, several steps are needed.

First of all, the number of biofuel producing countries has to increase (Kloss 2012, MRE 2010b:1f., Simões 2007a:23). As only a large number of producers can enhance a reliable supply, the reliance on a small number of producing countries would raise energy security concerns in consuming countries (Feres 2010:33, Simões 2007a:16).[302] Kloss (2012:11) and Casa Civil staff[303] add that in the Brazilian case, it was very important for the consumers' confidence that Brazil could import biofuels from the USA when its domestic production declined. Lula da Silva points out that cooperation, technology transfer and open markets can help other developing countries to develop their potential for biofuel production:

> "Brazil's ethanol is competitive because we have technology, fertile land, abundant sun, water and competent farmers. And we are not alone. Most African, Latin American and Caribbean countries, in addition to some in Asia, enjoy similar conditions. Through cooperation, technology transfer and open markets, they can successfully produce sugar-cane ethanol or biodiesel too, and generate jobs, income and progress for their peoples. So this "Golden Revolution," combining land, sun, labour and high technology, can also happen in other developing countries. The African savannahs, for example, are very similar to Brazil's Cerrado plains, where very high crop yields are obtained" (Lula da Silva 2008c).

Lula da Silva (2008a) emphasizes that biofuels facilitate science and technology cooperation with poorer countries because biofuel technology can be transferred at low costs. However, more important than technology transfers is to provide developing countries with access to markets in the developed world (Feres 2010:34). Kloss (2012:9,55) points out that the economic viability of ethanol production in

---

[300] See, for example, Amado 2010a, Amado 2010b, Amorim 2007b, Kloss 2012, Lula da Silva 2008a, Lula da Silva 2008c, Lula da Silva 2008d, Lula da Silva 2007a, Lula da Silva 2007c, MRE 2010b, Simões 2007a.
[301] Interview with Head of Energy Department, Itamaraty, April 2013.
[302] This point was reinforced in the interviews with the Secretary and team, Secretariat of Planning and Energy Development, MME, March 2013, and the team of Bioenergy Projects, FGV, February 2013. Climatic conditions, for example, that may prejudice the harvest in one country are unlikely to occur in a large number of countries at the same time.
[303] Interview with Biodiesel Federal Program Coordinator and staff member, Analysis and Following up of Government Policies, Casa Civil, March 2013.

developing countries often depends on access to industrialized countries' markets. Only if industrialized countries open up their markets, developing countries can produce biofuels on a large scale which is often a precondition for economic competitiveness (Kloss 2012:112).

The Brazilian government demands the elimination of the trade barriers on biofuels imposed by industrialized countries. Lula da Silva calls for an elimination of distortions in agricultural trade in general, to allow for free trade not only in biofuels but in all agricultural goods:

> "Enforcing fair rules in international agricultural trade is of fundamental importance not only in order to eliminate world hunger, but also to face another crucial challenge of our time – that of reconciling environmental protection with energy security" (Lula da Silva 2008d).

Kloss (2012:57) claims that it is necessary to assess if non-tariff trade barriers in industrialized countries, such as environmental requirements or technical standards, impede biofuel imports from developing countries. Simões (2007a:29) adds that it is of crucial importance to prevent the creation of technical, i.e. non-tariff, trade barriers on biofuels, which could take the form of "discriminatory sustainability criteria" or "restrictive norms and standards".

In order to transform biofuels into a commodity, a harmonization of biofuel standards and norms is needed (Feres 2010:59, MRE 2010b:4, Lula da Silva 2008a, Simões 2007a:23). As a first step towards harmonization, existing standards and norms should be assessed (MRE 2010b:4). In addition, transboundary policy-making should strengthen research on biofuels-related standards and measurements (Simões 2007a:29). Simões (2007a:29) emphasizes that norms and standards should not be too restrictive in order to avoid that they serve as non-tariff trade barriers. Instead, they should be balanced, simple and non-discriminatory with regard to the production, distribution and quality of biofuels. He adds that biofuels are a nascent industry competing with agricultural commodities and well-established energy industries, such as fossil fuels. Therefore, norms and standards should be carefully set so that they do not harm the competitiveness and expansion of biofuels (Simões 2007a:29).

In addition, international stock exchanges on biofuels are to be established to create a futures market and facilitate long-term contracts (Kloss 2012, Simões 2007a:23,31). Kloss (2012:48-52) highlights the benefits of futures trade. He states that it enhances transparency, as only one price serves as an international reference and enables hedging, making it possible to transfer risks to other market participants. While hedging also increases the risk of speculation, it fulfills important functions, particularly for less developed countries that export only a few primary products. With the help of futures trade, they can hedge against possible price fluctuations. Simões (2007a:31) adds that oil already enjoys such market conditions.

The Brazilian administration highlights that transboundary policy-making should also focus on information exchange to increase the acceptance of biofuels in the consuming markets of the developed world (Amorim 2008, Lula da Silva 2008a, MRE 2010b:3). Lula da Silva (2008a) emphasizes that there is a need for an open information exchange without prejudice. He claims that a vast number of developing countries can produce biofuels in a sustainable way. What is needed is a correct assessment of their production potential:

> "It is time for political and economic analysts to make a correct analysis of developing countries' capacity to help in food, energy and climate-change issues. Nearly 100 countries have a natural vocation to produce biofuels sustainably" (Lula da Silva 2008c).

Amorim (2008) also urges an "informed dialogue about biofuels" (translation S.R.).[304] An Itamaraty official reinforces that information exchange is needed to bridge the gap between the frontier of knowledge and that of discussion.[305] Information exchange shall be used to demonstrate "on rational and scientific bases" (translation S.R.)[306] that the promotion of biofuels is an efficient instrument to achieve sustainable development (MRE 2010b:4).

The joint development of sustainability standards for biofuels shall enhance their acceptance in industrialized countries and thus facilitate the opening of markets in industrialized countries for biofuels from developing countries. Kloss (2012:26,28,30) highlights that sustainability issues gained more and more relevance for the establishment of biofuels as a commodity, as concerns about the possible negative impacts of biofuels on food production and environmental protection increasingly affected consumer markets in the developed world. At the same time, Kloss (2012:12) acknowledges that sustainability requirements can be an important tool to promote the positive aspects of biofuels while preventing possible negative impacts. For the Brazilian government, it is important to assess how sustainability criteria influence the realization of the goal to create an international market for biofuels (Kloss 2012:28). It emphasizes that sustainability requirements should not create unjustified barriers to the dissemination of biofuels (Kloss 2012:12,30, MRE 2010b:4). As such, they should not be used in a discriminatory way and should be flexible enough to take into account the differences between different production systems (Kloss 2012:12, Simões 2007a:29). Kloss (2012:137) summarizes that the Brazilian government favors "the adoption of objective sustainability criteria which should be inclusive, transparent, scientific, multilaterally accepted and not constitute, in any way, unjustified trade barriers making the production of biofuels in developing countries impossible" (translation S.R.).[307] An official of the MDIC

---

[304] Original text of this quotation: "dialogo informado sobre os biocombustíveis".
[305] Interview with diplomat, Itamaraty, March 2013.
[306] Original text of this quotation: "em bases racionais e científicas".
[307] Original text of this quotation: "a atuação de critérios objetivos de sustentabilidade, os quais deveriam ser inclusivos, transparentes, científicos, aceitos multilateralmente, ademais de não constituírem, em nenhuma hipótese, em barreiras injustificadas ao comercio, sob pena de inviabilizarem a produção de biocombustíveis em países em desenvolvimento."

furthermore highlights the importance of achieving uniform sustainability requirements as potential biofuel exporters can have difficulties in complying with different requirements, which in turn could inhibit exports.[308] Kloss (2012:172) adds that the Brazilian government tries to avoid sustainability discussions evolving in a way that could stunt the development of biofuel industries in countries that have comparative advantages in the production of same. In discussions on sustainability, it is therefore important for the Brazilian government to reinforce the positive aspects of biofuels (Kloss 2012:172).

Before the start of its ethanol diplomacy, the Brazilian administration presented a different (and less elaborated) analysis of the tasks for transboundary policy-making. The Brazilian Energy Proposal for the WSSD in 2002 argues for the establishment of global targets on renewable energy usage and refers to EU targets as examples (Goldemberg 2002a, 2002b). It states that markets alone are not an effective tool to promote sustainable development; policies and political frameworks are required to enhance access to energy and to promote energy efficiency and renewable energy through the development of technologies, the creation of markets and the implementation of special policies that also incentivize demand (Goldemberg 2002b:9,11). Vargas (2004:3) adds that governments and international financial institutions should invest in renewable energy and that transboundary policy-making should furthermore focus on improving capacities for policy analysis and technology assessment.

5.2.5 Global Governors: Focusing on Biofuels and Trade Issues

The Brazilian government focuses on global governors dealing with trade issues and/or concentrating on biofuels. Thus, it particularly points at IBF, WTO, GBEP and ISO. However, it also refers to UN bodies and further organizations dealing with renewable energy.

IBF

During the existence of the IBF, from 2007-2010, the Brazilian government regarded this forum as the primary platform for transboundary cooperation on biofuels (Kloss 2012, Lula da Silva 2008a, Lula da Silva 2007a, Simões 2007a). As the Brazilian government had initiated the IBF, the composition and focus of this forum widely matched its ideas on transboundary biofuel cooperation. Kloss (2012:25,125) states that for Brazil, two criteria were central when deciding on the composition of the IBF member group: it wanted to unite the principal consumers and producers

---

[308] Interview with Foreign Trade Analyst, Secretariat for Innovation, MDIC, March 2013.

of biofuels and ensure equilibrium between developed and developing countries. IBF members should also be able to take concrete measures with immediate impact. Kloss (2012:125,127) adds that the government sought to establish an informal forum that respects the particularities of each country. Simões (2007a:23,30) emphasizes that the IBF was thought of as a coordination mechanism between the major producers and consumers of biofuels, aiming at the creation of an international biofuels market. According to Simões (2007a:30), IBF members shared the understanding that the creation of an international biofuels market contributed to increased efficiency in the worldwide production, distribution and use of biofuels. Kloss (2012:188) underlines that for the Brazilian administration, it was very important to enhance the participation of developing countries in the forum to make their voice heard in the creation of a new market. When sustainability concerns gained importance on the international agenda, the government favored the IBF over the GBEP to discuss these issues. It thought IBF's flexible and informal character and its composition of developed and developing countries would allow for a more "balanced" discussion of the issue (Kloss 2012:132f.).[309]

Kloss (2012:126) reports that the Brasília initially wanted the IBF to focus on the following issues: adoption of blending goals for ethanol, sustainable production increase of ethanol, development of internationally accepted standards and technical specifications, as well as quotation of ethanol on the main stock markets. However, the last aspect did not enter the IBF declaration. As the USA and the EU wished to include biodiesel next to ethanol, it was decided to focus IBF's work on biofuels in general (Kloss 2012:129). Simões (2007a:30) adds that the Brazilian administration expected the IBF to advance on the following issues: international biofuel standards and norms, infrastructure and logistics, international trade, and information exchange on scientific and technological advancements, particularly with regard to second and third generation biofuels. Lula da Silva (2008a, 2007a) emphasizes that the IBF should ensure the expansion of biofuels was not delayed by protectionist barriers. In its foreign policy review of the Lula da Silva government, the Itamaraty highlights two results of the IBF: the International Biofuels Conference and the White Paper on Internationally Compatible Biofuels Standards (MRE 2010b). Kloss (2012:134) underlines that the work on the White paper on internationally compatible biofuel standards created strong links of credibility between the metrology institutions involved.

---

[309] According to Kloss (2012:133), the Brazilian government initially argued that sustainability certificates should be left to the private sector. Yet, it noticed at the second IBF meeting that it was isolated in this demand.

WTO

In his statements on transboundary policy-making on biofuels, Lula da Silva (2008a, 2008c, 2008d) repeatedly underlines the role of the WTO. He regards trade liberalization in agricultural goods as a decisive step, not only for biofuel promotion, but for agricultural development in general. As such, he emphasizes the importance of concluding the Doha Round of trade liberalizations:

> "Overcoming today's hurdles depends on a successful conclusion, as soon as possible, of the WTO's Doha Round, with an agreement that will no longer treat agricultural trade as an exception to the rule, and that will allow the poorest countries to generate income with their own production and exports" (Lula da Silva 2008c).

Lula da Silva (2008d) claims that all agricultural products should be placed under the "multilateral discipline of the WTO". Kloss (2012:25) affirms that the Brazilian government expected the Doha Round to bring about new possibilities for a liberalized biofuels trade. As the Doha negotiations did not advance, the government tried to push for trade liberalization of biofuels by establishing them as environmental goods. Yet, it did not succeed in doing so. Kloss (2012:159f.) emphasizes that WTO negotiations on environmental goods were characterized by North-South cleavages. The Brazilian administration argued that the list of environmental goods had to comply with the spirit of the Doha development round. It wanted to include products with comparative advantages in developing countries and proposed that the WTO adopt the definition of environmentally preferable products developed by the UN Conference on Trade and Development (UNCTAD), which privileges products based on natural resources. According to Kloss (2012:159f.), many countries – particularly from the industrialized world – favored the inclusion of technology-intensive products – mainly made by companies in industrialized countries – such as wind turbines and water filtration equipment. He states that developing countries and the UNCTAD supported the Brazilian proposal, whereas the WTO rejected it due to possible inconsistencies with multilateral trade rules (Kloss 2012:161). However, Kloss (2012:164) admits that only a few developing countries had a direct interest in the biofuels industry, mentioning Colombia as the only example. In the discussions on biofuel sustainability, the Brazilian government emphasizes that WTO rules must have primacy over possible sustainability standards (Feres 2010:210, Kloss 2012:138f., MRE 2010b:10). It considers the WTO dispute settlement mechanism as key to challenging protectionist policies in potential export markets (Kloss 2012:25).

GBEP

When the IBF ceased to exist, the GBEP became the principal platform for the Brazilian government to advance transboundary cooperation on biofuels (MRE

2010b).[310] The Itamaraty highlights that GBEP structures the dialogue between countries that are the major producers and consumers of biofuels and emphasizes the voluntary and non-binding nature of this forum. It underlines that GBEP cooperates on biofuel research and development, the integration of bioenergy into energy markets, the substitution of fossil fuels for biofuels, and improved biomass utilization. It expects GBEP's work on sustainability and technical cooperation to have a decisive impact on the creation of an international biofuels market. Out of the GBEP tasks, Kloss (2012:171f.) highlights biofuels market development and the promotion of biofuels as substitutes to fossil fuels as especially important for the Brazilian government. He underlines that GBEP covers the main analytical elements that are relevant for the creation of an international biofuels market. First of all, GBEP focuses on policies that help to create biofuels markets in member countries. Secondly, it addresses barriers to trade and technology transfer. Moreover, a transboundary agreement on sustainability criteria can enhance public acceptance of biofuels. According to Kloss, such an agreement can be taken as a guarantee for public opinion that the positive impacts of biofuels prevail over possible negative consequences. The Head of Itamaraty's Energy Department reinforces the fact that reaching consensus on GBEP sustainability criteria was a big step forward. She adds that for the Brazilian government, it is important that GBEP not only works on sustainability standards but also offers a framework for technical cooperation and capacity building.[311]

The government initially viewed GBEP sustainability discussions with distrust. This was reinforced by the fact that GBEP has been the initiative of industrialized countries, launched by the G8 (Kloss 2012:172). The administration originally perceived GBEP's Task Force on Sustainability "as an exercise by developed countries to create obstacles to the growth of the biofuels industry" (Kloss 2012:180, translation S.R.).[312] It feared that sustainability discussions could pose barriers to the establishment of biofuel industries in countries with comparative advantages, and to the creation of an international biofuels market as a whole. As Brazil could not avoid that the G8 countries would treat the issue with a "restrictive perspective" (translation S.R.),[313] it decided to play a constructive role and actively engage in GBEP's work on sustainability indicators (Kloss 2012:180). Within GBEP, Brazil regards itself as the voice of developing countries, emphasizing their perspective on

---

[310] Interviews with Head of the New and Renewable Energy Resource Division, Itamaraty, March 2013; diplomat, Itamaraty, March 2013; Biodiesel Federal Program Coordinator and staff member, Analysis and Following up of Government Policies, Casa Civil, March 2013; Director of the Department of Renewable Fuels, MME, March 2013.
[311] Interview in April 2013.
[312] Original text of this quotation: "como um exercício dos países desenvolvidos a fim de criar obstáculos ao crescimento da indústria de biocombustíveis".
[313] Original text of this quotation: "ótica restritiva".

bioenergy and its benefits, and acting as a counterweight to the "exaggerated European precaution" (Kloss 2012:172, translation S.R.).[314] Kloss (2012:179,192) criticizes that with the exception of Brazil, the participation of developing countries in GBEP is too weak. As a consequence, sustainability discussions within GBEP have focused too strongly on environmental protection which is the main sustainability concern of industrialized countries. Against the strong opposition of European countries, the Brazilian government insisted on including the economic dimension into GBEP sustainability discussions as well. Kloss (2012:181f.) claims that European GBEP members wanted to discuss criteria for indirect land use change within GBEP in order to get an internationally accepted reference they could then use to justify the sustainability criteria foreseen in the EU directive on renewable energy. The Brazilian government, however, regarded GBEP as an inappropriate forum for such discussions. It claimed that issues such as indirect land use change should be treated under the UNFCCC because it not only relates to biofuels but to agricultural production in general (Kloss 2012:121). Kloss (2012:173) highlights that despite all these difficulties, the government continues to regard GBEP as a valuable forum, being the only intergovernmental forum where the sustainability of biofuels is discussed.

ISO

Kloss (2012) mentions the ISO as a further forum that is relevant to the Brazilian goal to establish an international biofuels market. In the expert interviews, government representatives from the Casa Civil and the MDIC also underline its importance.[315] Next to the IBF, the Brazilian government used the ISO to push for the harmonization of technical ethanol standards (Kloss 2012:184). As Kloss (2012:189) underlines, the development of technical ethanol norms continues to be the Brazilian priority in the ISO. He emphasizes that ISO standards are voluntary and not binding for governments, yet serve as a reference for the industry. In addition, ISO standards are difficult to challenge under the WTO as they are regarded as not constituting unjustified trade barriers (Kloss 2012:25,154).

The Brazilian government viewed the ISO negations on biofuels with some skepticism, because biofuels were discussed in a technical committee that deals with oil products and lubricants. According to Kloss (2012:144), this technical committee

---

[314] Original text of this quotation: "exacerbada precaução européia". According to the manager of MMA's Climate Change Department, GBEP discussions on sustainability issues can be an important counterweight to the sustainability discussions in the EU (interview in March 2013). The Head of Itamaraty's Energy Department even speaks of GBEP as a way to confront European criticism of biofuels (interviews in April 2013).
[315] Interviews with Biodiesel Federal Program Coordinator and staff member, Analysis and Following up of Government Policies, Casa Civil, March 2013, and Foreign Trade Analyst, Secretariat for Innovation, MDIC, March 2013.

is dominated by the US fossil fuel industry which is not interested in promoting biofuels. Besides, the administration was skeptical about discussing sustainability issues within the ISO. As the private standardization organization ABNT represented the country at the ISO, it feared that the ISO could decide about issues that were of primary importance for the Brazilian government without taking its interests into account. Kloss (2012:156-158) adds that the strong participation of European countries in ISO meetings gives a disproportional weight to EU positions. He criticizes that within ISO sustainability discussions, the EU focused on the environmental pillar only, favoring a precautionary approach. He furthermore charges that the EU sought to include controversial issues such as social aspects, and hard-to-measure criteria, such as indirect effects of biofuel production on biodiversity and greenhouse gas emissions. Kloss claims that the European countries try to legitimize the sustainability requirements of the European directive by establishing a corresponding ISO standard. To provide counterweight to the strong influence of the European perspective on sustainability debates within the ISO, the Brazilian government would like to strengthen the participation of developing countries in the ISO committees on biofuels.

UN Organizations and Agencies

The Brazilian government regards the UN system as an important platform to advance transboundary cooperation on biofuels.[316] At the International Biofuels Conference in São Paulo in 2008, Lula da Silva (2008a) states that he would like to discuss biofuels at various international forums, particularly within the UN. In its foreign policy review of the Lula da Silva government, the Itamaraty underlines the role of different UN bodies and treaties for transboundary policy-making on renewables, such as the General Assembly, Convention on Biological Diversity, CSD, UNFCCC, UNDP, FAO, Common Fund for Commodities, UNIDO and UNCTAD (MRE 2010b:1). Itamaraty officials furthermore highlight the importance of SE4All and the Rio plus 20 summit which included issues such as access to energy.[317] Kloss (2012:121) states that the Brazilian government regards the UNFCCC as the most appropriate forum to discuss climate-related issues, such as indirect land use change. Kloss (2012:192) claims that FAO and UNIDO work on biofuels in a partial way; he states that the Brazilian government tried to prevent the FAO from working on the relationship between biofuels and food production but did not succeed in doing so.

Kloss (2012:142) furthermore refers to the World Bank. He reports that the Brazilian government wanted the World Bank to support biofuel production in

---

[316] Interview with Head of Energy Department, Itamaraty, April 2013.
[317] Interviews with Head of Energy Department, Itamaraty, April 2013, and Head of the New and Renewable Energy Resource Division, Itamaraty, March 2013.

developing countries, particularly on the African continent. It pointed out that access to international finance was crucial for the establishment of a biofuels industry in African countries. World Bank analysts, however, questioned the possibility of replicating good biofuel policies on the African continent. An Itamaraty official admits that the Brazilian government used to have many problems with the World Bank. He states that the World Bank produced studies on biofuels that were not adequate and that it even had to redo a study after Brazil complained about the quality.[318]

Further Organizations Dealing with Renewable Energy

In its foreign policy review of the Lula da Silva government, the Itamaraty states that it closely follows the process of IRENA's creation. The Itamaraty underlines that the conditions are not yet given that joining IRENA would be advantageous for Brazil (MRE 2010b:1). The Head of Itamaraty's New and Renewable Energy Resource Division adds that Brazil has participated as an observer in all IRENA General Assemblies.[319] Itamaraty officials underline that the Brazilian government still views IRENA with caution; yet, they are now seriously thinking about joining IRENA. According to an Itamaraty official, the Brazilian government had the impression that IRENA was created in an undemocratic way and that the country was called to join when major decisions had already been taken.[320] The government's main point of criticism is that during the process of IRENA's creation, different sources of renewable energy were not treated equally. There were proposals on the table that differentiated between 'good' and 'bad' types of renewable energy. Whereas wind and solar energy were favored, biofuels and large hydropower – the renewable energy options that prevail in Brazil – were considered to be 'bad'. Some proposals did not even consider biofuels and large hydropower as renewable energy at all.[321] An Itamaraty official points out that the Brazilian government felt that IRENA was too strongly influenced by European perspectives and that the agency wanted to draw global lessons to be followed everywhere, without taking into account different circumstances.[322] Feres (2010:201) adds that the Brazilian government saw IRENA as a platform to advance German export interests with regard to solar and wind energy. The Head of Itamaraty's New and Renewable Energy Resource Division emphasizes that IRENA's approach has been changing since Adnan Amin has become its Secretary General. Now, IRENA advances a broader

---

[318] Interview with Head of the New and Renewable Energy Resource Division, Itamaraty, March 2013.
[319] Interview in March 2013.
[320] Interview with diplomat, Itamaraty, March 2013.
[321] Interviews with Head of Energy Department, Itamaraty, April 2013; Head of the New and Renewable Energy Resource Division, Itamaraty, March 2013; diplomat, Itamaraty, March 2013; professor (#1), Institute of International Relations, UnB, April 2013.
[322] Interview with diplomat, Itamaraty, March 2013.

view on renewable energy. In addition, he underlines that more developing countries have become IRENA members and that it now has close to universal participation.[323]

In the expert interviews, Itamaraty officials also point to the IEA.[324] Nevertheless, the Itamaraty does not mention the IEA in its listing of different multilateral forums that are relevant for global renewable energy governance (MRE 2010b:1). The Brazilian government participates in four IEA implementing agreements, two of which deal with hydropower and bioenergy. However, as the Head of Itamaraty's New and Renewable Energy Resource Division underlines, the administration's cooperation with the IEA is stronger on oil and gas issues than on renewable energy. He refers positively to the cooperation on creating an IEA handbook on hydropower as this enabled the Brazilian government to transmit its experience.[325] A MME official claims that in the IEA implementing agreement on bioenergy, around ninety percent of the participants are not interested in biofuels, but rather in heating and electricity generation based on biomass.[326] The Brazilian government furthermore criticizes IEA's limited participation as an OECD organization.[327]

Government officials refer to the CEM as another forum to advance transboundary cooperation on biofuels. They claim that CEM was created as part of a US attempt to take climate negotiations out of the UN. As such, the Brazilian administration views CEM's origin with caution.[328] Itamaraty officials also point to the role of REN21. Brazilian diplomats underline that REN21, in contrast to IRENA, has always been a very balanced forum which considers all renewable energy options and developments all over the world. They add that REN21 does not have a heavy structure and that it is very good in terms of sharing experiences and providing information.[329]

In its foreign policy review of the Lula da Silva government, the Itamaraty mentions further global forums that also work on biofuels. It explicitly refers to the following, yet without elaborating on their specific contribution: G20, Major Economies Forum, G8+G5, Ramsar Convention on Wetlands, G15, and OECD (MRE 2010b:1). In the appendix of the document, the Itamaraty names two Memoranda

---

[323] Interview in March 2013.
[324] Interviews with the Head of Energy Department, Itamaraty, April 2013 and the Head of the New and Renewable Energy Resource Division, Itamaraty, March 2013.
[325] Interview in March 2013.
[326] Interview with the Director of the Department of Renewable Fuels, MME, March 2013.
[327] Interview with the Head of the New and Renewable Energy Resource Division, Itamaraty, March 2013.
[328] Interviews with Head of the New and Renewable Energy Resource Division, Itamaraty, March 2013; diplomat, Itamaraty, March 2013; Director of the Department of Renewable Fuels, MME, March 2013.
[329] Interviews with the Head of Energy Department, Itamaraty, April 2013 and the Head of the New and Renewable Energy Resource Division, Itamaraty, March 2013.

of Understanding signed by IBSA members to cooperate on biofuels and wind energy (MRE 2010a). Kloss (2012:156) also points to non-governmental initiatives, namely the Roundtable on Sustainable Biofuels (RSB) and the Roundtable on Sustainable Palm Oil (RSPO).

5.2.6 Salient Features: North-South Conflicts, Power Asymmetries and Re-Balancing

For the Brazilian government, international discussions on biofuels are embedded in structural conflict between industrialized and developing countries. In its opinion, transboundary policy-making in general is, in its opinion, characterized by pronounced power asymmetries, with industrialized countries trying to impose their perspectives and preferences on the rest of the world. A press statement by the governmental Press Agency Agência Brasil on the G8+5 meeting in Hokkaido 2008 illustrates the Brazilian perspective of transboundary policy-making being dominated by industrialized countries:

> "Tired of being a luxury figurehead of the powerfuls' meeting, President Luiz Inácio Lula da Silva started to cogitate about not going to this year's meeting, yet he changed his mind: he decided to continue his crusade in defense of biofuels" (Fiori 2008, translation S.R.).[330]

The Brazilian government especially points to conflicts in international agricultural trade, condemning the agricultural protectionism of industrialized countries that negatively impacts or even impedes the development of agricultural markets in developing countries (see, for example, Amorim 2008, Kloss 2012, Lula da Silva 2008d). In addition, the administration claims that industrialized countries are the main culprits of environmental destruction, yet they try to impose environmental restrictions on developing countries, limiting their choices to pursue socioeconomic development (Kloss 2012:172, 180). Feres (2010:108f.) states that industrialized countries have imposed their view on science and nature in global environmental governance; he sees the risk that the same happens with regard to biofuels. Accordingly, Itamaraty officials highlight that the representatives of developed countries, especially from Europe, often act in an arrogant way vis-à-vis delegates from developing countries, assuming a lack of knowledge and incompetence.[331] Representatives from industrialized countries were convinced that they knew what was right and wrong, and as developing countries often had less scientific force, they were in a difficult position to challenge these perspectives.[332]

---

[330] Original text of this quotation: "Cansado de ser figurante de luxo do encontro de poderosos, o presidente Luiz Inácio Lula da Silva chegou a cogitar não vir ao encontro deste ano, mas acabou mudando de idéia: decidiu continuar sua cruzada em defesa dos biocombustíveis".
[331] Interviews with Brazilian diplomats.
[332] Interview with diplomat, Itamaraty, March 2014.

The Brazilian government emphasizes that industrialized and developing countries have different perspectives on biofuels and that there is a need to strengthen the perspective of the latter in transboundary policy-making on biofuels. The Itamaraty underlines that Brazil's perspective on biofuels is influenced by its condition as a developing country (MRE 2010b:2). At the International Biofuels Conference in Brussels in 2007, Lula da Silva claims that it is important to look at biofuels through the eyes of developing countries:

> "It is important to look at biofuels not so much through the eyes of a European citizen but through the eyes of a citizen of the world. It's necessary to look at the world map not only with the logic of the developed countries – where the conquests have already been made, where the economic issue has been resolved, and therefore, where importing oil at 60 or 70 dollars [a barrel] isn't a problem. Look at biofuels with an eye on the map of Africa, of South America, and Latin America" (Lula da Silva 2007a).

In his opinion, developing countries are more concerned with economic issues and more troubled than their industrialized counterparts with high expenses on oil imports. The Itamaraty also points to the need to strengthen the influence of developing countries in transboundary policy-making (Kloss 2012:190, MRE 2010b:2). It states that in contrast to industrialized countries, developing countries share concerns about how to provide their population with access to energy (MRE 2010b:2). Kloss (2012:126) adds that the benefits of biofuels production differ between developing and developed countries. According to Kloss, developing countries can profit from reduced dependency on fossil fuel imports, the liberation of financial resources that were formerly spent on fossil fuel imports and can then be invested in areas such as education and health, new export possibilities, and jobs and income generation in rural areas. Developed countries can increase their energy security, create new economic alternatives for the rural sector where competitiveness is only maintained due to high subsidies, and achieve environmental goals, such as reducing greenhouse gas emissions. In line with this argument, the Itamaraty points out that the Brazilian government organized the International Biofuels Conference in São Paulo to transmit a developing country's perspective on biofuel challenges and opportunities, in particular in relation to sustainability (MRE 2010b:2).

The administration underlines that industrialized countries dominate the worldwide debate on biofuels and accuses them of acting on the basis of double standards. Citing Embaixador Seixas Correa, Feres (2010:205) states that it is typical of the behavior of industrialized countries with regard to global issues that they demand developing countries practice what they say, but not what they do themselves ("façam o que eu digo, mas não o que eu faço"). At the UNFCCC meeting in Bali, Amorim (2007a) claims that the protectionist policies of industrialized countries vis-à-vis biofuels show that they do not take climate protection seriously. Whereas biofuel production in developing countries offers great potential for the worldwide reduction of greenhouse gases, this is still unexplored as the "big energy consumers of the developed world have put all types of barriers on biofuels from developing countries" (translation S.R.),[333] while spending a lot of money to subsidize their inefficient domestic production. Kloss (2012:191) emphasizes that the global debate on the sustainability of biofuels disadvantages the social and the economic pillars – those pillars that are especially important to developing countries. Kloss (2012:173) speaks of an "exaggerated" precaution by European countries regarding possible negative impacts on the environment in developing countries. He feels that European countries try to hide their true motives behind the argument of environmental protection; their true motive being to protect their domestic biofuels industries against more competitive producers from other countries. According to his view, industrialized countries try to concentrate sustainability discussions on environmental aspects only as this tends to be the weak point in developing countries, industrialized countries being typically less vulnerable to environmental concerns. Factors on indirect land use change and the loss of biodiversity weigh more strongly in developing countries. The weak point of biofuel production in industrialized countries, by contrast, is economic sustainability. This is because industrialized countries shield their domestic biofuel production with subsidies and trade restrictions against competition. Kloss furthermore states that European positions on biofuels are contradictory. Europeans say that they want to establish biofuels as a true alternative to fossil fuels, yet they define sustainability criteria in a way that practically impedes the global production and trade of biofuels (Kloss 2012:173f.).

The Brazilian government emphasizes that transboundary policy-making on renewable energy should primarily benefit developing countries (Amorim 2008, Goldemberg 2002a:4). It states that transboundary cooperation on renewable energy should follow the principle of common but differentiated responsibilities and underlines that developing countries need access to new and additional financial resources and technologies to effectively promote renewable energy within their countries (Goldemberg 2002b:2, Lula da Silva 2008a, Vargas 2004:2). The Brazilian administration clarifies that what developing countries need most are not North-South transfers, but competition on equal grounds by accessing the markets of industrialized countries. At the International Biofuels Conference in São Paulo in 2008, Lula da Silva (2008a) states that the previous G20 meeting was one of the most important of his presidency because he could make his point that rich countries do not need to talk about providing aid to emerging or poor countries. Instead, the best help they can provide to poorer countries is to fix their own economies without increasing protectionism. In an article for the Brazilian newspaper Valor Econômico, Amado (2010b) emphasizes that Brazil wants to share its biofuels technology with other developing countries. He contrasts this to the way developed

---

[333] Original text of this quotation: "grandes consumidores de energia no mundo desenvolvido têm colocado todo tipo de barreira aos biocombustíveis dos países em desenvolvimento".

countries have acted, claiming that developing countries could have advanced significantly in their fight against poverty and inequality if the former had been more willing to share their technological advancements.

According to the Brazilian government, the promotion of biofuels in poorer countries helps to re-balance economic and political power relations between developing and industrialized countries. Lula da Silva (2008a) presents biofuels as the response from developing countries to address the challenge of sustained growth. Biofuels help developing countries to achieve independence and autonomy by strengthening their economies and providing opportunities to sell their goods to industrialized countries:

> "The only 'bad' thing that could happen is that African countries will be able achieve their independence and autonomy, when the richer countries have to buy biodiesel from them in order to de-pollute the planet. And we should always remind ourselves that the poor countries are not responsible for this pollution but are actually its victims" (Lula da Silva 2007a).

Simões (2007a:15) claims that it is a characteristic of the worldwide transition towards a low-carbon energy supply that the BRICs gain more economic weight vis-à-vis the USA and Europe. Lula da Silva (2008a) underlines that transboundary cooperation on biofuels makes it possible to integrate developing countries into transboundary collaboration on science and technology, since biofuels involve technologies that can be transferred at low costs.

The Brazilian government emphasizes the importance of national sovereignty, underlining that each country experiences different circumstances and must find its own way. It further adds that no country should impose its model on others (Amorim 2008, Amorim 2007b, Lula da Silva 2008c, Lula da Silva 2008d, Vargas 2004). It emphasizes that there is no one-fits-all-model for biofuel promotion; instead, each country must assess its biofuel potential according to its own circumstances (Amorim 2008, Lula da Silva 2008d).[334] As a consequence, the government also emphasizes that it does not want to impose the Brazilian model on other countries (Lula da Silva 2008d).[335] Amorim (2007b) states that it is the sovereign decision of each administration to decide whether it wants to take advantage of the environmental, economic and social benefits the production of biofuels offers. Vargas (2004:1) relates the question of sovereignty to the promotion of renewable energy in general, stating that "the increase in the global share of renewable energy supply is closely linked to the sustainable use of natural resources that is a sovereign decision of countries." Lula da Silva reinforces the fact that national sovereignty should be respected; he claims that it is even disadvantageous for developing countries to be

---

[334] This point was also made by MAPA's Director of Sugarcane and Agroenergy and Itamaraty's Head of the Energy Department at the Bioenergy Week in Brasília 2013.
[335] Interview with Diplomat, Itamaraty, March 2013.

influenced by other countries or organizations, as these may advance the interests of agricultural and oil lobbies – perhaps without even being aware of it:

> "These [developing] countries will have to do their own studies and decide whether or not they can produce biofuels, and on how large a scale. They will need to decide which crops are the most appropriate and design their projects based on economic, social and environmental criteria.
>
> These are important decisions that they will have to make on their own, rather than leaving them to other countries or organizations that often echo - even in good faith - interests of the oil industry or of farm interests hooked on subsidies and protectionism" (Lula da Silva 2008c).

Vargas (2004:1) reinforces that developing countries need to increase their energy supply in order to achieve socio-economic development; as such, he claims that "the use of those resources, especially of their hydrological potential cannot be forgone or be subject to conditionalities."

Lula da Silva states that the manner in which international cooperation works must change in order to effectively tackle the challenges of climate protection, energy security and food security. He calls for more coordination, solidarity, transparency and democratic rules. While international cooperation should respect the necessity of all, it should particularly consider least developed countries (Lula da Silva 2008a). An Itamaraty official reinforces that it is very important for the Brazilian government that international organizations and forums on renewable energy function in a democratic way. He adds that the Brazilian administration felt this was not the case when IRENA was created: Brazil had been invited when most of the important decisions had already taken.[336]

5.3 The Brazilian Government's Action and Ideas at a Glance

The Brazilian government's first actions on global renewable energy governance were made in the realm of international conferences that placed renewable energy on their agenda. Against the backdrop of its clean energy matrix and early efforts to promote biofuels, it proposed a target for increasing the worldwide share of renewables at the WSSD and led the regional preparation process for the Renewables 2004 Conference. At that time, transboundary policy-making on renewable energy was still politically low profile for the Brazilian government. This changed significantly with the commencement of ethanol diplomacy.

Brazilian ethanol diplomacy started in light of an expanding domestic ethanol industry and worldwide growing interest in the promotion of biofuels. The government regarded this as an important window of opportunity to increase the export strength of its economy and to leverage its own influence in transboundary policy-making. Facilitating international biofuels trade became a priority in the dip-

---

[336] Interview with diplomat, Itamaraty, March 2013.

lomatic action of President Lula da Silva and the Itamaraty. When biofuels became the target of global sustainability concerns, the administration intensified its diplomatic action in support of biofuels. It placed biofuels on the agenda of international meetings, launched the IBF, organized an international conference on biofuels and engaged at the WTO, ISO and GBEP to push for these. Within the Brazilian government, the Itamaraty and Casa Civil closely coordinated their action with the MME and MAPA. The Brazilian Sugarcane Industry Association, UNICA, also became a central player in the government's ethanol diplomacy. Since 2010/11, biofuels are no longer a priority of Brazil's transboundary action. As its ethanol industry experienced economic difficulties, the prospects for export diminished significantly. Yet, Brazilian ethanol diplomacy continues despite its lower visibility.

According to the Brazilian government, global renewable energy governance should primarily enhance socio-economic development in developing countries. Economic competitiveness should guide the choice between different renewable energy options. In its deliberations on global renewable energy governance, it concentrates on biofuels, and, more specifically, on ethanol. It presents the protectionist trade policies and prejudices of industrialized countries as major barriers to the worldwide spread of biofuels. These prejudices relate to the supposed negative implications of biofuel production on global food security and environmental protection. According to the government, transboundary policy-making should focus on the creation of an international biofuels market. It underlines the importance of global governors dealing with trade issues and biofuels and particularly points to IBF, WTO, GBEP and the ISO. According to the Brazilian government, transboundary policy-making on renewables – and global governance in general – is shaped by conflicts and power asymmetries between industrialized and developing countries, with the former dominating transboundary policy-making to the detriment of the latter. It expresses its caution vis-à-vis global policy-making and presents biofuel promotion as a means to re-balance economic and political power relations between industrialized and developing countries.

6 Ideational Differences in Global Renewable Energy Governance

This chapter answers the main research question of the study: how and why do the ideas of the German and Brazilian governments on global renewable energy governance differ? As a first step, the chapter examines the ideational differences between both governments. This section is structured along the categories that have already guided the analysis of the case studies and thus focuses on ideational differences concerning (1) global challenges addressed by global renewable energy governance, (2) renewable energy options, (3) barriers to a worldwide promotion of renewable energy, (4) main tasks for transboundary policy making, (5) global governors on renewable energy, and (6) meta-aspects/salient features in global renewable energy governance. Whereas the ideational analysis in the case studies followed the logics of each government's deliberations and therefore only considered aspects that the administrations present as relevant, the comparison of ideas makes it possible to reveal blind spots in their deliberations. For the purpose of comprehensibility, the analysis of ideational differences refers to both governments as composite actors without differentiating between either entity. Although the main focus of this study is on ideational differences, this chapter also considers similarities and congruent ideas to allow for a comprehensive comparison of both governments' ideas.

The second part of the chapter identifies possible reasons behind the differing ideas. Following the theoretical-analytical framework of this study, the search for possible reasons focuses on the policy contexts and the self-interests of the government actors in charge of global renewable energy governance, as well as their efforts to achieve coherency between ideas. The policy contexts comprise the domestic policy context of renewable energy promotion, the embeddedness in global (renewable energy) governance and the institutional context due to the division of competencies in each government. To allow for more comprehensibility, the search for possible reasons starts with those aspects that can be identified by merely analyzing global governance ideas, i.e. which concern the governments' strive for coherency between ideas. Here, it is specified what ideational differences can be traced back to other ideational differences. The chapter concludes with a short overview of the ideational differences and the reasons behind them.

6.1 Identifying Ideational Differences[337]

6.1.1 Global Challenges: Environment Protection vs. Socio-Economic Development

The German and Brazilian governments refer, in general terms, to the same global challenges when speaking about the worldwide promotion of renewable energy. Both governments underline the importance of climate protection and mention improved air quality as a further environmental benefit. Both emphasize the importance of energy security in its economic sense, which means that they refer to the need of satisfying growing energy demand and cushioning energy price increases. Each stresses that rising energy demand is confronted with absolutely limited fossil fuel deposits. They particularly refer to rising oil prices which negatively affect the global economy as a whole, and poor and fossil fuel importing countries in particular. Both state that the promotion of renewable energy enhances economic development and growth, and they particularly underline positive effects on jobs creation. Further, both single out economic benefits for developing countries, such as economic growth and improved trade balances. In addition, each administration underlines that renewable energy promotion contributes to poverty reduction and worldwide conflict prevention, as an expanding supply of renewable energy sources reduces rivalry over scarce energy resources.

However, both attach different priorities to the global challenges mentioned above. The German government clearly prioritizes environmental protection, particularly climate change mitigation. References to the need to combat climate change are never missing in its statements on the worldwide promotion of renewable energy. The Brazilian government recognizes that for most policy actors who promote renewables around the globe, climate protection and energy security are the major concerns. However, it points out that its own priority lies in enhancing socio-economic development in developing countries. It particularly refers to the need to combat poverty and hunger. It argues that for poor countries, socio-economic development is more important than climate protection. In line with this, the Brazilian government claims that biofuels should not be seen as a mere clean energy solution; instead, socio-economic benefits should also be taken into account.

In keeping with the prioritization of divergent global challenges, both governments draw a different relationship between renewable energy and conventional forms of energy. While the Brazilian government regards renewable energy and conventional energy as complementary sources, its German counterpart wants renewable energy to replace conventional energy. As the German administration

---

[337] The subsequent analysis builds on the assessment of government ideas in the case studies on Germany and Brazil. For references to specific sources, please see chapters IV.2 (Germany) and V.2 (Brazil).

focuses on environmental protection, it sees renewable energy promotion as a tool to prevent negative environmental externalities that go along with the production and use of conventional energy. The Brazilian government, by contrast, prioritizes socio-economic development and presents renewable energy and conventional energy as alternative options that both contribute to an expanding energy supply.

The German and Brazilian administrations establish divergent causal links between the promotion of renewable energy and poverty reduction. The German government emphasizes the importance of enhancing access to energy by providing energy infrastructure in remote areas. It states that access to energy is crucial for satisfying basic needs: access to energy not only facilitates the processing of clean drinking water, but also provides the electricity needed to access information, communications and education, and engage in economic activities. For the Brazilian government, the main link between the promotion of renewable energy and poverty reduction is income generation. With regard to poverty reduction, it particularly highlights the need to combat hunger. It states that lack of income is the main cause for hunger and thus presents income generation as the most important tool for tackling poverty and hunger. In the Brazilian opinion, the promotion of renewable energy should directly generate income for poor people by involving them in the economic process of generating renewable energy services. The German administration, in contrast, addresses the promotion of renewable energy as an input/precondition for income-generating activities for poor people, not as an income-generating activity in itself.[338] Nor does it explicitly address hunger.

In addition, both governments partly advance distinct assessments on how the promotion of renewable energy can contribute to economic development in developing countries.[339] As with poverty reduction, the German administration presents renewable energy as an input/precondition for economic activities and macroeconomic growth. It states that access to energy is needed to process raw materials and thus scale up economic activities. It furthermore emphasizes that a reliable energy supply is an important precondition for macroeconomic stability. The Brazilian government, by contrast, again refers to renewable energy provision as an economic activity in itself which enables developing countries to take economic advantages of their natural resources and produce more technologically advanced products. Accordingly, both governments' assessments on the trade effects of renewable energy promotion for developing countries also vary. Whereas the German administration focuses on positive trade effects by replacing imports, its Brazilian counterpart brings export opportunities in as well.

---

[338] An exception to this is a BMZ publication dealing with biofuels (BMZ 2011:4).
[339] Again, BMZ (2011) is an exception to the prevailing line of argument in the German government's deliberations.

The comparison between the government statements also reveals some blind spots on both sides. In the Brazilian deliberations, agricultural development in developing countries is of central importance, whereas the German administration does not pay special attention to the agricultural sector.[340] The Brazilian government sees the lack of strong agricultural sectors as a structural cause of hunger. Thus, it claims that renewable energy promotion should aim at strengthening agricultural sectors in developing countries, and as such their capacity to locally produce food. It states that the production of biofuels enables developing countries to improve the living standards of poor people in rural areas by generating income and employment. In addition, it also alleviates negative impacts of rising oil prices on the agricultural sector, as it provides an alternative fuel for transport and fertilizers. The German government, by contrast, underlines that the promotion of renewable energy should also combat deforestation and loss of biodiversity whereas their South American counterpart does not refer to these challenges.

6.1.2 Renewable Energy Options: Electricity vs. Transport Sector

While the German government mostly speaks about 'renewable energy' in general without differentiating between different renewable energy options, the Brazilian government concentrates on biofuels, especially sugarcane ethanol, and rarely pronounces on other renewable energy options. Only the deliberations before the commencement of ethanol diplomacy generally refer to 'renewable energy'. Yet, biomass already prevails when examples of specific renewable energy sources are given. After the beginning of ethanol diplomacy, the term 'renewable energy' is rarely used. One notable exception is the denomination of Itamaraty's "Division for New and Renewable Energy", which deals with biofuels and other renewable energy options. Its section in the foreign policy review of the Lula da Silva Government is also entitled 'renewable energy' (MRE 2010b). Despite this broad title, biofuels are the only renewable energy option that is explicitly mentioned. The Itamaraty explicitly states in this section that it regards biofuels as the most relevant option of renewable energy deployment. The Brazilian government often uses 'biofuels' and 'ethanol' as synonyms. Yet, it also pays some attention to biodiesel and biomass for electricity generation. Whereas it clearly focuses on the transport sector, its German counterpart concentrates on the electricity sector.[341] The latter often uses 'energy' and 'electricity' as interchangeable terms and if it gives examples for specific renew-

---

[340] Exceptions to this are the German government's statements that concentrate on biofuels (see, for example, BMZ 2011, BMZ 2013b).
[341] The third energy sector, heating and cooling, is outside of the focus of both governments' deliberations.

able energy options, the electricity sector is clearly prevailing. Within the electricity sector, wind and solar energy receive most attention.

Both governments refer to economic competitiveness, affordability and environmental friendliness as important criteria for assessing different renewable energy options. These are the criteria that are commonly regarded as composing the 'triangle' of energy policy-making (see, for example, Lesage, Van de Graaf & Westphal 2010:37f.). Yet, they clearly attribute different weights to these criteria: the German government emphasizes environmental sustainability, whereas the Brazilian administration prioritizes economic competitiveness. According to the Brazilian administration, economic feasibility should guide the choice between different renewable energy options. It makes clear that market forces, not government subsidies, should decide what kind of renewable energy options win the race. The German government, by contrast, presents sustainability as the precondition for supporting renewable energy. It argues that transboundary policy-making should only promote renewable energy if these are produced and used in a sustainable manner. Although the German government commonly claims that all three dimensions of sustainability must be taken into account equally, its deliberations clearly show that it prioritizes the ecological dimension. With regard to the affordability of renewable energy, both governments underline the need to achieve the lowest possible prices and state that affordability is a particularly important issue in developing countries.

A closer look at the deliberations of both governments on economic competitiveness reveals that they use different time horizons when assessing economic feasibility and link economic competitiveness to distinct production scales. The Brazilian government argues that large scale production is an important precondition for economically competitive biofuels production, whereas its German counterpart claims that small-scale options are particularly suited for achieving economically competitive renewable energy deployment. In addition, the Brazilian government focuses on the status quo of renewable energy deployment, pointing at options that are already competitive, whereas the German administration emphasizes future potentials and predicted cost reductions.

Applying their core criteria to the various renewable energy options, both governments present different forms of renewable energy as worthy of support. The Brazilian government praises the economic competitiveness of biofuels. It furthermore argues that sugarcane ethanol is more competitive than ethanol based on other feedstock, such as corn and sugar beet, and that developing countries have more favorable conditions for economically competitive biofuel production than industrialized countries. The German administration presents wind and solar energy as 'win-win options', and to a lesser degree small hydropower and geothermal energy as well. However, it raises sustainability concerns and thus trade-offs with regard to large hydropower and biofuels. As such, it particularly presents solar and wind energy as renewable energy options that are worthy of support, whereas it makes clear that it sees large hydropower and biofuels with caution. With regard to solar and wind energy, it underlines the growth potential and predicted cost reductions. If it mentions negative aspects, such as supply variability, it quickly points out how these can be overcome. With regard to large hydropower and biofuels, the German government particularly points at environmental risks. In its opinion, a careful risk assessment is needed before deciding on whether biofuels and large hydropower should be employed.

The diverging assessments of the pros and cons of biofuels are particularly striking.[342] Whereas the Brazilian government speaks about the promotion of biofuels in very positive terms, its German counterpart expresses its caution and doubts. In contrast to the Brazilians, the Germans emphasize the various risks of biofuels production that they are not willing to accept. They state that these risks particularly affect developing countries. Above all, the German government refers to increasing pressure on natural resources which may have several negative implications, such as environmental destruction, loss of biological diversity, food price increases and (forced) displacement of small famers. It furthermore points at poor working conditions on biofuel feedstock plantations and ambiguous effects on emissions reduction. The Brazilian administration, by contrast, does not mention the negative consequences of biofuels deployment – at least with regard to sugarcane ethanol.[343] The German government also recognizes positive effects of biofuel production. It mentions positive aspects that the Brazilian administration also brings forward, such as economic opportunities for rural regions in developing countries, positive trade effects (replacing imports and generating export opportunities), income generation and expanding energy supply. However, it does not refer to the economic competitiveness of biofuels vis-à-vis fossil fuels and other renewable energy options, as the Brazilian government does.

Both governments even reach contrary conclusions on the effects of biofuel production on hunger. Whereas the Brazilian government states that biofuel production reduces hunger, the German administration regards it as posing a risk to food security (see Figure 8). The two base their assessment on divergent assumptions about the underlying causes of hunger. Whereas the Brazilian administration explicitly states that hunger is the result of a lack of income and weak agricultural sectors in developing countries which are not able to produce their own food, the German government implicitly argues that hunger is a result of an overall food

---

[342] On the different perspectives on biofuels in Germany and Brazil see also Acosta, Zilla 2011 and Röhrkasten, Zilla 2012.
[343] In light of the worldwide debate on a possible competition between biofuels and food production, Lula da Silva (2009b) states that these concerns might be justified with regard to some types of biofuels. However, he affirms that Brazilian sugarcane ethanol does not pose a risk to food security. The Head of Itamaraty's New and Renewable Energy Resources Division adds that in relation to palm oil, some of the global sustainability concerns on biofuels are justifiable (interview, March 2013).

scarcity, i.e. insufficient food production and food prices that are too high. Taking the scarcity of arable land as a starting point, the German government argues that biofuel production competes with food manufacture. If land is used for biofuel production, there is less land available for food production which leads to a declining food supply and rising food prices. Rising food prices make it more difficult for poor people to afford food.

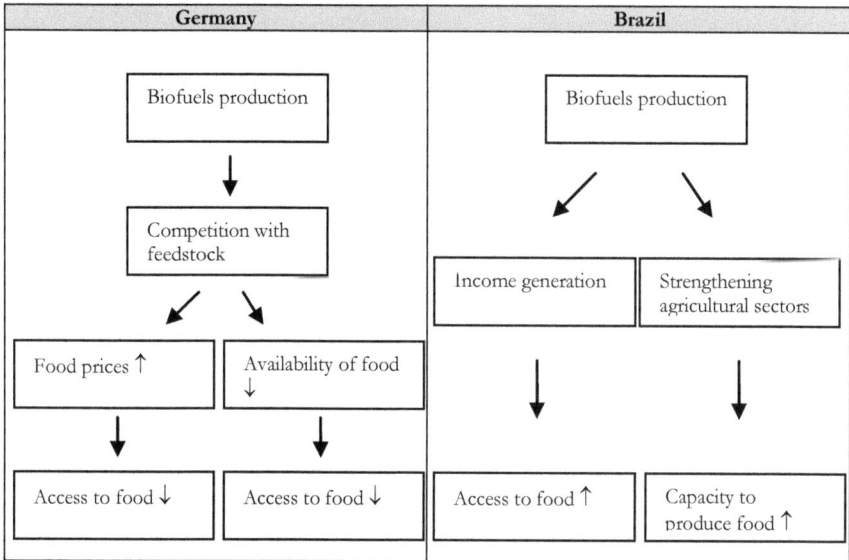

Figure 8: Different Causal Links between Biofuels Production and Food Security

The Brazilian government, on the contrary, states that there is enough arable land for producing both biofuels and food. It argues that hunger is not a result of insufficient overall food production, but of an unequal distribution. People suffer hunger because they lack the income to buy food. It stresses that biofuel production reduces hunger as it provides poor people with the opportunity to generate income, thus improving their capacity to buy food. It adds that biofuel production also helps to strengthen agricultural sectors in developing countries, thus improving their capacity to produce sufficient food.

6.1.3 Barriers: Discriminatory Energy Policies vs. Discriminatory Trade Policies

Both governments concentrate on different forms of discrimination when elaborating on the barriers to worldwide promotion of renewable energy. The German government points at discrimination of renewable energy vis-à-vis conventional forms of energy, whereas the Brazilian administration directs attention to discriminatory trade policies of industrialized countries vis-à-vis developing countries which not only affect biofuels but agricultural products in general. The comparison of the government statements furthermore reveals that the Brazilians attribute much more space to the barriers than the Germans do. Whereas the German administration's line of argument remains in the realm of energy policy, the Brazilian government links the promotion of biofuels to an unequal trade between industrialized and developing countries. It states that agricultural subsidies and trade barriers in industrialized countries impede developing countries from exploiting their comparative advantages in the production of agricultural goods. If developing countries do not have access to large consumer markets in industrialized countries, it is difficult for them to produce at a scale large enough for economically competitive production. In addition, developing countries are deprived of potential export revenues and related economic benefits. As biofuels are agricultural goods, these are also affected by such protectionist policies. The Brazilian government criticizes the agricultural protectionism of industrialized countries *per se* but also underlines that these policies disadvantage biofuels vis-à-vis fossil fuels, as the latter are globally traded and not confronted with import barriers. It furthermore underlines that protectionist barriers imposed by industrialized countries do not receive sufficient attention in international debates on biofuels.

In the German deliberations on barriers to the worldwide promotion of renewable energy, trade policies are not mentioned. The German government, by contrast, underlines factors that prize conventional forms of energy over renewable, particularly at the domestic level. It states that conventional energy counts with established technologies, markets and market players; as such, market requirements and structures are typically geared towards these energy options whereas the features of renewable energy are not (sufficiently) taken into account. It furthermore points to additional disadvantages renewable energy is confronted with. Energy prices, for example, do not include the external costs of energy production and consumption, such as environmental damage caused by fossil fuels, and fossil fuel subsidies still prevail in many parts of the world. The German administration furthermore states that adequate policies to promote renewable energy are lacking both at the national and international level. It identifies a lack of willingness to address renewable energy in the international community and claims that national governments do not do enough to implement renewable-friendly policies and support the development of a renewable energy industry. Before commencing its active ethanol diplomacy, the Brazilian government advanced a very similar problem assessment, pointing at inappropriate regulatory and legal frameworks, market barriers for renewable energy and market distortions caused by subsidies and prices that do not cover external costs. However, once it had begun its ethanol diplomacy, its problem assessment changed, focusing on the trade aspects mentioned above.

Whereas the German government is cautious at blaming political actors for imposing barriers to the worldwide promotion of renewable energy, speaking about such barriers in rather vague terms, the Brazilian administration explicitly names actors who try to obstruct the promotion of biofuels. It furthermore claims that these actors successfully manipulate global debates about biofuel sustainability by also spreading misinformation. The German government states only in very general terms that lacking political will and public awareness prevents policy-makers from taking the necessary steps to improve political and market frameworks for renewable energy. It also complains that policy actors often ignore renewable energy. According to the Brazilian administration, unjustified concerns about possible negative impacts of biofuels are a major barrier to their worldwide promotion. These concerns relate to a possible competition between biofuels and food production (food vs. fuel) and to environmental risks, such as deforestation and biodiversity loss. They focus not only on direct relationships but also take into account the indirect effects on food production and environmental protection (indirect land use changes, iLUC). It claims that the Brazilian experience with sugarcane ethanol shows that these concerns are not justified, and repeatedly underlines that the production of sugarcane ethanol has no negative consequences on food security. On the contrary, in the Brazilian opinion, it enhances food security and does not lead to deforestation. The Brazilian government states that these concerns are raised by political actors in industrialized countries, particularly Europe and dominate the global debate about biofuel sustainability without being based on an objective analysis of reality. It claims that misinformation, distortions and prejudices prevail in the global debate about biofuel sustainability. Political actors not only look for easy answers, blaming others instead of re-thinking their own actions, but also defend their own political interests. The government explicitly names two lobbying groups that obstruct the promotion of biofuels: the agricultural lobby in industrialized countries which seeks to impede competition from developing countries, and the oil lobby which wishes to maintain worldwide oil dependency. It insinuates that the global debate about food vs. fuel mirrors the interests of such lobby groups. Instead of ascribing rising food prices to rising oil prices and instead of addressing distortions in agricultural trade and related socio-economic inequalities as a major cause for hunger, biofuel production is blamed for rising food prices and hunger. Interestingly, the German government also complains about prejudices. However, it refers to prejudices concerning a different causal relation, claiming that sustainability and environmental protection are often seen as impeding economic growth and job creation, while the promotion of renewable energy demonstrates that the opposite is true.

Both governments advance a similar assessment on another barrier. They both emphasize that developing countries often lack the capabilities that are needed for effective promotion of renewable energy. They specifically refer to the lack of knowledge and information, technological know-how and financial resources.

6.1.4 Tasks: Improving Domestic Policies vs. Strengthening International Markets

Whereas the German government regards the improvement of domestic regulatory frameworks as the most important step for the worldwide promotion of renewables, its Brazilian counterpart calls for the unleashing of market forces at the international level: it suggests that the most effective means for the worldwide promotion of biofuels is to create an international biofuels market. Thus, transboundary cooperation should concentrate on the different steps necessary for the enhancement of free trade. The German administration, by contrast, states that transboundary cooperation shall enhance the commitment of governments to promote renewable energy within their countries and support them to achieve their commitments. It regards international agreements on time-bound targets for increasing the worldwide share of renewable energy as the best way to enhance government commitments for the promotion of renewable energy.[344] Voluntary commitments, combined with a monitoring process, are regarded as a second best option. According to the German government, global renewable energy governance should focus on policy advice in order to improve domestic regulatory frameworks. It emphasizes that governments in both developing and industrialized countries require advice on issues such as renewable energy potentials, technologies, business models and policy options. In addition, transboundary cooperation should support developing countries through capacity-building, financing and technology transfer.

The Brazilian administration also states that transboundary cooperation should support developing countries through financing, technology transfer and science cooperation. It emphasizes that increasing the number of biofuel-producing countries in the developing world is a crucial step to counter energy security concerns in biofuel-consuming countries. However, it makes clear that more important than providing support to developing countries is to open up markets. Thus, transboundary policy-making should eliminate trade barriers on biofuels operated by industrialized countries, harmonize biofuel standards and norms to facilitate trade, and create an international stock market for biofuels.

---

[344] The German government several times tried to establish such targets within the UN framework. It only chose other options when it realized that this strategy did not succeed (see chapter 4.1.)

In line with various perspectives on the main tasks for transboundary policy-making, both governments point to a different thematic focus when elaborating on the need for information exchange, knowledge and research. The German government focuses on aspects that are relevant for improving domestic regulatory frameworks, whereas the Brazilian government refers to elements that are instrumental for creating an international biofuels market. The former states that more information is needed on renewable energy potentials, available resources, existing activities that promote renewables, effective policies and technological expertise. It adds that transboundary policy-making should also promote best practice and lessons learned. The Brazilian government, by contrast, underlines the necessity of assessing existing biofuels standards and norms and to strengthen research on biofuel-related standards and measurements. In addition, it seeks transboundary policy-making to engage in information exchange in order to increase the acceptance of biofuels in consuming markets of the industrialized world.

Before ethanol diplomacy began, the deliberations of the Brazilian government on tasks for transboundary policy-making were very similar to those of the Germans. Here, the Brazilian government also called for the establishment of global targets on renewable energy deployment, improved policies and political frameworks. It furthermore emphasized that transboundary cooperation should improve capacities for policy analysis and technology assessments. After the commencement of ethanol diplomacy, the Brazilian government shifted its attention to the creation of an international biofuels market, no longer mentioning these aspects. Trade issues, by contrast, never played a prominent role in the German deliberations on global renewable energy governance.

Both governments emphasize the need to establish internationally agreed sustainability standards on biofuels. However, they endow these with distinct purposes: whereas for the German administration, the main aim of sustainability standards is to prevent negative side effects, particularly in developing countries, the Brazilian government regards these as an instrument for ensuring biofuel demand in industrialized countries. The Brazilian government recognizes that these standards are instrumental in enhancing the acceptance of biofuels in industrialized countries, thus facilitating the opening of markets in industrialized countries. It states that sustainability standards should be designed in a way that facilitates trade and repeatedly underlines that it seeks to prevent sustainability standards being employed as non-tariff barriers to biofuels trade, impeding the production of biofuels in developing countries. In the view of the Brazilians, standards should be based on uniform requirements that developing countries can easily adapt to. The German government, by contrast, calls for ambitious sustainability requirements that prevent the negative side-effects of biofuel production. According to its view, sustainability standards on biofuels should ensure that human rights, social aspects and environmental protection are taken into account when producing them.

6.1.5 Global Governors: Diverging Relevance and Assessments

Both governments present distinct actors as the main global governors on renewable energy. The Brazilian government points to IBF, GBEP, WTO and ISO whereas the German government underlines the importance of IRENA, IEA and REN21. As such, the Brazilian administration focuses on those bodies that exclusively deal with biofuels and/or focus on trade issues. The bodies highlighted by the German government, by contrast, belong to the realm of energy policy. It particularly refers to the analytical work and policy advice of these bodies. It is interesting to note that the actors that are considered as being most relevant by one government are barely mentioned in the other's public statements on global renewable energy governance.

Both administrations apply different criteria when assessing the work of global governors dealing with renewable energy. For the Brazilian government, the relative weight of developing countries vis-à-vis industrialized serves as a decisive criterion, whereas the German government primarily refers to the importance that global governors attribute to renewable energy – either in comparison to conventional energy or in relation to their overall portfolio. The German government praises IRENA and REN21's exclusive focus on renewable energy, but criticizes that global governors such as IEA and the World Bank are still tend to favor conventional energy. It furthermore criticizes that the different organizations within the UN systems bestow on renewable energy only a subordinate stance. The Brazilian administration, by contrast, does not refer to the relative weight of renewable energy vis-à-vis fossil fuels. However, it pays attention to the comparative importance that is given to different renewable energy options, particularly to biofuels and hydropower. It points out that all global governors dealing with renewable energy are affected by North-South cleavages. It presents the IBF as the only forum in which developing countries and industrialized countries have equal participation, whereas it regards global governors such as IRENA, GBEP and ISO as being dominated by industrialized countries. The Brazilian government particularly criticizes the strong influence of European countries and their environmental concerns on these global governors. The German administration also pays attention to the involvement of developing countries in the work of global governors. It repeatedly emphasizes that global governors should have a global scope. It positively underlines that IRENA and REN21 are global governors comprising industrialized and developing countries alike and criticizes that IEA membership is limited to OECD countries. However, in contrast to the Brazilian government, it only focuses on membership issues and does not scrutinize the de facto influence of developing countries on global governors.

The comparison of government statements reveals that both governments draw a diverging picture of GBEP's thematic focus. Whereas for the German government, GBEP is about environmental risk prevention, for the Brazilians it is about the promotion of biofuel production and trade. The German administration concentrates on the development and implementation of international sustainability standards within GBEP and claims that these should prevent environmental damage caused by bioenergy production. The Brazilian government, by contrast, emphasizes that GBEP covers all analytical elements that are important for creating an international biofuel market: it promotes policies that help to increase biofuels deployment in member countries, addresses barriers to trade and technology transfer and works on sustainability standards to enhance public acceptance of biofuels by demonstrating that the positive aspects overweigh the negatives.

Both government assessments on IRENA vary as well. The German government presents it as the body that is best suited to foster transboundary cooperation on renewable energy. It particularly highlights IRENA's exclusive focus on renewable energy, its institutional structure as an intergovernmental organization, its global membership and focus, and its mandate to advise governments. The Brazilian government, by contrast, views IRENA with caution. It claims that IRENA's creation was not democratic and that the agency is too strongly influenced by European countries. It furthermore criticizes that IRENA's initial approach towards different renewable energy options was biased, favoring wind and solar energy while disadvantaging large hydropower and biofuels.

6.1.6 Salient Features: Mutual Benefits vs. Conflicts and Power Asymmetries

The North-South divide appears as a defining feature in both governments' deliberations on global renewable energy governance. If they differentiate between different groups of countries, they typically refer to industrialized countries on the one hand and developing countries on the other. Sometimes, emerging economies such as those of the BRICS appear as a third category. Both governments underline the importance of the principle of common but differentiated responsibilities. They each characterize industrialized countries as being the main culprits of global environmental problems and underline the special needs of developing countries due to their lower level of socio-economic development. Both governments point out that developing countries are more vulnerable to oil price increases and recognize that the promotion of socio-economic development is a political priority for these countries. With regard to energy policy, both governments consider access to energy and energy affordability as issues that are particularly important for developing countries. Both emphasize that transboundary policy-making on renewable energy should respect the right to development of developing countries and enhance North-South transfers of technology and finance. In addition, developing countries should be the primary beneficiaries of transboundary cooperation on renewable energy. Yet, their assessment of how this can be achieved varies significantly.

Both governments agree that global renewable energy governance is led by industrialized countries. Yet, they advance different assessments on the leadership of industrialized countries and on the interaction between industrialized and developing countries. The Brazilian government explicitly addresses conflicts and power asymmetries between industrialized and developing countries in global renewable energy governance, while the German administration insinuates that transboundary policy-making is to the mutual benefit of all involved without advantaging one group over the other. While the Brazilian administration criticizes the dominance of industrialized countries, its German counterpart claims that industrialized countries have a special responsibility to take the lead. The comparison of government deliberations furthermore reveals that the Brazilians endow developing countries with an active role in global renewable energy governance, whereas the Germans attribute developing countries a rather reactive role in global action to promote renewable energy.

The German government characterizes industrialized countries as frontrunners in the promotion of renewable energy, with regard to both technological development and policy implementation. It furthermore argues that their historical contribution to global environmental problems and their economic strength endows them with a duty to lead. It suggests that industrialized countries should implement renewables-friendly policies that can later on serve as positive examples in developing countries. In addition, they should develop appropriate renewable energy technologies that they can later transfer to developing countries as well. Thus, the German government locates economic, technological and political dynamics of renewable energy deployment in industrialized countries. These also benefit developing countries, it states.

For the Brazilian government, global renewable energy governance is embedded in deeper conflicts between industrialized and developing countries, which affect global governance in general. It particularly refers to inequalities in international trade. In its opinion, agricultural protectionism as practiced by industrialized countries deprives developing countries from economic development opportunities. Instead of being able to compete on equal grounds and thus to promote socio-economic development by themselves, developing countries depend on the aid that is provided by industrialized. The Brazilian administration states that the power asymmetries between aid providers and aid receivers are also mirrored in global governance, with industrialized countries attempting to impose their perspectives and preferences on the rest of the world. It complains about the arrogance of industrialized countries in global governance and points out that representatives of industrialized countries are convinced that they know what is right and wrong, assuming incompetence and a lack of knowledge on the other side. As developing countries are scientifically in a weaker position, challenging the perspectives of industrialized countries is a difficult undertaking. At the same time, industrialized countries, particularly in Europe, base their action in global governance on double standards, trying to hide their true motives. For the Brazilian government, global environmental protection is a case in point. While industrialized countries argue that they wish to protect the environment, their main interest is to shield their markets against competition. They try to impose environmental restrictions on developing countries, limiting even further the possibilities for developing countries to pursue socio-economic development. The Brazilian government suggests that this also applies to transboundary policy-making on biofuels. Due to the dominance of industrialized countries, global debates about biofuel sustainability primarily focus on environmental protection and disadvantage the pillars of social and economic development. As such, global debates concentrate not only on the priorities of industrialized countries, they also focus on the 'weak points' of developing countries. The 'weak points' of industrialized countries, such as their protectionist barriers, are merely addressed. The Brazilian administration adds that industrialized countries use the argument of environmental protection as a pretext to protect their domestic biofuel production. It furthermore states that European countries only claim that they wish to establish biofuels as a real alternative to fossil fuels, yet define sustainability criteria in a way that actually impedes the worldwide expansion of biofuels.

In line with what has been previously stated, both governments advance divergent views on how global renewable energy governance best benefits developing countries. While the German government presents developing countries as beneficiaries of North-South transfers of knowledge, technology and finance, its Brazilian counterpart criticizes transboundary cooperation that is merely based on North-South transfers. Instead of merely providing aid to developing countries and thus maintaining the dominance of industrialized countries, these countries should open up their markets and allow for equal competition so that developing countries can strengthen their economies and achieve independence and autonomy. The Brazilian government emphasizes that developing countries should be the main producers and exporters of biofuels and thus the place where the economic and technological development of biofuels takes place. It states that it is crucial to strengthen the influence of developing countries within global renewable energy governance. Moreover, it presents the promotion of biofuels in developing countries as an instrument to strengthen the economic capacities of developing countries and thus to tackle power asymmetries between industrialized and developing countries in global governance.

Whereas a positive framing of global governance prevails in the deliberations of the German government, the Brazilian administration clearly expresses its caution vis-à-vis global governance. Accordingly, the German administration favors the strengthening of global governance on renewable energy while the Brazilian government emphasizes the importance of national sovereignty. The German government states that global problems demand global solutions; it favors global commitments and wants to achieve as much bindingness as possible. The Brazilian government, by contrast, does not want to transfer too much decision-making power to the global level and emphasizes that each country has different circumstances and must find its own way. No country should impose its model on others. It argues that developing countries should not accept any conditions that limit their freedom to pursue socio-economic development. It adds that for developing countries, it might even be disadvantageous to enable the external influence of other countries or international organizations, as these may be captured by the interests of economic groups, such as the agriculture lobbies of industrialized countries. Whereas the Brazilian government seeks to prevent external influence on domestic policy-making, its German counterpart regards such an influence as an important task for transboundary policy-making.

6.2   Reasons Behind Ideational Differences

6.2.1   Coherency Between Ideas

Some ideational differences result from the governments' efforts to draft ideas in a consistent way, i.e. they are caused by other ideational differences. Thus, some reasons behind ideational differences can be identified by merely connecting the differences across analytical categories.

First of all, the differences concerning the main barriers to the worldwide promotion of renewable energy and the main tasks for transboundary policy-making are closely linked. The German government presents inadequate policy frameworks at the domestic level as a major barrier to the worldwide spread of renewable energy; in line with this, it states that transboundary policy-making should aim at improving domestic regulatory frameworks, by enhancing government commitments to renewable energy promotion and by supporting administrations to implement effective policies. Out of the different options to support governments, the German government particularly highlights policy advice. Its Brazilian counterpart presents trade barriers as a major obstacle to the worldwide spread of biofuels and seeks transboundary policy-making to focus on the creation of an international biofuels market in order to unleash market forces at global level. Whereas it is apparent that each government's interpretation of the main tasks for transboundary policy-making is consistent with its interpretation of the main barriers, it is not possible to determine whether the interpretation of the main tasks is a function of the interpretation with regard to the barriers or if the contrary is the case.

Several ideational differences can be traced back to the diverging views on salient features in global renewable energy governance. That the Brazilian government endows developing countries with an active role in global renewable energy governance while its German counterpart observes a rather reactive role is reflected in the different assessments of how renewable energy promotion enhances economic development in developing countries. The German government does not present the provision of renewable energy as an economic activity in itself, but rather refers to it as an input or pre-condition. Thus, in the deliberations of the German government, developing countries appear as consumers and not producers of renewable energy. The Brazilian administration, by contrast, presents the provision of renewable energy as an economic activity that directly creates economic value for developing countries. Thus, it speaks of developing countries as producers of renewable energy. The divergent assessments of trade benefits are in line with this. Whereas the Brazilian government speaks of developing countries as exporters of renewable energy, the Germans refer to developing countries as importers.

The diverging views on salient features also contribute to ideational differences with regard to the main barriers to a worldwide promotion of renewables and the main tasks for transboundary policy-making. In the Brazilian government's deliberations on barriers and tasks, the dominance of industrialized countries takes up much space, whereas it is not problematized in the German statements. While the Brazilian government raises confrontational issues, the German one follows a more cautious approach. It insinuates that global governance is to the mutual benefit of all involved. As it frames transboundary cooperation as a win-win option, it tries to concentrate on consensual issues. Therefore, it is cautious about blaming actors for obstructing the expansion of renewable energy but rather mentions barriers without linking them to specific actors. The Brazilian government, in contrast, claims that industrialized countries are those that pose the main obstacles to the worldwide promotion of biofuels. The Brazilian administration points out that developing countries have competitive advantages in the production of biofuels and agricultural goods in general. However, they cannot fully exploit these advantages as industrialized countries erect trade barriers to shield their domestic agricultural production. According to the Brazilian government, this policy seriously harms economic development in developing countries and also prevents biofuels from playing a greater role in worldwide energy supply. The Brazilian administration emphasizes that trade barriers to biofuels not only take the form of tariffs, sustainability criteria may also serve as non-tariff trade barriers. It points out that industrialized countries furthermore obstruct the worldwide spread of biofuels by circulating unjustified concerns about possible negative impacts of biofuel production, undermining their public acceptance. The German government, by contrast, mentions less confrontational barriers such as lacking political will and public awareness. In accordance with its problem assessment, the Brazilian administration claims that global renewable energy governance should primarily focus on the elimination of biofuel trade barriers operated by industrialized countries.

The focus on distinct forms of discrimination in each government's deliberations on the barriers is not only due to their differing perspectives on the salient features, but also due to the divergent relationship that is drawn between renewable and conventional energy. As the German administration regards the promotion of renewables as a means to replacing conventional energy, it assesses the regulative environment and the market opportunities of renewable energy in comparison to conventional energy. Thus, the relative (dis)advantages of renewable vis-à-vis conventional energy are of central importance for the German government. This is not the case for its Brazilian counterpart. As it regards renewable and conventional energy as complementary, it does not focus on the relative (dis)advantages. Thus, the German government concentrates on the discrimination of renewable vis-à-vis conventional energy, while the Brazilians rather point to discriminatory trade policies of industrialized countries.

A number of ideational differences with regard to global challenges can be traced back to the concentration on distinct renewable energy options: biofuels in the case of Brazil and renewable electricity in the case of Germany. Their concentration on different renewable energy options explains why both governments draw divergent causal links between renewable energy promotion and poverty reduction, with the Brazilian government focusing on income generation and the Germans concentrating on access to energy. The production of biofuels is less technology-intensive and sophisticated than the production of solar panels, wind turbines etc. Thus, poor people can relatively easily participate in the production of biofuels, for example, by cultivating their land or working as cane cutters. And if poor people are involved in the production process, they can directly generate income from it. As opposed to this, it is much more difficult for poor people to generate a direct income out of the production or provision of renewable electricity. The manufacture of solar panels, wind turbines etc. is technology-intensive and thus less accessible for the poor. Solar and wind energy, by contrast, facilitates the provision of electricity in remote areas without access to grids and thus contributes to poverty reduction by providing access to energy.

The different assessments of how renewable energy promotion enhances economic development in developing countries follow a similar logic. As the production of biofuels is not very technology intensive but rather requires input factors which are relatively abundant in many developing countries, such as fertile land and cheap labor, it is an economic activity in which poor countries can easily engage. Thus, the Brazilian government refers to biofuel production as an activity that directly generates economic value for developing countries. The provision of renewable electricity, by contrast, relies on more sophisticated technologies. Thus, it is less likely that countries with a low degree of development will directly engage in the production of these technologies. However, if the employment of renewable electricity improves the country's power supply, it can have important indirect effects on economic development. Consequently, the German administration presents the promotion of renewable energy as an input for, or precondition of, economic value creation in developing countries and not as an activity that directly generates economic value.

The focus on the dinstinct trade benefits for developing countries springs from the same reasoning. In this context, it is important to note that the tradability of biofuels and renewable electricity varies significantly. Whereas biofuels are easy to transport and thus to trade, renewable electricity is not easily traded as it requires grid infrastructure for transport. As a consequence, trade relations are typically limited to neighboring countries and renewable electricity is not a globally traded good. In this case, the globally tradable good is the technology involved in the production of renewable electricity. Industrialized and developing countries enjoy different competitive advantages in the production of these tradable goods: whereas developing countries have comparative advantages in the production of biofuels, the comparative advantages for the production of renewable electricity technologies typically lie in more technically advanced countries. As such, most developing countries have only limited opportunities to export renewable electricity technologies, although they can exploit their comparative advantages to become biofuels exporters. As a consequence, the German government refers to trade benefits through import substitution, whereas the Brazilian administration emphasizes two ways of improving trade balances: substituting imports and generating export opportunities.

The focus on different renewable energy options also contributes to a better understanding of why the Brazilian government draws a close link between renewable energy promotion and agricultural development while its German counterpart does not pay special attention to the agriculture sector. Biofuels have agricultural goods as feedstocks, thus directly involve agricultural production. Electricity based on renewable resources only comprises agricultural production in the case of bioenergy. Renewable electricity such as solar and wind energy can be employed in the agricultural sector, yet it is not necessarily linked to that sector as it does not build on agricultural production.

Some ideational differences with regard to the global governors are due to ideational differences on the tasks for transboundary policy-making. Both governments only rank global governors as important if these engage in what they regard as the most important task for transboundary policy-making. Two of the global governors that the Brazilian government presents as being most relevant – IBF and WTO – explicitly deal rsp. dealt with trade issues. The global governors that the German government considers as most relevant – IRENA, IEA and REN21 – all engage in policy advice. The divergent view on GBEP's thematic focus is an interesting case in point. For the Brazilian government, GBEP is mainly about promoting biofuel production and trade. It points out that GBEP deals with all analytical aspects that are crucial for building an international biofuels market. The German government, by contrast, does not link GBEP's work to trade liberalization but rather emphasizes the development and implementation of biofuel sustainability standards – which can be one tool to influence domestic policy-making on biofuels.

The divergent view on GBEP's thematic focus is also due to the different assessment of biofuel pros and cons which not only affect the view of both governments on GBEP, but also their general perspective on internationally agreed sustainability standards for biofuels. The Brazilian administration presents biofuel promotion as a socio-economic development opportunity for developing countries, with positive effects on poverty reduction and the fight against hunger. The German government, by contrast, connects the promotion of biofuels with a number of environmental and social risks. It fears that biofuel production could pose a threat to food security. As such, it regards sustainability standards as a means to preventing the negative side effects of biofuel production. The Brazilian administration, meanwhile, sees sustainability standards as a means to increase worldwide production and use of biofuels and to facilitate the creation of an international biofuels market. In line with this, it presents GBEP as a global governor that helps to promote biofuels, while the German government speaks of GBEP as a global governor that helps to prevent the negative side effects of biofuels.

The different criteria to assess global governors dealing with renewable energy mirror the varying forms of discrimination that prevail in the governments' deliberations on the barriers to a worldwide promotion of renewable energy. Whereas the German government focuses on the discrimination of renewable vis-à-vis conventional energy and accordingly uses the relative weight of this as a central criterion to assess global governors, the Brazilian government points to the relationship between developing and industrialized countries. It presents the trade policies of industrialized countries as a major barrier to the worldwide spread of renewable energy and uses the relative influence of developing countries vis-à-vis industrialized countries as a core criterion for assessing global governors.

6.2.2 Domestic Policy Context of Renewable Energy Promotion

Different domestic policy contexts of renewable energy promotion are one important reason behind the ideational differences on global renewable energy governance. The political motivations behind the domestic promotion of renewables, the policy interventions and structure of the renewables sector differ considerably between both countries.

Germany

In Germany, environmental concerns are the main driving force behind the domestic promotion of renewable energy. These primarily relate to the risks of using nuclear energy. Climate change concerns also come into play. Since the 1970s, Germany has experienced the emergence of several civil society initiatives that called for the phasing out of nuclear energy, climate protection and the replacement of fossil fuels by renewable energy (Hirschl 2008:127, Jacobsson, Lauber 2006:261, Laird, Stefes 2009:2621, Schreurs 2013:88-91). Two of the principal actors who pushed for the domestic promotion of renewable energy originate from these environmental movements. Participants of the environmental movements founded the Green party and Hermann Scheer, the SPD parliamentarian who lobbied intensively for IRENA's creation, was closely embedded in the anti-nuclear movement. After the nuclear accident in Chernobyl in 1986, which by then belonged to the Soviet Union, public opinion turned against nuclear energy (Hirschl 2008:127, Jacobsson, Lauber 2006:261-263, Laird, Stefes 2009:2621). The German government, then comprising a Coalition of CDU/CSU and FDP, continued to support nuclear energy. However, it established environmental protection as one important goal of the country's energy policies and created the BMU (Hirschl 2008:96,124). During the 1980s, climate concerns gained political importance in Germany (Hirschl 2008:128f., Jacobsson, Lauber 2006). The quest for environmental protection through renewable energy promotion united parliamentarians with different party affiliations (Jacobsson, Lauber 2006:271, Laird, Stefes 2009:2622). Due to the political pressure of the German Bundestag, the government finally introduced its first policy measure for an electricity feed-in of renewable sources in 1990 (Hirschl 2008:19).

However, the breakthrough for renewable electricity only came with the government takeover by SPD and the Greens in 1998 who established the 'Energiewende' as a central building block of their government program. They presented concerns about nuclear risks and climate protection as the main motivations behind this step (SPD, Bündnis 90/Die Grünen 1998:17-20). When the Greens called for the transfer of competencies for renewable energy from the BMWi to the BMU in 2002, they further reinforced the link between renewable energy promotion and environmental protection in Germany. CDU/CSU and FDP continued to support nuclear energy and when they regained power in 2009, they decided to expand the lifespan of nuclear power plants. This decision met strong public opposition. After the nuclear accident in Fukushima, Japan, in 2011, the government coalition (CDU/CSU and FDP) withdrew from its decision and started the second 'Energiewende'. Since then, the phasing out of nuclear energy is borne by an all-party consensus (Röhrkasten, Westphal 2012a:333, Schreurs 2013). Hirschl (2008:177,184) underlines that the widespread skepticism against nuclear power and environmental awareness in German society endows renewable energy with very broad public support.

Renewable energy contributes to 12.7 percent of German energy supply (BMU 2013c:5).[345] The renewable energy share is highest in the electricity sector. Here, it supplies 23.5 percent, while it only makes up 10.2 percent of the heating sector and 5.7 percent of transport (BMU 2013c:10). Political and public debates on renewable energy promotion commonly concentrate on electricity generation.[346] Initial policy measures to promote renewable energy aimed at the transformation of the electricity supply, and the electricity sector still has the most ambitious and publicly visible policy measures to support renewables.[347] This concentration on the electricity sector can be understood against the background that the phasing out of nuclear energy has been a major driving force behind the efforts to promote renewable energy in Germany.

The promotion of renewable energy is a rather recent phenomenon in Germany. It is still considered a 'newcomer' on the market that needs government support in order to compete with conventional energy. Thus, in Germany the expansion of renewable energy is led by policy interventions. The "flagship legislation" (Laird, Stefes 2009:2624) of the German 'Energiewende', the EEG, was adopted in 2000.[348] It provides feed-in tariffs for electricity generated from renewable resources. With the introduction of the EEG, the dissemination of renewable electricity increased significantly in Germany (Hirschl 2008:188).[349] Figure 6 on page 122 illustrates the rising contribution of renewable energy sources to German electricity supply.

It is interesting to note that the EEG was not initiated by the government itself. The BMWi, which by then was in charge of renewable energy, argued against market intervention to promote renewables. It rather wanted market forces to guide the choice between different energy options (Hirschl 2008:195f., Laird, Stefes 2009:2623). On the contrary, parliamentarians of the ruling parties, SPD and the Greens, agreed that renewable energy could not become competitive in the market without state support. They thus drafted the bill that initiated the EEG (Jacobsson, Lauber 2006:264, Laird, Stefes 2009:2622f.). The BMU also supported feed-in measures for renewable energy. However, by then, it did not have the core compe-

---

[345] If not stated otherwise, the data in the section on the domestic policy context in Germany refer to the year 2012.

[346] The concentration on the electricity sector is not unique to political and public debates on renewables, but concerns German energy policy-making in general (Röhrkasten, Westphal 2012a).

[347] The electricity sector is likely to continue as the area with the highest share of renewable energy. The government goals establish that the renewable share in the electricity sector is to be raised to at least 35 percent by 2020 and to at least 80 percent by 2050, while the renewable share of overall energy consumption is set at 18 percent by 2020 and 60 percent by 2050 (BMU 2013c:9).

[348] In addition, the government set ambitious targets for increasing the renewable share in the domestic electricity supply, and launched a program providing low-interest loans for investments in photovoltaics (the 100,000 roof program) (Laird, Stefes 2009:2624).

[349] In comparison to the preceding legislation, the Act on the Sale of Electricity to the Grid (Stromeinspeisungsgesetz, Streg) of 1991, the new tariffs significantly increased the premium for most renewable energy sources, particularly solar energy (Laird, Stefes 2009:2624).

tency for renewable energy (Hirschl 2008:188f.). The EEG is still the main instrument for promoting renewable energy in German electricity supply.

When the German government initiated the 'Energiewende' in 1998, the transport and heating sector were not the focus of political attention. Yet, the ecological tax reform in 1999, which increased the fiscal burden on fossil fuel consumption, led to a first expansion of biodiesel (Brand-Schock 2010:290). In 2002, the German government exempted the blending of biofuels from the mineral oil tax, and in 2005 it introduced quotas for biofuel blending (Beneking 2011:73, Brand-Schock 2010:301-306,381f.). In the heating sector, the German government only endorsed an act in 2009 promoting renewable energy sources. The Heat Act establishes that new buildings must cover part of their heat supply with renewable sources.[350]

Since 2008, biofuels have met strong criticism in Germany.[351] When the government sought to increase the blending quotas for biofuels from five to 10 percent in 2008, it withdrew the decision due to strong public opposition. While NGOs and churches feared negative implications on global food security and environmental protection, representatives of the car industry questioned the compatibility of the blending requirements with car engines (Beneking 2011, Brand-Schock 2010:313-317, Kohlhepp 2010:231, Selbmann, Kaup 2010:30). In the same year, the German Bundestag introduced a bill to prevent negative environmental impacts of biofuel usage.

The expansion of renewable energy in the electricity sector has primarily built on decentralized and small-scale applications. This stands in clear contrast to the centralized structures of the conventional energy system (Hirschl 2008:21,71, Jacobsson, Lauber 2006:261). Pioneering work on renewable energy was carried out by small-scale businesses, municipalities and civil society initiatives, while large utilities and conventional energy suppliers strongly opposed the promotion of renewable energy (Beneking 2011:58, Brand-Schock 2010:373f., Laird, Stefes 2009:2622). As a result, the renewable energy sector emerged with new economic players (Hirschl 2008:555f.,579). For many of its advocates, achieving a decentralized energy supply without monopolies was an important motivation (Beneking 2011:102). Keller (2010:4741) points at the bottom-up structure of German support policies for renewable electricity, as mechanisms for direct compensation favor small investors. The government support for photovoltaics in particular followed a highly decentralized model and the photovoltaic applications at household level turned energy consumers into energy producers (Hirschl 2008:184). This was also instrumental for increasing public support for renewables.

---

[350] BMU, Act on the Promotion of Renewable Energies in the Heat Sector (Heat Act, EEWärmeG), http://www.erneuerbare-energien.de/en/topics/acts-and-ordinances/heat-act/ (accessed January 14, 2014).
[351] Biofuels are also strongly contested at the EU level. See Fischer, Röhrkasten 2013.

A closer look at the support polices for renewable energy in Germany reveals that the German government has particularly aimed at expanding the use of wind and solar energy. The initial research projects that it launched in 1988 to promote renewable energy focused on photovoltaics and wind energy (Hirschl 2008:127-129, Jacobsson, Lauber 2006:264). An important goal of the first feed-in measures in 1990 was to expand the use of wind energy (Laird, Stefes 2009:2622).[352] The coalition agreement of the SPD/Greens government of 1998 foresaw not only legislative means to promote renewable electricity in general, but also a special program in support of solar cells (100,000 roofs program) (SPD, Bündnis 90/Die Grünen 1998:17-20). Jacobsson & Lauber (2006:272) underline that these measures were thought to provide financial incentives for investments in solar cells and to further improve the investment conditions for wind energy. As a result of the support policies, photovoltaics for the first time became a lucrative investment option in Germany (Jacobsson, Lauber 2006:268). Solar and wind energy are furthermore the renewable energy options that commonly enjoy the highest support rates in German public opinion polls (Hirschl 2008:184).[353]

The contribution of different renewable energy sources for energy supply and economic value creation provides a mixed picture. As Figure 9 illustrates, bioenergy by far makes up the largest share in the renewables-based final energy supply (64.6%),[354] followed by wind energy (15.9%), photovoltaics (8.3%), hydropower (6.9%), geothermal energy (2.2%) and solar thermal energy (2.1%) (BMU 2013c:14). In the electricity sector, wind energy is the dominant renewable energy source. It contributes to 35.6 percent of the renewables-based electricity supply. Different biomass uses provide another 30.[355] Significant shares are furthermore provided by photovoltaics (18.5%) and hydropower (15.5%) (BMU 2013c:14). In the heating sector, biomass makes up the bulk of renewables-based supply. It share amounts to more the 90 percent, the rest being divided almost evenly between geothermal and solar thermal energy. The renewables-based motor fuel supply is dominated by biodiesel (72.2%). Bioethanol provides another 26.1 percent (BMU 2013c:15). The bulk of economic revenues from the operation of renewable energy installations are generated by bioenergy (72.5%). By a wide margin follow wind energy (9.7%), photovoltaics (8.2%) and geothermal energy (5.3%) (BMU 2013c:32).[356] Bioenergy,

---

[352] Another aim was to maintain the use of small hydropower.
[353] Agentur für erneuerbar Energien, Umfrage 2013: Bürger befürworten Energiewende und sind bereit, die Kosten dafür zu tragen, http://www.unendlich-viel-energie.de/themen/akzeptanz2/akzeptanz-umfrage/umfrage-2013-buerger-befuerworten-energiewende-und-sind-bereit-die-kosten-dafuer-zu-tragen (accessed January 22, 2014).
[354] The bulk of bioenergy is employed in the heating sector (39.8%). Biomass for electricity generation contributes to 13.7 percent of the renewables-based final energy supply, biofuels to 11.1 percent.
[355] The different biomass uses cover biogas (17.4%), biogenic solid fuels (8.1%), biogenic fraction of waste (3.4%) and biogenic liquid fuels (0.3%).
[356] Hydropower only accounts for 2.6 percent of the revenues, solar thermal energy for 1.7 percent.

solar and wind energy are the renewable sources that generate most jobs (BMU 2013c:32).[357] In the realm of investments, photovoltaics is by far the leading renewable energy source, accounting for 57.6 percent of overall investments in the construction of renewable energy installations. Wind energy and bioenergy follow with a wide margin (BMU 2013c:30).[358]

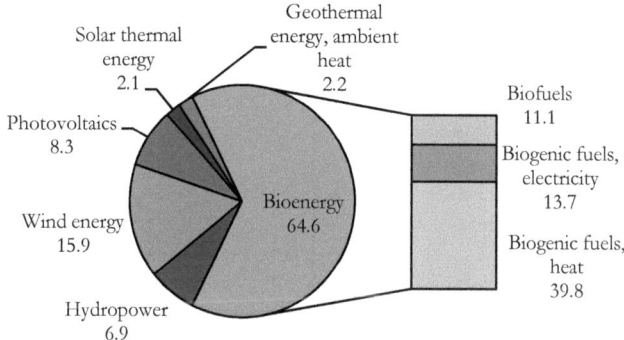

Figure 9: German Renewables-Based Final Energy Supply, 2012[359]

Brazil

The Brazilian electricity supply has always relied on renewable energy sources.[360] The country's territory offers vast hydropower potential and exploiting this has been the most economic way of meeting the rising electricity demand. The IEA therefore states that "Brazil's early determination to press ahead with alternatives to fossil fuels was a natural choice" (OECD/IEA 2013:304). As hydropower has been the most competitive electricity source in Brazil, there has been no need for specific policies promoting its use (Schutte, Barros 2010:34).[361] Hydropower is still by far

---

[357] Between 2010 and 2012, they each employed more than 100,000 people (bioenergy: 125,100 employees, solar energy: 115,500 employees, wind energy: 105,000 employees. During the same period of time, only 13,800 people were employed in the geothermal sector, while hydropower only accounted for 7,400 employees.
[358] Out of the overall investments, 19.3 percent were made in wind energy and 13.1 percent in bioenergy. Solar thermal heat and geothermal heat only reached a share of around 5 percent, while hydropower accounted for even less than 0.5 percent of overall investments.
[359] Source: illustration S.R. with data from BMU (2013c:14).
[360] Interview with professor (#1), Energy Institute, USP, April 2013.
[361] Interview with professor, Energy Planning Program, UFRJ, February 2013.

the most important basis of Brazilian electricity supply. In 2011, hydropower provided 81.9 percent of the Brazilian electricity supply (Empresa de Pesquisa Energética 2012:16).[362] The bulk of Brazilian hydropower is generated by large dams. Biomass such as firewood and sugarcane bagasse[363] is the second most important source of electricity generation in Brazil, contributing another 6.6 percent. Wind energy only makes up a share of 0.5 percent, while further renewable energy sources do not make significant contributions to the electricity supply. Conventional energy sources only make up a share of 11 percent of Brazilian electricity supply (Empresa de Pesquisa Energética 2012:16).[364]

Brazil is confronted with a sharply rising electricity demand. Since 1990, electricity supply has more than doubled (OECD/IEA 2013:309). The Brazilian Energy Plan for the year 2021 foresees annual increases in the domestic electricity supply of 4.8 percent. Hydropower will continue to provide the bulk of the supply increases. Yet, its relative contribution will decrease (MME 2012). Brazilian hydropower potential is far from exhausted; yet expanding its use is becoming more difficult. The remaining hydropower potential is located in remote areas such as the Amazon and the north, and the construction of new dams in these areas is becoming controversial in the country. Protests are raised with regard to environmental issues and the rights of the local population (OECD/IEA 2013:311, Viola 2013:8).[365] In addition, the Brazilian government seeks to reduce the vulnerability of the electricity supply vis-à-vis droughts and other adverse hydrological conditions. After droughts had led to a severe electricity crisis in 2001/02, the government introduced a system of contract auctions to foster other sources of electricity generation. Here, the most competitive bid decides on long-term power supply contracts. Some bids were held for renewable energy sources excluding large hydropower. Wind power has been a major profiteer of these auctions (OECD/IEA 2013:311f.). The Brazilian Energy Plan 2021 foresees that the contribution of wind power will rise significantly. The relative importance of biomass is also set to increase (MME 2012).

In the transport sector, fossil fuels have been dominant. Until the end of the 1970s, the transport sector relied on fossil fuel imports. After the oil price shocks of the 1970s, the government not only began to invest in domestic oil production, but also introduced policy measures to foster alternatives to fossil fuels.[366] Energy secu-

---

[362] If not stated otherwise, all data cited in this section refer to the year 2011.
[363] Sugarcane bagasse is a byproduct of sugar or ethanol production. It is also used for heating (Coelho, Goldemberg 2013:464, Oliveira 2012:6, Sanchez Badin, Godoy 2013:20).
[364] This share is divided as follows: natural gas 4.4 percent, nuclear energy 2.7 percent, oil products 2.5 percent and coal and coal products 1.4 percent.
[365] Interviews with diplomat, Itamaraty, March 2013; Vice-President, CEBRI, April 2013; staff member, Reporter Brasil, April 2013.
[366] Since 2006, Brazil is self-sufficient in petroleum (Kohlhepp 2010:224).

rity concerns were the major driver behind these policy measures, as the administration sought to reduce its import dependence (Galli 2011:48-53, Schutte, Barros 2010:34). It opted for the promotion of sugarcane ethanol, which offered positive side-effects for the country's socio-economic development. At that time, the Brazilian sugarcane industry was hit by low international sugar prices and the promotion of ethanol provided new market opportunities for this traditional and economically important industry (Kloss 2012:140, Simões 2007b:17-20). When President Lula da Silva began again to promote biofuels, he also aimed at fostering socio-economic development. With regard to biodiesel, he particularly underlined social issues, such as promoting agricultural development in remote areas and supporting small and family farmers (Simões 2007b:21-23). His support for ethanol primarily aimed at improving Brazil's economic strength through increased export activities (Galli 2011:55, Lula da Silva 2006:19).

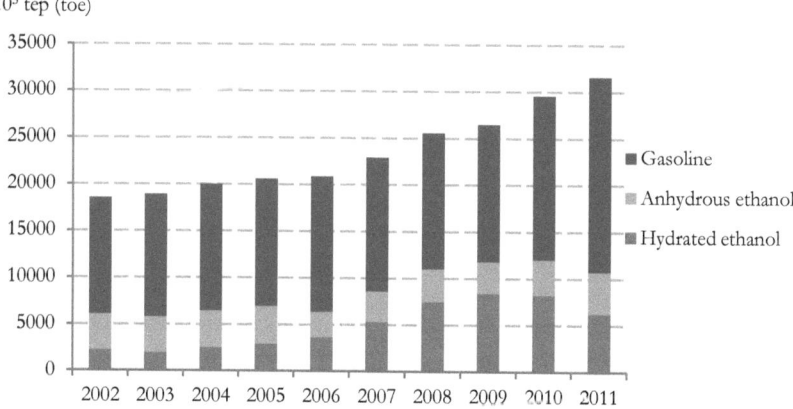

Figure 10: Brazilian Consumption of Gasoline and Ethanol in Road Transport, 2002-2011[367]

The government supports the use of biofuels through obligatory blending requirements. Gasoline contains an obligatory blend of anhydrous ethanol that ranges between 18 and 25 percent. The obligatory blending quota for biodiesel is set at five percent. Yet, the use of ethanol in Brazil is not only driven by policy measures, but also by market forces. Since the market introduction of flex fuel cars in 2003, consumers can choose flexibly between any mixture of gasoline (containing the obligatory blending of anhydrous ethanol) and pure (hydrated) ethanol. Figure 10 illus-

---

[367] Source: illustration S.R. with data from Empresa de Pesquisa Energética (2012:79).

trates the contribution of anhydrous and hydrated ethanol to Brazilian fuel consumption in road transport. Thus, hydrated ethanol directly competes with gasoline. As it has a lower energy content than gasoline, it is an economically attractive choice if its price does not exceed 70 percent of the gasoline price. Yet, the competition between gasoline and hydrated ethanol is not free from government interventions. Since 2006/07 the government has frozen the gasoline price in order to fight inflation.[368] As the gasoline price sets an indirect price cap for hydrated ethanol, this policy measure negatively affects the competitiveness of hydrated ethanol. Representatives of the Brazilian government often tend to play down the role of government interventions in the Brazilian transport fuel market. In his article in a Brazilian newspaper Amado (2010a), for example, claims that in Brazil ethanol prices are determined by market forces and that the government does not distort the market with subsidies.[369] As this section shows, this is only half the story. Only hydrated ethanol competes with gasoline on the market, while the consumption of anhydrous ethanol is determined by the blending quota. In addition, not only market forces determine the price of hydrated ethanol, but also the government-induced price cap for gasoline.

Brazil has an established and internationally competitive ethanol industry. Its ethanol is made of sugarcane, a product that has traditionally been of central importance for the economy. The ethanol industry is part of its agribusiness and one of the most industrialized agricultural sectors in the country. As sugarcane in Brazil has traditionally been a large-scale crop (Bastos Lima 2012:346), the ethanol industry builds on large-scale land holdings. The Brazilian Development Bank BNDES considers the sugarcane industry as strategic in national development because its whole value chain is dominated by Brazilian producers and it is a sector where the country can turn its natural resource endowment into global technological leadership (Herrera, Wilkinson 2010:752).

From 2003 to 2008, the Brazilian ethanol industry experienced times of expansion, with high levels of private investment and technological modernization (Herrera, Wilkinson 2010:751, Simões 2007b:20). Its exports increased significantly during this period of time (Empresa de Pesquisa Energética 2012:67). In the following two years, exports declined significantly and since 2010/11 the ethanol industry has passed through economic difficulties (see also Figure 7 on page 189 for an illustration of the Brazilian ethanol production and trade balance between 2002 and 2011). This has been due to a variety of factors, such as the freezing of the domestic

---

[368] Interviews with Secretary and team, Secretariat of Planning and Energy Development, MME, March 2013; professor (#1), Energy Institute, USP, April 2013; professor, Economics Institute, UFRJ, March 2013; professor, FFLCH, USP, April 2013; staff member, Solidaridad, April 2013.
[369] At the GBEP Bioenergy Week in Brasília in April 2013, Brazilian government representatives advanced similar assessments (participant observation by the author of this study).

gasoline price, the international financial crisis and bad harvests (Kloss 2012:10).[370] In 2011, Brazil even had to import significant amounts of ethanol to close its domestic output gap. Nevertheless, it continues as a net exporter of ethanol (Empresa de Pesquisa Energética 2012:67).

In contrast to the ethanol industry, the biodiesel industry is emergent and concentrated on the domestic market only. Biodiesel production only began after the launch of the Biodiesel Program in 2004. Yet, both industries resemble each other in one respect: they each form part of agribusiness and thus rely on large-scale farming. This stands in clear contrast to the initial goal of the Biodiesel Program. While the government presented the support of small-scale and family farming as an important goal, according to (Herrera, Wilkinson 2010:764) "the main feature of the biodiesel program (…) has been its appropriation by the soy complex". Soy has been the main feedstock of Brazilian biodiesel production. As such, the country's biodiesel production is closely linked with the soy agribusiness (Herrera, Wilkinson 2010:765).

In Brazil, ethanol enjoys strong domestic support, both in the media and the political sphere (Giersdorf 2011:153, Viola 2013:7).[371] One academic based at a Brazilian university even claims that ethanol forms part of the national pride.[372] Domestic critics do not argue against the use of biofuels but rather call for improved regulatory frameworks to minimize negative social and environmental side-effects. Several domestic NGOs and academics claim that the ethanol industry is more responsive to sustainability concerns than other sectors of agribusiness, as sustainability issues are crucial for its global reputation and export possibilities.[373] The working conditions on sugarcane plantations have been a major focus of domestic conflicts on ethanol production (Herrera, Wilkinson 2010:754).[374] In addition, the local environmental implications of sugarcane burning have met strong public opposition. In order to tackle both concerns, sugarcane harvesting is increas-

---

[370] Interviews with Secretary and team, Secretariat of Planning and Energy Development, MME, March 2013; Biodiesel Federal Program Coordinator and staff member, Analysis and Following up of Government Policies, Casa Civil, March 2013; Director of the Department of Renewable Fuels, MME, March 2013; diplomat, Itamaraty, March 2013; Vice-President, CEBRI, April 2013; professor, CPDA, UFRRJ, March 2013; professor, Energy Planning Program, UFRJ, February 2013; team, Bioenergy Projects, FGV, February 2013; staff member, Solidardidad, April 2013; former President, UNICA, April 2013; professor, Luiz Queiroz College of Agriculture, USP, April 2013; professor (#1), Energy Institute, USP, April 2013.
[371] Interviews with Head of Energy Department, Itamaraty, April 2013; professor, CPDA, UFRRJ, March 2013; professor (#2), Energy Institute, USP, April 2013; professor, Luiz Queiroz College of Agriculture, USP, April 2013.
[372] Interview with professor, CPDA, UFRRJ, March 2014
[373] Interviews with policy adviser, Brazilian Office, Oxfam, April 2013; staff member, Reporter Brasil, April 2013; staff member, Solidaridad, April 2013.
[374] Interviews with policy adviser, Brazilian Office, Oxfam, April 2013; staff member, Reporter Brasil, April 2013; staff member, Solidaridad, April 2013.

ingly mechanized (Galli 2011:71-84).[375] Some domestic NGOs and academics furthermore criticize the negative environmental impacts of monocultural farming and expanding agricultural frontiers (Herrera, Wilkinson 2010, Giersdorf 2011:117f.).[376]

The major sustainability concerns that non-Brazilian policy actors raise with regard to the country's ethanol production – deforestation of the Brazilian Amazon and displacement of food production – are not shared by domestic actors. It is generally argued that these concerns do not mirror the realities of Brazilian ethanol production (Herrera, Wilkinson 2010:763-765, Johnson 2010:45, Kohlhepp 2010:232-241, Schutte, Barros 2010, Viola 2013:7). On the food versus fuel debate, Herrera & Wilkinson (2010:765) add that in Brazil, "food politics are seen primarily in terms of income and access", while "Brazil's exceptional availability of land and water in favorable climatic conditions permits the continued growth of both fuels and food."

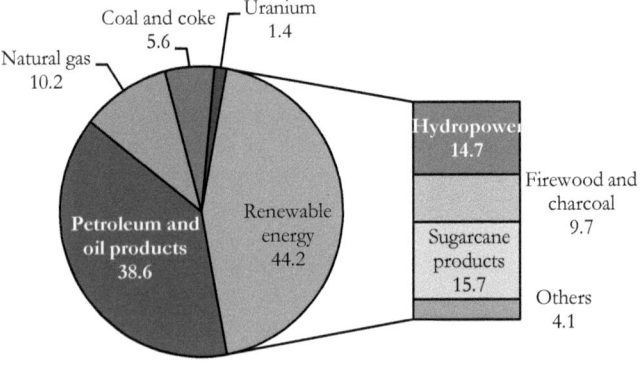

Figure 11: Brazilian Energy Supply, 2011[377]

Of the country's total energy supply, sugarcane products make up a slightly bigger share than hydropower (see Figure 11). Out of the renewables-share of 44.2 percent, sugarcane products provide 15.7 percent, while the contribution of hydropower amounts to 14.7 percent. Another 9.7 percent is provided by firewood and charcoal (Empresa de Pesquisa Energética 2012:22).

---

[375] Mechanized harvesting does not rely on pre-harvest burning and substitutes manual harvesting.
[376] Interviews with policy adviser, Brazilian Office, Oxfam, April 2013; staff member, Reporter Brasil, April 2013; staff member, Solidaridad, April 2013.
[377] Source: illustration S.R. with data from Empresa de Pesquisa Energética (2012:22).

Connecting the Domestic Policy Context to the Global Governance Ideas

The motivations behind the domestic promotion of renewable energy strongly influence the view of both governments on the global challenges addressed by transboundary policy-making. In the German case, the emphasis on global environmental protection is in line with the environmental concerns that motivated the domestic promotion of renewable energy. Yet, there is one important difference between the domestic motivations and the administration's pronouncements on global renewable energy governance. At the global level, the German government does not touch the environmental risks of using nuclear energy, although concerns about it were the main driving force behind the promotion of renewable energy at the domestic level. In its statements on global renewable energy governance, it rather concentrates on climate protection. Concerns about climate change also motivated the domestic promotion of renewables, yet were much less prominent than nuclear risks. The Brazilian government's emphasis on the promotion of socio-economic development in developing countries corresponds to the motivations behind the domestic promotion of biofuels, at least in recent times. When the Brazilian government began to promote ethanol in the 1970s, energy security concerns were dominant. Yet, since the 2000s socio-economic development has been the prevailing motivation.

As the views of both governments on the global challenges correspond to the domestic motivations behind the promotion of renewable energy, they also draw the same relation between renewable and conventional energy at the domestic and global level. The German administration presents renewable energy as a substitute for conventional energy, helping to prevent the latter's negative side effects both at the domestic and the global level. The Brazilian administration, by contrast, presents renewable energy and conventional energy as complementary sources that enhance socio-economic development, both domestically and globally.

The blind spots in the deliberations on global challenges also relate to the different domestic motivations behind renewable energy promotion. Whereas the Germans point to global environmental challenges that the Brazilians neglect, such as protecting forests and preserving biodiversity, the Brazilians emphasize agricultural development as an essential component of socio-economic development in developing countries.

In the Brazilian case, the domestic policy context is also mirrored in the government deliberations on the link between renewable energy promotion and economic development in developing countries. When the Brazilian administration elaborates on the economic gains of producing biofuels and related export opportunities, it clearly draws on the Brazilian experience.

The concentration on distinct renewable energy options is also embedded in the domestic policy context. In their deliberations on global renewable energy governance, both administrations only focus on renewable energy sources that provide significant shares to their domestic energy mix. This helps to understand why solar and wind energy only appear in the German statements. While wind and solar energy provide significant shares to the German energy supply, their shares in the Brazilian energy supply are tiny. Wind energy provides only 0.5 percent, while solar energy does not even appear in the Brazilian energy balance (Empresa de Pesquisa Energética 2012). It is important to note that there is no direct relationship between the emphasis placed on specific renewable energy options at the global level and their predominance in domestic energy supply. On the contrary, renewable energy sources with major shares in the domestic energy supply – biomass in the German case and hydropower in the Brazilian – do not play a major role in government deliberations on global renewable energy governance. Moreover, there is no direct link between the concentration on different energy sectors in the deliberations and the prevalence of these sectors in the domestic usage of renewable energy. In both countries, the renewable energy share is highest in the electricity sector. Yet, while the electricity sector plays a central role in the German deliberations on global renewable energy governance, this is not the case in the Brazilian statements.

In the German case, the focus on the electricity sector at global level corresponds to domestic prioritization. There, political and public debates on the promotion of renewable energy concentrate on the electricity sector. The electricity sector also has the most ambitious policies to promote renewables. As in the government's deliberations on global renewable energy governance, solar and wind energy are also in the focus of domestic political attention. Policy action to promote renewable electricity has particularly aimed at expanding wind and solar energy. Even though biomass provides a major share of the renewables-based electricity supply, it receives much less political attention than wind and solar energy. Considering the German energy supply in general (i.e. not only the electricity supply) and the economic value creation of the renewables sector, the gap between the actual contribution of bioenergy and its scarce political and public attention becomes even wider. A reason for this might be the diverging public support rates of different renewable energy options in Germany. While solar and wind energy are the renewable sources that typically receive the highest public support rates in public opinion polls, bioenergy enjoys less public backing. In particular, the use of biofuels is strongly contested at the domestic level.

The fact that hydropower and the electricity sector in general do not appear in the Brazilian government's deliberations on global renewable energy governance cannot be understood by looking at the domestic policy context only. Almost 90 percent of the electricity supply relies on renewable energy sources, the bulk of which is provided by hydropower. The government's deliberations on transboundary policy-making instead concentrate on the transport sector. Here, the domestic policy context helps to understand the administration's focus on (sugarcane) ethanol. At the domestic level, ethanol clearly prevails over biodiesel – in terms of supplied volumes, economic importance and deployment experience. Sugarcane is the dominant feedstock of Brazilian ethanol production.

The domestic policy context also contributes to a better understanding of why both governments advance a divergent assessment of the pros and cons of biofuels. In Germany, biofuels are highly contested and there has been strong opposition against domestic biofuel blendings. The risks that the government emphasizes at the global level are also voiced by NGOs and media at the domestic level. In Brazil, by contrast, ethanol enjoys strong public support. When the administration elaborates on the advantages of worldwide deployment of biofuels, it clearly draws on the Brazilian experience. However, it does not mention those aspects that meet criticism at the domestic level. It is generally argued that the sustainability concerns that prevail at the global level do not apply to the Brazilian context of ethanol production. When the government elaborates on the relationship between biofuels and food production, it transfers the Brazilian situation to the global level. The country's territory offers enough fertile lands to produce both biofuels and food. If people suffer hunger, it is not the result of an overall food scarcity, but rather due to income poverty. Thus, income generation is regarded as a central tool for fighting hunger.

The different criteria for assessing renewable energy options can also be traced back to the domestic policy context. In Germany, the promotion of renewable energy has been led by policy interventions, not by market forces. The major motivation behind these policies has been environmental protection. The policies were enacted precisely because of the recognition that without government support, renewable energy could not compete with its conventional counterpart. As a consequence, the German government argues against basing the choice between different energy options on economic competitiveness only, both domestically and globally. It rather presents environmental sustainability as the most important criterion. In Brazil, environmental protection has not been a driver behind the domestic usage of renewable energy. Here, economic aspects have been prevailing. Thus, the government argues that economic competitiveness should be the key criterion that guides the choice between different renewable energy options – both domestically and globally.

The domestic policy context furthermore explains why both governments advance diverging views on renewable energy competitiveness. The varying degrees of renewable energy competitiveness at domestic level are in line with the different time horizons both administrations use when elaborating on the economic feasibility of renewable energy options. In Germany, renewable energy is not yet competitive on the market and therefore, the government points to future potentials. In the Brazilian case, the bulk of renewable energy sources are deployed because of their economic competitiveness. This particularly applies to hydropower and hydrated ethanol. Thus, the Brazilian government refers to the status quo. The different scales of renewable energy deployment at the domestic level also affect the views of both governments on economic competitiveness. As large scale deployment of renewable energy prevails in Brazil, its government links economic competitiveness with large-scale deployment. The German administration, by contrast, points to the competitive advantages of decentralized, small-scale renewable energy options, as these have been dominating the country's renewable energy landscape.

The different balance between state interventions and market forces at the domestic level is also reflected in the ideational differences on the barriers to the worldwide promotion of renewable energy and on the tasks for transboundary policy-making. The German administration focuses on policy interventions and government action, whereas its Brazilian counterpart underlines the need to strengthen market forces. In keeping with its domestic experience, the German government argues that inadequate domestic regulatory frameworks form a major barrier to the worldwide spread of renewable energy, and it presents the improvement of domestic regulatory frameworks as the most important step in the worldwide promotion of renewables. Thus, the German administration particularly points at the need to provide policy advice to governments. The Brazilian government, by contrast, points to market interventions as a major barrier to the worldwide spread of biofuels. It presents free trade, without government intervention, as the best way to promote biofuels at a global scale.

6.2.3 Embeddedness in Transboundary Policy-Making

The ideational differences on global renewable energy governance are also due to the governments' different involvement and self-positioning in transboundary policy-making. Interestingly, the general global governance context turns out to be more relevant for the understanding of the ideational differences than the renewables-specific context of transboundary policy-making.

With regard to the global context, the different self-positioning within global power structures appears as the most important reason behind differing ideas on global renewable energy governance. Both governments divide the world up into two main groups: industrialized countries and developing ones. Both recognize that industrialized countries have been the leading force in shaping global governance. Yet, whereas the German administration ascribes itself to the group of industrialized countries, i.e. to the country group with the greatest say in global governance, the Brazilian government sees itself as a leading actor within the group of developing countries. According to the Brazilian administration, these countries have been deprived from influencing global governance.

Such different self-positioning strongly influences how both governments characterize the functioning of global governance and its 'division of labor' between industrialized and developing countries. These differences are mirrored in their perspectives on the salient features in global renewable energy governance. As the Brazilian government positions itself on the weaker side in global governance, i.e. within the country group that has been deprived from influence, power asymmetries between industrialized and developing countries are a central analytical category in its view on global governance. It claims that industrialized countries dominate global governance to the detriment of developing countries, shaping transboundary policy-making in a way that benefits primarily their own self-interests. For the Brazilian government, increasing the influence of developing countries is a central precondition for the improved functioning of global governance. In its view, global renewable energy governance is closely embedded in general power asymmetries between industrialized and developing countries. Thus, it explicitly addresses conflicts and power asymmetries between industrialized and developing countries in transboundary policy-making on renewable energy. Moreover, it assesses how transboundary policy-making on renewable energy may affect general power asymmetries in global governance. In line with this, it regards the relative weight of developing countries vis-à-vis industrialized countries as a central criterion to assess global governors on renewable energy.

The German administration locates itself on the stronger side in global governance, i.e. within the country group with most influence. Therefore, it is not negatively affected by power asymmetries between industrialized and developing countries. It might be for this reason that its deliberations on global renewable energy governance do not contain explicit references to power asymmetries between industrialized and developing countries. It rather justifies the leadership of industrialized countries, underlining that these have a duty to lead and support developing countries through North-South transfers of knowledge, technologies and finance. As such, it presents the leadership of industrialized countries in global renewable energy governance as part of a transboundary division of labor that is also beneficial to developing countries. In line with this, the German government presents transboundary policy-making on renewable energy as a win-win option which is to the mutual benefit of all involved. The Brazilian government, by contrast, expresses its caution vis-à-vis global action on renewable energy.

In the Brazilian case, another aspect comes in. The government has experienced conflicts between global governance and its national sovereignty. This particularly applies to global environmental governance. Here, demands formulated at the global level have interfered with its domestic policy priorities, as the following statement illustrates:

> "Since it has vast natural resource reserves—including the largest reserves of fresh water—and is the biggest repository of biodiversity on the planet, Brazil is the target of constant attention. The focus of international public opinion, concentrating on the preservation of natural resources, collided with the Brazilian emphasis on industrial and agricultural development. Since Stockholm, the international perception that Brazil is unable to preserve its extraordinary herit-

age has been consolidated. This perception became stronger in the following years, and was aggravated in the second half of the 1980's due to the repercussions of the intensifying forest burnings in the Amazon" (Corrêa do Lago 2009:19).

With regard to the Brazilian energy system, the construction of large hydropower plants has met much international environmental criticism. Pimentel (2011:81) claims that powerful attacks by transboundary environmental NGOs targeted Brazilian hydropower plants, exacerbating their negative aspects in worldwide public opinion. In recent times, the country's biofuel production has also been in the spotlight of international criticism. As a consequence of these experiences, the government does not wish to transfer too much decision-making power at the global level. It rather emphasizes that global renewable energy governance must respect the national sovereignty of states. As such, it should not regulate domestic matters. The administration particularly emphasizes that global renewable energy governance should not impose conditionalities on developing countries which limit their freedom to pursue socio-economic development. The German government, by contrast, has not experienced global policy-making interfering in its domestic policies on renewable energy. Consequently, it does not raise sovereignty concerns.

Different involvement in transboundary policy-making on renewables helps to understand the varying perspectives on global governors dealing with renewable energy. There is a clear relationship between the relevance attached by both governments to certain global governors and their own involvement in the work of these governors. Both governments attach high priority to only those global governors which they join as members. Those with whom they are not directly involved are seen to have limited or no relevance. This applies to IRENA in the Brazilian case and IBF in the German case. In addition, one of the main points of criticism that the Brazilian government raises with regard to IRENA is that the founding process was not democratic. In its opinion, the process was dominated by European countries and did not offer enough opportunities for involvement for countries such as Brazil. The divergent weight that the IEA receives in the deliberations of both governments also reflects the different involvement of each administration in IEA's work. The German government, which is an IEA member, ranks the IEA among the most important global governors on renewable energy, whereas the Brazilian government, a non IEA-member that only participates in some IEA-implementing agreements, rarely mentions this global governor.

Considering the global policy context furthermore contributes to a better understanding of why the German government's deliberations on global environmental protection concentrate on climate protection, yet do not mention the domestically dominant topic of phasing out nuclear energy. Here, it clearly adapts itself to the political priorities prevailing globally. Climate change has been the dominant issue in global environment governance. The phasing out of nuclear energy, by contrast, is not on the global environment governance agenda, as government assessments on nuclear energy differ widely.

The global policy context also affects the German administration's view on the pros and cons of biofuels and the relationship between biofuel promotion and food security. Here, the German government's main concerns do not relate to possible negative impacts in Germany, but to problems that might arise in developing countries.

6.2.4 Institutional Context: Division of Competencies

The division of competencies for global renewable energy governance within both governments also contributes to ideational differences. In Germany, the competencies on global renewable energy governance are divided between various ministries. From 2002 to the end of 2013, the BMU took the lead in transboundary policy-making on renewable energy. It was also tasked with the domestic promotion of renewables. The BMZ is responsible for German development cooperation on renewable energy. It also participates in most of the government's actions on global renewable energy governance, though with a lower degree of engagement than the BMU. The AA only plays a minor role in the administration's action on global renewable energy governance. An exception to this was the German initiative for IRENA's creation. In this, the AA was actively involved on the diplomatic side but did not engage much on substantive issues. With regard to transboundary cooperation on bioenergy, the BMELV comes in as well. It holds the core responsibility for the domestic promotion of bioenergy. As such, bioenergy is the only renewable energy option in which the BMU did not take the lead, neither in domestic policy-making nor in transboundary cooperation (Brand-Schock 2010:249,253 Hirschl 2008:93). From 2002 to the end of 2013, the BMWi was only in charge of renewable energy technology exports. It did not directly engage in the German initiatives on global renewable energy governance. In general terms, global renewable energy governance has not been a political priority for the German government. Consequently, the Chancellor has rarely participated in the administration's action on global renewable energy. A notable exception was Chancellor Schröder's involvement in the WSSD 2002 and the Renewables 2004 Conference.

In Brazil, the Foreign Ministry Itamaraty is in charge of global energy governance. This distinguishes the Brazilian government from many other that often have divided responsibilities for transboundary energy relations. During the Lula da Silva Government, the president's office (Casa Civil) was actively engaged in biofuels diplomacy. Lula da Silva took a very active stance in Brazilian foreign policy in general. His presidential diplomacy was shaped by the foreign policy agenda of the Brazilian Workers' Party PT (Zilla 2011b:11). Dilma Rousseff, by contrast, engages less in foreign policy-making and barely makes statements on transboundary renewable energy cooperation. The Itamaraty closely coordinates its action on renewables with the MME and MAPA. Further ministries are consulted as well but only play a minor role.[378]

The division of competencies influences how transboundary policy-making on renewable energy is connected to other policy issues, both domestically and globally. It both expresses and reinforces how governments relate the promotion of renewable energy to other issue areas. In the German case, the fact that the BMU was in charge of renewable energy is an expression of the close relationship that the government draws between renewable energy promotion and environmental protection. Yet, this allocation of competencies also reinforced further the perspective of renewable energy promotion being an environmental policy. As a consequence, environmental issues are of central importance in the German government's statements on global renewable energy governance. This is particularly apparent with regard to global challenges and the criteria to assess different renewable energy options. The German government points to climate protection as the main global challenge that renewable energy promotion shall address, referring to further environmental challenges that the Brazilian government blocks out (combating deforestation and preserving biodiversity) and presents environmental sustainability as the main criterion to assess different renewable energy options. In this context, it is important to note that all representatives of the German government, independent of their institutional affiliation, emphasize the importance of environmental protection and particularly climate protection when speaking about global renewable energy governance. However, it is probable that a different division of competencies would have reduced the relative weight of environmental issues in the government deliberations. As the competency for renewable energy was re-transferred from the BMU to the BMWi in December 2013, it is likely that in the future, economic issues will gain more importance in the German government's deliberations. In the Brazilian case, the environment ministry has only a minor influence on the government's engagement in global renewable energy governance. Again, the minor role that the MMA plays in Brazilian ethanol diplomacy can be interpreted as both a cause for, and a result of, the low priority given to environmental issues.

As the Itamaraty has the sole competency for transboundary policy-making on renewable energy, for the Brazilian government global renewable energy governance is closely embedded in foreign policy-making. Thus, the administration's statements on global renewable energy governance are strongly influenced by the background and experience of its diplomats. As Lula da Silva actively engaged in Brazilian ethanol diplomacy, his government deliberations were furthermore shaped by the foreign policy background of his party PT. This is particularly visible in the

---

[378] Among these are the MCTI, MDIC, MMA and MDA.

deliberations on the salient features. Itamaraty's approach to foreign policy-making has been characterized by suspicion about the dominance of industrialized countries, and particularly the USA, in global politics (Hurrell, Narlikar 2006:430). Under the presidency of Lula da Silva, North-South conflicts and global power asymmetries became dominant issues in the country's foreign policy (Christensen 2013:274, Plagemann, Röhrkasten 2013:104, Soares de Lima, Hirst 2006:22, Zilla 2011b:21-26). In addition, Brazilian foreign policy has traditionally placed a strong emphasis on national sovereignty (Hurrell, Narlikar 2006:430). Viola (2013:5) states that within the Brazilian government, traditional conceptions of national sovereignty particularly prevail in the Itamaraty.

With regard to global challenges, the Brazilian government's emphasis on socio-economic development in developing countries and agricultural development corresponds to the developmentalist tradition in Brazilian foreign policy (Dauvergne, Farias 2012:903,906-908, Hirst 2006:22, Soares de Lima, Vieira 2013:383f., Zilla 2011b:5f.). The importance of trade issues in its foreign policy, particularly with regard to agricultural goods, is also mirrored in its deliberations on global renewable energy governance. This is particularly the case in Lula da Silva's statements which put trade issues at the top of his presidential diplomacy agenda (Christensen 2013:274, Viola 2013:7, Zilla 2011b:27-30). Trade issues are also of central importance in the government's deliberations on barriers and tasks for transboundary policy-making on renewable energy.

The division of competencies seems to reinforce the focus on distinct forms of discrimination in the government deliberations on the barriers to a worldwide promotion of renewable energy. Between 2002 and 2013, the German government split competencies for renewable and conventional energy. While the BMU was in charge of renewables, conventional energy was the responsibility of the BMWi. As such, the BMWi acted as BMU's main counterpart and opponent in German energy policy. This institutional competition might have intensified the perceived rivalry between renewable and conventional energy.[379] This strong rivalry is also reflected in criteria used by the German government for assessing different global governors. Here, it focuses on the relative strength of renewable vis-à-vis conventional energy. BMWi staff reveals in interviews that it does not share this perspective. This becomes particularly apparent in the assessment of IRENA and IEA. While official government declarations praise IRENA's exclusive focus on renewables and criticize the IEA for providing insufficient support for renewables, BMWi staff does not share this criticism but rather questions IRENA's value added vis-à-vis IEA. In the Brazilian administration, competencies for different energy sources are not split.

---

[379] Within the German government, the BMU had been an early advocate of renewable energy (Brand-Schock 2010:55, Hirschl 2008:188f., Laird, Stefes 2009:2623). Its establishment after the nuclear accident in Chernobyl and its responsibility for climate protection endowed the BMU with a strong institutional interest in substituting conventional energy (Brand-Schock 2010:55).

The MME is responsible for domestic policies on both renewable and conventional energy, while Itamaraty's Energy Department is in charge of transboundary policy-making on all kinds of energy sources. Thus, in the Brazilian government there is no institutional competition between renewable and conventional energy. In its deliberations on the barriers to the worldwide promotion of renewables, it instead points to the discriminatory trade policies of industrialized countries, an aspect that is generally criticized by Brazilian diplomats. According to Christensen (2013:274), in Brazilian foreign policy the actions of industrialized countries "are largely treated as barriers to the realization of its (…) aims."

In the German case, the division of competencies on renewable energy has further implications on the government's perspective of global renewable energy governance, particularly with regard to biofuels. The ministries that took the lead in global renewable energy governance, BMU and BMZ, are more critical about the impacts of biofuels than ministries such as the BMELV and the BMWi. In addition, BMU and BMZ are ministries with relatively close NGO interaction. This facilitated NGO access to the German government when they began to campaign against biofuels. Thus, the division of competencies might have reinforced the administration's perspective of biofuels posing threats to global food security and environmental protection. In addition to that, the BMU and BMZ departments that were in charge of transboundary policy-making on renewable energy, particularly of the German initiatives IRENA and REN21, did not deal with biofuels. Therefore, they were not in charge of GBEP.[380] This might explain the minor attention that biofuels and the GBEP receive in government statements on global renewable energy governance.

The division of competencies not only influences how the two governments construct their context of action, it also affects the self-interests that each pursue in global renewable energy governance. The next subsection elaborates further on these interests.

6.2.5 Self-Interests

The ideational differences on global renewable energy governance are also influenced by the self-interests both governments pursue in transboundary policy-

---

[380] In the BMU, the Division "International and European Affairs of Renewable Energy" was in charge of transboundary policy-making on renewables in general. However, the Division "Solar energy, biomass, geothermal energy, market stimulation programmes for renewable energies" dealt with GBEP. In the BMZ, the Division "Water; Energy; Urban development; Geosciences sector" has the main responsibility for global renewable energy governance, while the Division "Rural Development, Agriculture and Food Security" is in charge of GBEP.

making on renewable energy. The following sections present these self-interests, building on the governments' deliberations on global renewable energy governance.

Germany

The German government seeks to position itself as a pioneer in the worldwide promotion of renewable energy. To demonstrate its pioneering role, it primarily highlights its political achievements at the domestic level. Here, it particularly refers to the EEG.[381] In recent statements, it also presents the 2011 'Energiewende' decision (Altmaier 2012).[382] To underline the success of its policies to promote renewables, it points to the rapidly increasing share of renewable energy in German electricity generation, economic benefits such as jobs creation, cost reductions for renewable energy and reduced greenhouse gas emissions. It furthermore emphasizes that the promotion of renewables has led to the emergence of an important and internationally competitive economic sector in Germany (Altmaier 2012, BMU 2009a, BMWi, Dena 2010:4,9, BMZ 2008:16, Westerwelle 2012, Wieczorek-Zeul 2009 [2001]:39).[383] To demonstrate its pioneering actions, the government also underlines its transboundary engagement. Here, it particularly refers to its development cooperation (BMU, BMZ 2012a, BMZ 2008:15f., Schröder 2004, Wieczorek-Zeul 2009).[384] In addition, it underlines its contribution in strengthening multilateral cooperation on renewable energy. Here, common examples are the organization of the Renewables 2004 Conference and the founding of IRENA (BMZ 2008:16, Trittin 2004c).[385]

The German government seeks to serve as a positive example of policymaking – primarily for other governments but also for international organizations. It repeatedly underlines that other governments can learn from its policies to promote renewable energy (Altmaier 2012, BMU, BMZ 2012b:6, Merkel 2012a,

---

[381] See, for example, (Schröder 2002c, Schröder 2004, Trittin 2004a, Wieczorek-Zeul 2009 (2001):39, BMWi, Dena 2010:4, Trittin 2004c, Schröder 2002b) as well as AA, energy security, http://www.auswaertiges-amt.de/EN/Aussenpolitik/GlobaleFragen/Energie/Energiesicherheit_node.html (accessed January 10, 2014).

[382] AA, energy security, http://www.auswaertiges-amt.de/EN/Aussenpolitik/GlobaleFragen/Energie/Energiesicherheit_node.html (accessed January 10, 2014).

[383] AA, energy security, http://www.auswaertiges-amt.de/EN/Aussenpolitik/GlobaleFragen/Energie/Energiesicherheit_node.html (accessed January 10, 2014).

[384] BMZ, promoting renewable energies and energy efficiency, http://www.bmz.de/en/what_we_do/issues/energie/index.html?PHPSESSID=ac135f430d4aa1ca857cb47dbc8f5e6e (accessed January 10, 2014).

[385] AA, energy security, http://www.auswaertiges-amt.de/EN/Aussenpolitik/GlobaleFragen/Energie/Energiesicherheit_node.html (accessed January 10, 2014); BMZ, international energy policy, International Renewable Energy Agency (IRENA), http://www.bmz.de/en/what_we_do/issues/energie/international_energy_policy/irena/index.html (accessed January 10, 2014).

Wieczorek-Zeul 2009 [2001]:39).[386] The following statement by Chancellor Angela Merkel is a case in point:

> "If we in Germany manage – and I'm convinced we can – to shift from fossil fuels and nuclear energy supplies first by phasing out nuclear energy and then by reducing the proportion of fossil fuels in the coming decades by moving towards renewable energies, then we will make a contribution insofar as other countries can learn and benefit from these experiences" (Merkel 2012a).

The AA emphasizes that there are already many countries that follow the German example, having introduced feed-in tariffs and adopted EEG principles. It adds that it is often merely a lack of knowledge that prevents further countries from following on the same path.[387] Chancellor Merkel claims that if the German energy transition is successful, "it can point the way towards a more sustainable energy supply for other countries too".[388] Wieczorek-Zeul (2009 [2001]:39) argues that the World Bank should promote the German model of feed-in tariffs instead of following the US American example of quota systems for renewable energy. Hirschl (2008:415,545,588) underlines that for the German government (and particularly for the BMU), establishing itself as an international pioneer of renewables promotion and as a positive example of policy-making also serves important domestic functions. As it sheds positive light on domestic policies, it can contribute to stabilizing or even strengthening these policies. Hirschl (2008:588-590) adds that engaging in transboundary policy-making on renewable energy also enabled the BMU to expand its institutional competences.

The government also refers to economic self-interests. It emphasizes that its economy benefits from the worldwide expansion of renewables, as it increases the demand for German exports of renewable energy technologies (Altmaier 2012, BMWi, Dena 2010:9, BMU 2009a, BMU 2009b, Wieczorek-Zeul 2009 [2001]).[389] However, it is somewhat hesitant in explicitly linking economic self-interest to its engagement in global renewable energy governance. This might be reinforced by the division of competencies, as the strengthening of the German economy is not the major institutional goal of BMU and BMZ. The following statement by Chancellor Angela Merkel serves as an illustrative example. Merkel refers to economic self-

---

[386] AA, energy security, http://www.auswaertiges-amt.de/EN/Aussenpolitik/GlobaleFragen/Energie/Energiesicherheit_node.html (accessed January 10, 2014).
[387] AA, energy security, http://www.auswaertiges-amt.de/EN/Aussenpolitik/GlobaleFragen/Energie/Energiesicherheit_node.html (accessed January 10, 2014).
[388] The Federal Government, Petersberg Climate Dialogue, Our shared responsibility for the global climate (Tuesday, 17. July 2012), http://www.bundesregierung.de/Content/EN/Artikel/2012/07/2012-07-13-petersberger-dialog-2.html (accessed January 10, 2014).
[389] AA, energy security, http://www.auswaertiges-amt.de/EN/Aussenpolitik/GlobaleFragen/Energie/Energiesicherheit_node.html (accessed January 10, 2014); BMWi, Internationale Energiepolitik, http://www.bmwi.de/DE/Themen/Energie/Energiepolitik/internationale-energiepolitik.html (accessed January 10, 2014).

interests, but explicitly links them to moral duties, emphasizing the government's global responsibility:

> "Let me say very clearly that my vision of Germany and Europe taking a leading role also has an ethical dimension. Of course, taking that role is partly about safeguarding our own standard of living. But it is also our moral duty to conduct test phases, to learn how best to deal with the complex of new energy supplies, resource efficiency and efficient technology, and to subsidize progress. After all, while other countries did not yet have the wherewithal to pursue the same prosperity as we enjoyed, we spent many years and decades overexploiting the world's resources. With that in mind, we have a duty to redress the balance somewhat. I feel that we should step up to that duty and, what's more, turn it to our advantage" (Merkel 2012b).

According to the expert interviews, economic self-interests did not play a major role in the German government's main initiatives to shape global renewable energy governance. A case in point is that these initiatives did not meet much interest in the German renewables industry (Hirschl 2008:482,526).[390] As interviewees underline, the administration has other channels that facilitate a more direct support for the transboundary expansion of the industry, such as its export initiative on renewable energy.[391] Government representatives that were involved in IRENA's founding process stress that it was very important for the success of this initiative to convince other governments that Germany did not initiate IRENA's creation in order to advance its own economic interests.[392] In the rare statements on global renewable energy governance that primarily target the German public, economic self-interests receive a more prominent place. In her Government Declaration in 2005, Chancellor Merkel states that her engagement for global climate protection not only serves the purpose of mitigating climate change, but also contributes to exporting German energy technologies.[393] In its election program of 2002, the SPD directly links the call for an international renewable energy agency with the promotion of German energy technologies (SPD 2002:35). However, in the coalition agreements of 2002 and 2005, the call for such an agency is not linked to German economic self-interests (SPD, Bündnis 90/Die Grünen 2002:37, CDU, CDU & SPD 2005:42).

Brazil

The Brazilian government also seeks to establish itself as a frontrunner in the worldwide promotion of renewable energy. Representatives of the Brazilian gov-

---

[390] Interviews with with former IRENA special envoy (#1), AA, March 2013; staff member, international cooperation, German Solar Industry Association, November 2013; board member, German Renewable Energy Federation, December 2012.
[391] Interviews with with former IRENA special envoy (#1), AA, March 2013, and staff member, international cooperation, German Solar Industry Association, November 2013.
[392] Interviews with with former IRENA special envoy (#1), AA, March 2013
[393] Die Regierungserklärung der Bundeskanzlerin Angela Merkel vom 30. November 2005 im Volltext, provided by Institut1, http://www.institut1.de/702.html (accessed January 10, 2014).

ernment repeatedly point out that the country has one of the cleanest energy matrixes of the world.[394] The statement by President Lula da Silva at the UN General Assembly is symbolic for this:

> "Brazil's energy blend is in one of the cleanest in the world. 45% of the energy my country consumes is renewable. In the rest of the world, only 12% is renewable, while no OECD country has a rate higher than 5%" (Lula da Silva 2009b).

The Brazilian government also highlights its pioneering activity in the promotion of biofuels. Here, it particularly refers to its long-lasting experience, economic competitiveness and technological capacities (Amorim 2007b:6, Kloss 2012:27, Simões 2007a:16). Simões (2007a:16) states that the country's technological capacity in the production of biofuels is superior to any other international player. Amado (2010b) even claims that "Brazil is nowadays the only country in the world where gasoline – and not ethanol – is the alternative fuel" (translation S.R.).[395]

By establishing itself as a frontrunner in the worldwide promotion of renewable energy, the Brazilian administration seeks to leverage its influence in transboundary policymaking and promote socio-economic development in Brazil (see, for example, Feres 2010:35, MRE 2003a, Pimentel 2011:23,194, Simões 2007b:16). The former Head of Itamaraty's Energy Department, Antônio Simões, highlights the importance of the country's comparative advantages in the production of biofuels for the pursuit of these two goals:

> "Brazil's comparative advantage with regard to biofuels can be crucial for consolidating the development of the country and its new role in the world" (Simões 2007a:11, translation S.R.).[396]

Several experts reinforce the fact that during the strongest phase of Brazilian ethanol diplomacy, ethanol was seen as an important foreign policy tool for the government (Bastos Lima 2012, Johnson 2010:37-45).[397]

For the Brazilian administration, its clean energy matrix and particularly its position as an international pioneer in the production of biofuels are important assets to consolidate and strengthen its influence in global governance.[398] It empha-

---

[394] See, for example, Amado 2010b, Lula da Silva 2007c, Lula da Silva 2008b, Simões 2007a:17; Speech by Ambassador Luis Alberto Figueiredo Machado, Itamaraty's Under-Secretary for Environment, Energy, Science, and Technology at the Bioenergy Week in Brasília, March 2013; Speech by Marcos Sawaya Jank, Ex-president of UNICA and ICONE, at Embrapa, Brasília, March 2013.

[395] Original text of this quotation: "o Brasil é, hoje, o único país do mundo em que a gasolina - e não o etanol - é o combustível alternativo".

[396] Original quotation in Portuguese: "A vantagem comparativa do Brasil em relação aos biocombustíveis pode ser fundamental para a consolidação do desenvolvimento do País e do seu novo papel no mundo."

[397] Interviews with Head of the New and Renewable Energy Resource Division, Itamaraty, March 2013; team, Bioenergy Projects, FGV, February 2013; Researcher on applied economics, FGV, March 2013; professor, CPDA, UFRRJ, March 2013.

[398] According to a FGV researcher on applied economics, energy and natural resources are the major asstes that Brazil enjoys in the international arena (interview, March 2013).

sizes that Brazil's clean energy matrix qualifies the government to play a leading role in global governance on energy and climate issues (MRE 2010b:1, MRE 2003a, Pimentel 2011:209f., Simões 2007b:17).[399] Simões (2007a:16) claims that its pioneering activity in the production of biofuels can help Brazil to transform itself into a "potência energética de primeira grandeza".[400] At the 2009 meeting of the UN General Assembly, President Lula da Silva (2009b) reinforces that the Brazilian government will consolidate its "role as a world power in green energy", notwithstanding its fossil fuel discoveries. Pimentel (2011:181) states that if Brazil exploits its pre-sal discoveries and bioenergy potentials, it can assume the position of energy superpower and as such amplify the scope of its diplomacy. He adds that assuming the role of energy superpower would not only upgrade the country's image in the world but also endow the government with more influence in important international processes, such as the UN Security Council and the Bretton Wood institutions (Pimentel 2011:184, 187f.). The administration argues that its domestic deployment of renewable energy demonstrates that Brazil assumes international responsibility and contributes to tackling global challenges (Amado 2010b, Lula da Silva 2008b, Lula da Silva 2008e, Lula da Silva 2007c). It furthermore presents the Brazilian experience as a positive example for other countries (Amorim 2007b:6, Lula da Silva 2008b, Lula da Silva 2007a, Pimentel 2011:209f.). According to Lula da Silva (2008a, 2008c) those factors that make its biofuel production competitive – such as technologies and favorable climatic and geographic conditions – are not a Brazilian privilege; biofuel production rather offers more than 100 developing countries a promising development alternative.

Kloss (2012:183, 186) emphasizes three distinctive features that transboundary policy-making on biofuels has on Brazilian diplomacy. First of all, it is an area where the country is recognized as a reference point by a number of governments and international organizations. This aspect is reinforced by Lula da Silva (2008c) in his speech at the FAO meeting in 2008. Secondly, it is an area in which Brazil can assume the role of donor country, promoting technologies and know-how to interested developing countries. This aspect is also stressed by Lula da Silva (2008a) and further government representatives (Feres 2010:101).[401] They underline that many developing countries are interested in the Brazilian experience of biofuel production as well as in its political, technological and economic know-how. Thirdly, it is one of the few areas where the country can interact with the USA and EU on equal grounds. Thus, it is an area in which the Brazilian government can chal-

---

[399] Interview with Secretary and team, Secretariat of Planning and Energy Development, MME, March 2013.
[400] In his government program, Lula da Silva also refers to Brazil as "energy power" (potência energética) (Lula da Silva 2006:18f.).
[401] Interview with General Coordinator for Sugar and Ethanol, MAPA, March 2013; Speech by the Head of the Energy Department, Itamaraty, at the Bioenergy Week in Brasília, March 2013.

lenge the OECD-led global governance architecture, demonstrating that it is no longer OECD countries that decide on economic issues and later transfer these decisions to other countries (Kloss 2012:58). In line with this, representatives of the Brazilian government usually compare the country's renewables share to the much lower OECD average (see, for example, Amado 2010b, Lula da Silva 2009b, Lula da Silva 2008b, Lula da Silva 2007c, Simões 2007a:17).[402] In addition, the administration considers the promotion of biofuels as an area that is particularly suited to establish itself as a leading actor within the group of developing countries. It even declares itself as the voice of developing countries in transboundary policy-making on biofuels (Feres 2010:35, Kloss 2012:172, MRE 2010b:2).

A MME representative[403] adds that for the Brazilian government, emphasizing its own achievements and presenting itself as a frontrunner at the global level also serves the purpose of communicating the success of its policies at the domestic level (see, for example, Amado 2010a, Amorim 2008, Amorim 2007b:6, Feres 2010:108, Kloss 2012:27). In keeping with this, Amado (2010a) even claims that "the bet on sugarcane ethanol is one of the most successful Brazilian public policies of all times" (translation S.R.).[404]

The Brazilian government explicitly communicates its economic self-interests in global renewable energy governance: it seeks to open up export markets for its ethanol production and biofuel technologies (see, for example, Kloss 2012:53,160f., Lula da Silva 2006:19, MRE 2010b:2).[405] An important goal of its ethanol diplomacy is to turn the country into a major exporter of ethanol.[406] The government furthermore regards biofuels as an area in which it can incentivize innovation and industrial development and thus strengthen the technological capacities of the economy (MRE 2007:325). Thus, representatives of the Brazilian government emphasize the importance of maintaining international leadership in biofuel technologies (Feres 2010:108, Kloss 2012:193). The importance of economic self-interest in the country's ethanol diplomacy is reinforced by the close interaction between the government and the Brazilian Sugarcane Industry Association, UNICA (Sanchez Badin, Godoy 2013:29). It is also underlined in the academic literature on Brazilian ethanol diplomacy (Johnson 2010, Sanchez Badin, Godoy 2013, Schutte,

---

[402] Speech by Ambassador Luis Alberto Figueiredo Machado, Itamaraty's Under-Secretary for Environment, Energy, Science, and Technology at the Bioenergy Week in Brasília, March 2013; Speech by Marcos Sawaya Jank, Ex-president of UNICA and ICONE, at Embrapa, Brasília, March 2013.
[403] Interview with Director of the Department of Renewable Fuels, MME, March 2013.
[404] Original text of this quotation: "a aposta no etanol de cana-de-açúcar está entre as mais acertadas políticas públicas brasileiras em todos os tempos".
[405] Kloss (2012:160f.) for example affirms that the Brazilian government argued for the inclusion of biofuels into the WTO list of environmental goods because these are goods that Brazil exports and produces with comparative advantages.
[406] Interviews with Head of the New and Renewable Energy Resource Division, Itamaraty, March 2013; Vice-President, CEBRI, April 2013; professor (#1), Energy Institute, USP, April 2013.

Barros 2010,). Several representatives of the government state that biofuels lost influence within Brazilian diplomacy when the country's ethanol industry passed through economic difficulties and the opening of new export markets became less of a priority for its producers.[407]

In addition, the Brazilian government also seeks to prevent external interference in its internal matters. In its foreign policy review of the Lula da Silva government, the Itamaraty clarifies that an important aim of Brazilian energy diplomacy is to ensure that international norms do not reduce or question the country's opportunities to pursue socio-economic development (MRE 2010b:2).

Connecting the Self-Interests to the Global Governance Ideas

Taking self-interests into account is of crucial importance for understanding the concentration on distinct renewable energy options. This particularly applies to the Brazilian case. Its government concentrates on biofuels and, more precisely, on ethanol because it wishes to open up export markets for its ethanol industry and regards ethanol production as an important asset to strengthen its own influence in transboundary policy-making. As opposed to this, it does not want transboundary policy-making to focus on hydropower. Its hydropower dams have met much international criticism and the Brazilian government therefore fears that global governance on hydropower could interfere with its domestic policy priorities. As such, it does not touch on hydropower in its statements on global renewable energy governance (Corrêa do Lago 2009:171f., Hirschl 2008:544, Vargas 2004:1f.). In the German case, self-interests are less decisive. Here, they rather seem to reinforce the focus that is already prevailing at the domestic level. The concentration on the electricity sector is in line with the government's political self-interests as the political 'showcase' for its international profiling is renewable energy promotion within the electricity sector, particularly the EEG. Considering the German export interests contributes to a better understanding of why the government's deliberations on global renewable energy governance particularly focus on wind and solar energy. Solar energy provides the bulk of the country's exports of renewable energy technologies. In 2011, solar energy accounted for 73.9 percent of its renewables-related export volume. Wind energy followed by a wide margin, its share amounting to 16.9 percent. All other renewable energy technologies (such as hydropower, thermal heat pumps, biomass and biogas) together only account for 9.2 percent of the export volume (Gehrke, Schasse 2013:16).[408] According to Groba & Kemfert (2011), Ger-

---

[407] Interviews with Secretary and team, Secretariat of Planning and Energy Development, MME, March 2013; Biodiesel Federal Program Coordinator and staff member, Analysis and Following up of Government Policies, Casa Civil, March 2013; Head of Energy Department, Itamaraty, April 2013; Manager, Climate Change Department, MMA, March 2013.

[408] In 2002, the German renewables-related export volume was divided up as follows: solar energy (71.9%), wind energy (11.2%), other renewable energy technologies (16.9%).

man producers of wind and solar technologies are among the world's leading exporters.

The self-interests of both governments also contribute to a better understanding of the ideational differences concerning barriers and tasks for transboundary policy-making. As the Brazilian government has a direct economic interest in the liberalization of biofuel trade and particularly in access to the markets of industrialized countries, it points to the protectionist policies of developed countries as a major barrier to the worldwide spread of biofuels and presents the creation of an international biofuels market as the major task for transboundary policy-making. The German government's assessments of the main barriers and tasks conform to its political self-interest, particularly the endeavor to present itself as a positive example of policy-making. It points to inadequate policy frameworks at the domestic level as being the main barrier to the worldwide spread of renewable energy and states that transboundary policy-making should focus on improving these frameworks, particularly by providing policy advice. The Brazilian administration, by contrast, does not want transboundary policy-making to focus on policy advice, as it does not want to receive external policy advice itself. In the Brazilian case, the emphasis placed by the government on the promotion of socio-economic development in developing countries and particularly on their export activities is also in line with its self-interests.

Taking self-interests into account also enables to understand the divergent assessments of the pros and cons of biofuels and the contrary relationships that are drawn between biofuel production and hunger. The influence of self-interests is particularly salient in the Brazilian case. As the government seeks to open up new export markets for its ethanol producers, it is interested in spreading a positive image of biofuel production. Thus, it does not mention negative consequences of biofuel deployment, but only refers to positive impacts. In the German case, there is no explicit causal relationship between economic self-interests and the assessments of the pros and cons of biofuels. However, there is a certain congruence (see also Acosta, Zilla 2011). Biofuel production in Germany is not internationally competitive, and there are trade barriers in place that help to shield domestic biofuel producers against competition from other countries. Competition from developing countries with comparative advantages in the production of biofuels could threaten Germany's domestic biofuels production. The emphasis that the government places on the risks of biofuels deployment helps to justify the trade barriers that are in place, in particular because the risks that it refers to – such as loss of biodiversity, food price increases, displacement of small farmers and poor working conditions on plantations – tend to weigh more strongly in developing countries. While it might be difficult or even misleading to sustain the thesis that the policy actors within the government that underline the risks of biofuel production – particularly the BMU and BMZ – do so with the *purpose* of shielding domestic producers against competition, it cannot be neglected that the emphasis on risks actually helps to justify trade barriers. However, the emphasis on the risks of biofuels production also undermines the general acceptance of biofuels in Germany which in turn also harms domestic biofuel producers.

With regard to the relationship between biofuel production and hunger, another aspect comes in, namely blaming others. The Brazilian government ascribes hunger to a lack of income and weak agricultural sectors in developing countries which – according to its view – are caused by protectionism of industrialized countries. It is striking that the Brazilian government blames industrialized countries as responsible for the underlying causes of hunger while its German counterpart refers to factors that lie in the responsibility of developing countries, such as land use regimes.

Similarly, the blind spots in government deliberations on global challenges might also be a result of the reluctance to address these issues in transboundary policy-making. The Brazilian government is not interested in placing deforestation and biological diversity on the agenda as it wants to avoid external influence on these matters. The German administration, by contrast, might not want to address agricultural development in developing countries and international agricultural trade in order to avoid discussions about industrialized countries' agricultural protectionism.

The different perspectives concerning the salient features of global renewable energy governance are also in line with self-interests. Again, this particularly applies in the Brazilian case. As the government seeks to strengthen its influence in global governance and establish itself as a leading actor within the group of developing countries, it calls for this group to have more influence and thus challenges the leadership of industrialized countries. Therefore, it explicitly addresses conflicts and power asymmetries between industrialized and developing countries in global renewable energy governance. The German government, by contrast, has no direct self-interest in addressing this as it belongs to the group of countries that enjoy more influence in global governance. Moreover, the Brazilian government's emphasis on national sovereignty corresponds to its own desire to avoid external demands restricting its freedom to promote socio-economic development. The German government, by contrast, does not fear such restrictions. It rather regards the strengthening of global governance as a means to expand political influence beyond the national border.

With regard to the different priority given to global governors, it can be noticed that both governments rank those global governors founded by them as among the most relevant: IRENA and REN21 in the case of the German government and IBF in the case of the Brazilian. Here, emphasizing own political achievements might be an important motivation.

6.3 Overview of Ideational Differences and Reasons Behind

The following table (on page 276) summarizes the results of this chapter. It lists the identified ideational differences and relates them to the analytical categories that guided the quest for the reasons. These categories comprise the coherency between ideas, the different contextual factors (the domestic policy context, embeddedness in transboundary policy-making and the institutional context due to the division of competencies) as well as self-interests. The ideational differences that can be traced back to other ideational differences are marked by numbers. The analytical category that is considered most relevant for understanding an ideational difference is highlighted in grey. Analytical categories that provide additional insights are shaded with diagonal lines if they apply to both governments, and marked with (G) or (B) where they apply to the ideas of the German or the Brazilian government only.

Ideational differences	Reasons behind				
	Coher. betw. ideas	Dom. policy context	Emb. in trans. policy-m.	Div. of comp.	Self-interests
Global challenges					
Environment (climate) protection vs. socio-economic development (1)			(G)	//////	(B)
Relationship between renewable energy (RE) and conventional energy: replacement vs. cohabitation (2)		(1)			
Link to poverty reduction and economic development in developing countries: *access to energy vs. income generation *RE as an input/precondition vs. RE as an economic activity in itself *merely replacing imports vs. enhancing exports as well		(3), (8)	(B)		(B)
Blind spots: agricultural development vs. deforestation and biodiversity		(1), (3)		//////	
Renewable energy options					
*Electricity sector vs. transport sector *Solar and wind energy vs. ethanol/biofuels (3)				(G)	
Criteria for assessing RE options: environmental sustainability vs. economic competitiveness				(G)	
Perspective on economic competitiveness: *time horizon: long-term vs. status quo *production scale: small-scale vs. large scale					
*Different assessments of the pros and cons of biofuels *Contrary relationship between biofuels production and hunger (4)			(G)	(G)	//////
Barriers					
Different forms of discrimination: RE/conventional energy vs. developing/industrialized countries (5)		(?), (8)		//////	
Inadequate dom. policy frameworks vs. trade barriers		(8)			(B)
*Caution in blaming actors vs. explicitly naming actors *Lacking political will and public awareness vs. unjustified concerns		(8)			
Tasks					
*Most important step for RE promotion: improving dom. regulatory frameworks vs. unleashing global market forces *Focus of transb. cooperation: committing and advising governments vs. opening up markets (6)		(8)			(B)
Function of internationally agreed sustainability standards on biofuels: risks prevention vs. ensuring demand		(4), (6)			

Global governors				
Relevance: IRENA,IEA and REN21 vs. IBF, GBEP, WTO and ISO	(6)		(G)	/////
Assessment: relative importance of RE vs. relative weight developing/industrialized countries (7)	(5),			/////
GBEP's focus: sustainability standards for environmental risk prevention vs. creating an international market	(4), (6)			
Different views on IRENA	(7)		(G)	
Salient features/meta-aspects (8)				
*Transboundary cooperation to the benefit of all involved vs. conflicts and power asymmetries *Leadership of industrialized countries: responsibility to lead vs. domination and pursuit of own benefits *The role of developing countries: reactive vs. active *How to benefit dev. count.: transfers vs. opening markets *Strengthening global governance vs. national sovereignty			(B)	/////

■ most relevant   ///// providing additional insights   (G)/(B) applying to the German / Brazilian gov. only

Table 7: Ideational Differences and the Reasons Behind

7 Conclusion

The present study demonstrates that the ideas of the German and Brazilian governments on global renewable energy governance differ considerably. The differences involve all elements that structured the analysis of global governance ideas: it includes the perspectives on the global challenges addressed by renewable energy promotion, the specification of and differentiation between different renewable energy options, the barriers to worldwide promotion of renewables, the tasks for transboundary policy-making, the relevant global governors and the salient features/meta-aspects in global renewable energy governance.

This study reveals that these ideational differences can be traced back to the policy contexts of the government actors in charge of global renewable energy governance, their self-interests and their efforts to draft ideas in a coherent way. With regard to the policy contexts, three different levels are important: the domestic policy context of renewable energy promotion, the embeddedness in transboundary policy-making and the institutional context resulting from the governments' division of competencies. The relevance of the reasons varies across the identified ideational differences. While Table 7 on pages 275 provides a comprehensive overview of the ideational differences and the reasons behind each of them, this chapter presents the main empirical findings and elaborates as to what these imply for academic research and policy-making in the realm of global renewable energy governance. It finally provides suggestions for further research.

7.1 Ideational Differences in Global Renewable Energy Governance

Global challenges. The German and the Brazilian governments both link the worldwide promotion of renewable energy to the global challenges that also prevail in academic discussions – energy security, environmental protection and poverty reduction (see chapter 3.2) – and add economic development to the picture. However, they clearly attach different priorities to these challenges. The German government prioritizes environmental protection, particularly climate change mitigation, whereas its Brazilian counterpart assigns priority to the promotion of socio-economic development in developing countries. The blind spots in the deliberations of both governments on the global challenges are in line with this. The German administration overlooks agricultural development which, according to the Brazilian government, is a central element of socio-economic development in developing countries, while the latter does not mention the environmental challenges of forest and biodiversity protection. The prioritization of different global challenges is furthermore mirrored in the relationship that each government draws between renewable and conventional energy: the German government refers to replacement, the Brazilian to complementarity and cohabitation.

The German focus on environmental protection and the Brazilian emphasis on socio-economic development are reflected in divergent approaches towards sustainable development. Both governments repeatedly point to sustainable development as an important goal for global renewable energy governance. However, when the German administration speaks about sustainability, it almost exclusively refers to the ecological dimension while its Brazilian couterpart clarifies that it prioritizes socio-economic development. The research results thus confirm the assertion of Bernstein (2012:368f.) that sustainable development is a "compromise [which] rests on an uncertain foundation." Despite more than twenty years of global governance efforts to align environmental protection and socio-economic development under the common roof of 'sustainable development', and thus bridge a decades-old North-South cleavage in global governance on environment and development (see, for example, Biermann 2007, Biermann 1998, Corrêa do Lago 2009), the different dimensions of sustainability are still not approached in an integrated manner. Even though declarations relating to the common goal of 'sustainable development' suggest that an agreement on the general direction of global governance efforts exists, it seems that they rather mask than merge different priorities.

Both governments furthermore draw different causal pathways as to how the promotion of renewables contributes to poverty reduction and economic development in developing countries. While the Brazilian government presents renewable energy provision as an economic activity in itself and underlines export opportunities for developing countries, the Germans regard it as an input or precondition for economic activities and merely refer to the trade benefits of replacing imports. With regard to poverty reduction, the German government points to access to energy, the Brazilian to income generation. The latter's assessment of the link between renewable energy promotion and poverty reduction contrasts not only with the German assessment, but also with the prevailing perspective in academic research on global energy governance. Academic discussions concentrate on the fight against *energy* poverty by providing energy infrastructure in remote areas (see chapter 3.2.2). The Brazilian government, by contrast, brings *income* poverty in and argues that the promotion of renewable energy should directly generate income for poor people by involving them in the economic process of generating renewable energy services.

Renewable energy options. In their deliberations on global renewable energy governance, the governments concentrate on renewable energy options in different energy sectors. The German administration focuses on the electricity sector and pays most attention to wind and solar energy. A case in point for the predominance of the electricity sector is that it often uses 'energy' and 'electricity' as synonyms. The Brazilian government, by contrast, concentrates on the transport sector and here on biofuels, particularly ethanol. It commonly uses 'biofuel' and 'ethanol' as interchangeable terms. The third energy sector, heating and cooling, does not receive much attention by either administration.

Both governments refer to the criteria of the energy policy triangle (see, for example, Lesage, Van de Graaf & Westphal 2010:37f.) to assess different renewable energy options –economic competitiveness, affordability and environmental friendliness – but regard different criteria as most important: economic competitiveness in the Brazilian case and environmental friendliness in the German. Moreover, the governments link economic competitiveness to different time horizons and production scales. The Germans refer to future potentials and small-scale options, the Brazilians to the status quo and large scale production.

Both present different renewable energy options as being worthy of support. The Brazilians underline the benefits of biofuels, above all sugarcane ethanol. The German government presents wind and solar energy as 'win-win options', while it raises trade-offs and thus caution and doubts with regard to biofuels and large hydropower. Small hydropower proved to be much less salient in the German deliberations than suggested in the author's previous research (Röhrkasten 2009).

The assessments on biofuels differ considerably. The governments even draw contrary relations between biofuels production and hunger, basing their arguments on different underlying causes of hunger. For the Brazilian government, the lack of income and weak agricultural sectors in developing countries are the main causes for hunger which the production of biofuels helps to tackle. For the German government, hunger is primarily a result of food scarcity, which biofuels production aggravates even further. While the Brazilians explicitly address the underlying causes of hunger, the German assumption of hunger caused by food scarcity serves as a good example for a problem definition that is taken for granted and enters political discussions only implicitly (Mehta 2011:34f.).

Barriers. Both point to different forms of discrimination that hinder the worldwide spread of renewable energy. The Germans refer to discrimination of renewables vis-à-vis conventional energy whereas the Brazilians point to the trade policies of industrialized countries which discriminate against developing countries. As such, they link the barriers to different areas of policy-making: trade in the Brazilian case and energy policy in the German. The barriers are furthermore located at different policy levels. While the Brazilian government addresses obstacles at the global level that directly involve transboundary interactions, its German counterpart primarily refers to domestic policy-making. Thus, the barriers mentioned by the German government do not necessarily comprise transboundary interdependencies.

This means from a theoretical point of view that they do not necessarily belong in the realm of global governance.

It is interesting to note that the Brazilian government explicitly defines *who* obstructs the worldwide promotion of biofuels, while the German administration does not hold specific policy actors accountable but rather speaks about barriers in vague terms. It merely complains about the lack of political will and public awareness, whereas the Brazilian government blames policy actors in industrialized countries, particularly those in Europe, for spreading unjustified concerns about biofuels which, in its opinion, are based on misinformation, distortions and prejudice. It further names two types of lobby groups that obstruct the promotion of biofuels: the agricultural lobbies of industrialized countries in addition to oil lobbies. As such, the Brazilian government clearly points to the politics behind the definition of global governance issues and associated struggles over the prerogative of interpretation. It claims that some policy actors who participate in this struggle behave in a dishonest way, purposely spreading false ideas.

Tasks. As in the deliberations relating to barriers, both governments choose different policy levels when they present leverage points for the worldwide promotion of renewables. They furthermore attach different weights to government interventions vis-à-vis market forces. The German administration points to the need for government interventions at the domestic level, its Brazilian counterpart to the unleashing of market forces at global level. According to the German government, the improvement of domestic regulatory frameworks is key for the worldwide promotion of renewables. It argues that transboundary policy-making should commit governments to the promotion of renewables within their own countries and support them to achieve their commitments. Here, it particularly refers to policy advice. The Brazilian government, by contrast, suggests that free trade is the most effective way to promote biofuels around the world. It wants transboundary policy-making to focus on the creation of an international biofuels market.

These differences touch a central question in global governance research, namely what types of actors are relevant in global governance. Following the claim of the Commission on Global Governance (1995:3) that it must involve all "actors who have the power to achieve results", global governance scholars explicitly open their analysis for different types of actors, comprising both state and non-state actors (see chapter 2.1.3). While they agree that governments are not the only relevant actors in global governance, their relative importance is highly contested in global governance research. In their deliberations on the subject, the German and the Brazilian governments opt for different assessments. While the Germans suggest that governments are primary actors for the worldwide promotion of renewables, the Brazilians present private actors as key. The latter insinuate that these dispose of the resources necessary to promote biofuels around the world and thus tackle the global challenges mentioned above.

Moreover, they present diverging perspectives on the purpose of internationally agreed sustainability standards for biofuels. Whereas the Brazilian government regards them as a means to ensure biofuels demand in industrialized countries, its German counterpart presents them as a means to tackle the negative side-effects of biofuels production, particularly in developing countries. As such, for the Brazilian government, sustainability standards primarily influence action in industrialized countries, while the German government points at action in developing countries.

Global governors. In line with academic discussions on global (renewable) energy governance (see chapter 3.4), both governments primarily refer to organizations with a transboundary set-up as the main global governors for the worldwide promotion of renewables. However, the relevance that each attach to individual global governors differs widely. The global governors presented by each administration as the most relevant – IRENA, IEA and REN21 in the German case and IBF, GBEP, WTO and ISO in the Brazilian – receive almost no attention in the public statements of the other. While the global governors highlighted by the Brazilian government focus on trade issues and/or deal exclusively with biofuels, those emphasized by the German administration belong to the realm of energy policy. It is interesting to note that the global governors presented by the Germans as most relevant also receive substantial attention in academic discussions on global (renewable) energy governance. Out of those mentioned by the Brazilians, this only applies to GBEP. Academic discussions on global (renewable) energy governance rarely cover trade issues[409] and therefore do not pay much attention to bodies such as the WTO and ISO. The scarce attention of IBF in academic discussions may be due to the informal character of this forum and its relatively short time in existence (2007-2010).

Moreover, both governments use different criteria to assess global governors. The Brazilians refer to the relative influence of developing countries vis-à-vis industrialized countries, while the Germans underline the relative importance attached by global governors to renewable energy in their overall portfolio or in comparison to conventional energy. Their diverging assessments of IRENA are partly in line with these criteria. The German government presents IRENA as the main global governor on renewable energy and particularly highlights its exclusive focus on renewable energy, while its Brazilian counterpart views IRENA with caution and criticizes that it has been influenced too strongly by European countries.

The governments furthermore characterize GBEP's thematic focus differently. The German government refers to environmental risk prevention, whereas the Brazilians point to the promotion of biofuels production and trade. While the Brazilian description matches more with GBEP's declared purpose, the German

---

[409] A notable exeption of this is Gosh (2011).

description relates to the main area of GBEP's activities, namely biofuel sustainability (see chapter 3.4.5). Here, the German focus on the environmental pillar of sustainability again becomes obvious.

Salient features. In their deliberations on global renewable energy governance, both governments reveal diverging assessments on the way global governance works. The German administration suggests that global governance is to the mutual benefit of all involved. The Brazilian government, by contrast, explicitly addresses conflicts and power asymmetries. Thus, their perspectives match different strands in global governance research. The German perspective is more compatible with global governance research that approaches global governance in terms of win-win options. The Brazilian view fits those approaches that criticize the power blindness of prevailing global governance research (see chapter 2.2.2).

Interestingly, the North-South divide is a defining feature in both perspectives on global renewable energy governance. If the governments differentiate between country groups, they typically refer to industrialized countries on the one hand and developing countries on the other. Thus, they use general country classifications that have been common in a wide range of areas in transboundary policymaking but do not take into account the specifics of the issue area. While the Brazilian government explicitly refers to, and strongly criticizes, the North-South divide, it enters the German statements rather implicitly. Yet, this implicitness does not mean that it is less dominant in the German perspectives on global renewable energy governance – it is just more subtle.

Both agree that industrialized countries are the *de facto* leading actors in global renewable energy governance. However, they judge this leadership differently. The German government justifies the *status quo* whereas its Brazilian counterpart denounces it. The German administration presents industrialized countries as frontrunners in the worldwide promotion of renewables in terms of economic and technological developments, and policy implementation. It argues that industrialized countries have a responsibility to lead and thus suggests that they are legitimate leaders in global renewable energy governance. The Brazilian government, by contrast, claims that industrialized countries dominate transboundary policy-making to the detriment of developing countries. It demonstrates a deep-rooted mistrust about the intentions of industrialized countries and sharply criticizes their dominance and arrogance in global renewable energy governance. Itamaraty officials complain that representatives from developing countries are not taken seriously in transboundary policy-making and are confronted with severe difficulties if they want their voice to be heard.

Both administrations furthermore attribute different roles to developing countries in transboundary policy-making. In the deliberations of the Brazilian government, they are assigned an active role, while in the German deliberations they are merely endowed with a reactive remit. In line with this, each reveals different perspectives on how transboundary cooperation best benefits developing countries. In the view of the German government, global governance on renewable energy is mainly about North-South transfers – not only of finance and technology, but also of ideas. The Brazilian government criticizes transboundary policy-making based on North-South transfers and claims that industrialized countries should rather open up their markets to allow for equal competition between industrialized and developing countries. As such, it argues for horizontal rather than vertical cooperation.

This study's analysis of global renewable energy governance (see chapter 3) seems to confirm that the North-South direction is the dominating logic of global action on renewables. While several global governance initiatives explicitly focus on the worldwide promotion of renewables, as for example IRENA and SE4All do, developing countries have clearly been the main target group of global governance efforts. Again, the North-South direction not only relates to the transfer of finance and technology but also involves the flow of ideas. As a consequence, global renewable energy governance runs the risk of falsely generalizing the perspectives of industrialized countries as 'global perspectives' without taking advantage of the know-how available in developing countries. The Renewables Global Futures Report 2013 of the transboundary policy network REN21 is a case in point. The report's assessments of the future of worldwide renewable energy deployment are based on policy scenarios and interviews with decision-makers and renewable energy experts. Out of the 170 interviews, almost 80 percent were conducted in Europe, USA and Japan (REN21 2013b:65f.). It is important to note that this publication concentrates on people in industrialized countries as generators of ideas, even though REN21 is a network that involves members from both industrialized and developing countries. The IEA, which publishes renowned and leading information sources on world energy markets such as the World Energy Outlook, even has its membership restricted to the OECD world. As its publications require approval by its member states, these have considerable influence on the ideas spread by the IEA.

This North-South direction, with industrialized countries as 'leaders' and developing countries as 'followers', stands in stark contrast to the prevailing 'partnership-rhetoric' in transboundary policy-making. This might also explain its implicitness in the German statements. Explicit affirmations would make the contradiction all too visible – both for the audience and the speaker. During the research for this study, the author repeatedly got the impression that many policy actors whose statements imply that they attribute developing countries a mere reactive role are not fully aware of this matter. If explicitly asked, they would presumably argue that this is not the case. However, these implicit assessments are major impediments for cooperation on an equal footing.

Last but not least, the German government argues for a strengthening of global renewable energy governance, whereas the Brazilian administration expresses its caution vis-à-vis global governance and emphasizes the importance of national sovereignty. This ideational difference is in line with the cleavage that Lesage, Van de Graaf & Westphal (2010:85f.) and Kuik, Bastos Lima & Gupta (2011:627) identify between the group of industrialized countries and that of developing and emerging countries: while the former tend to favor the expansion of global energy governance, the latter hesitate due to sovereignty concerns.

7.2 The Reasons Behind the Ideational Differences

Domestic policy context. The domestic policy context is of central importance in order to understand the ideational differences with regard to the global challenges and renewable energy options. It furthermore helps to understand the different perspectives on the main barriers to a worldwide promotion of renewables and tasks for transboundary policy-making.

The prioritization of different global challenges corresponds to the political motivations behind the domestic promotion of renewables in both countries. In the German case, environmental concerns have been the main driver, whereas in the Brazilian case, the priority has been given to socio-economic development, particularly in the agricultural sector. As such, the different relationships drawn between renewable and conventional energy at the global level also prevail domestically. In Germany, the promotion of renewables is seen as a means to replace conventional energy and thus to mitigate environmental damages caused by the latter. In Brazil, renewable and conventional energy are regarded as complementary sources that both contribute to socio-economic development.

It is interesting to note that the German statements on global renewable energy governance do not refer to the environmental concern that has been dominating the domestic promotion of renewables. The German government avoids addressing the risks of using nuclear energy as this is a politically sensitive and highly contested issue at global level. Instead, it concentrates on a topic that is more compatible with the agenda of global environmental governance. Here, concerns about climate change are the center of attention. While these also motivated the domestic promotion of renewables, they are much less prominent than nuclear risks.

The results of this study, to a certain degree, contradict conventional wisdom that governments support the worldwide promotion of those renewable energy sources that prevail in their domestic energy mix. While both administrations do indeed only argue for the global spread of renewable energy sources that also contribute to their domestic energy mix, not all the renewable energy sources that provide significant shares receive much attention in the deliberations on global renewable energy governance. This applies to hydropower in the Brazilian case and bioenergy in the German. Moreover, there is no direct link between the concentration on different energy sectors in the statements and the prevalence of these sectors in domestic renewable energy deployment. While the electricity sector has the highest renewable energy shares in both countries, it only plays a central role in the German deliberations.

In the German case, the domestic policy context is decisive in order to understand the discrepancy between the predominance of bioenergy in domestic renewables-based energy supply[410] and the scant attention given to it in the statements on global renewable energy governance. Bioenergy is also at the sidelines of domestic efforts to promote renewables – at least from a declaratory point of view. Government statements on the domestic promotion of renewables strongly concentrate on the goal of expanding wind and solar energy use. Both sources commonly serve as symbols for renewable energy deployment in public debates and also receive the highest support rates in public opinion polls. Bioenergy enjoys less public backing and domestic biofuels blendings have even met strong public and political opposition. However, the gap between the actual contribution of bioenergy to the renewables-based energy supply and the declared political support for this energy source remains intriguing. In the realm of this study, it remains an open question as to whether bioenergy has competitive advantages that facilitate its expansion despite relatively weak policy support or whether there is a mismatch between political declarations and action. This also touches on a major limitation of ideational research that is based on statements: ideas that are expressed in words are not necessarily converted into action.

In the Brazilian case, considering only the domestic policy context is insufficient in order to understand why the government does not issue statements relating to the electricity sector and hydropower, even though the latter has the second-largest share in the Brazilian renewables-based energy supply.[411] However, the domestic level helps to understand why the government concentrates on sugarcane ethanol within the transport sector. In this sector, sugarcane ethanol has been the dominant renewable energy source. It provides the bulk of the renewables-based energy supply and builds on established economic structures.

The different assessments of the 'pros' and 'cons' of biofuels and the relationship between biofuels production and food security generally prevail in the public and political debates in both countries. While biofuels, and particularly sugarcane ethanol, enjoy widespread public support in Brazil, the use of biofuels is strongly contested in Germany. Here, NGOs and the media repeatedly warn about negative impacts on global food security. Next to the different assumptions regarding the underlying causes of hunger (see also Figure 8 on page 231), the choice of different reference points seems to be the major reason behind the diverging assessments of the relationship between biofuel production and food security. The

---

[410] In 2012, its share amounted to 64.6 percent (BMU 2013c:14).
[411] In 2011 hydropower had a share of 33.3 percent. Sugarcane products provided with 35.5 percent the largest share. Own calculations based on Empresa de Pesquisa Energética (2012:22).

Brazilian debate takes the Brazilian experience of biofuels and particularly ethanol production as a reference point, which has not posed risks to domestic food production. The German debate draws on the statements of NGOs, international organizations and the media that refer to risks that biofuels production can cause in developing countries. As such, the Brazilian debate apparently generalizes the Brazilian 'success story' while the German debate generalizes 'problematic cases' that appeared (or are considered likely) in some developing countries.

In this context, it is interesting to note that NGO assessments of biofuels also differ considerably between Brazil and Germany. In Germany, many NGOs campaign against biofuels deployment. Among their Brazilian counterparts, it is not common to argue against the use of biofuels *per se*. They rather call for improved regulatory frameworks to minimize negative social and environmental side-effects. Moreover, they point out that the ethanol industry is more responsive to sustainability demands than other sectors of agribusiness. Brazilian NGOs have particularly criticized the working conditions on sugarcane plantations and the local environmental implications of sugarcane burning. The sustainability concerns that are commonly raised by non-Brazilian policy actors with regard to Brazilian ethanol production – deforestation of the Brazilian Amazon and displacement of food production – do not prevail in the Brazilian NGO community. It is commonly argued that these do not apply to the Brazilian context of ethanol production. Cleavages even arise between international headquarters and Brazilian chapters of transboundary NGOs. While Oxfam International, for example, denounces biofuels production in Brazil and argues against EU biofuels imports from Brazil, Oxfam Brazil does not want the EU to close its market for Brazilian biofuels exports.[412] As such, the ideas that NGOs advance on global renewable energy governance also prove to be highly context-dependent.

The structures of domestic renewables sectors and the differing balance between state interventions and market forces explain the different perspectives on economic competitiveness, and also contribute to a better understanding of the ideational differences with regard to the main barriers and tasks. In Germany, renewable energy receives government support to compete in the market. In Brazil, the bulk of renewable energy (hydropower and hydrated ethanol) is used because it is the most competitive option. As such, the Brazilian government underlines the importance of economic competitiveness at present, whereas the German administration refers to future potentials and argues that the choice between different energy options should not be based on competitiveness only. While the German deliberations on barriers and tasks focus on government action, the Brazilian statements concentrate on market forces. Moreover, the governments link economic

---

[412] Interview with policy adviser, Brazilian Office, Oxfam, April 2013.

competitiveness to the deployment scales that dominate at domestic level: large scale in Brazil, and small scale in Germany.

Embeddedness in transboundary policy-making. The involvement and self-positioning in transboundary policy-making is key for understanding the ideational differences on the meta-aspects/salient features in global renewable energy governance and global governors. Interestingly, the renewables-specific context of transboundary policy-making turns out to be less relevant for the understanding of ideational differences than the general global governance context.

The differing perspectives on salient features in global renewable energy governance can be traced back to the governments' self-positioning in global governance. The German government ascribes itself to the group of industrialized countries, i.e. the countries that both governments regard as most influential in global governance; whereas the Brazilian government positions itself within the group of developing countries. According to the former, these countries have been deprived from influence in global governance. As such, it locates itself on the weaker side of global governance and considers itself negatively affected by existing power asymmetries. While power structures and asymmetries are a central analytical category in the Brazilian view on global governance, the German statements do not contain explicit references to power asymmetries between industrialized and developing countries. In the realm of this study, it remains open whether the German government considers power asymmetries as an analytical category of limited relevance (since it is not negatively affected) or whether it purposely avoids addressing these asymmetries to maintain the *status quo*.

In this context, it should be noted that there is often a mismatch between Brazilian self-portrayal as a developing country and its outer ascription. Often, Brazil is referred to as an emerging power, i.e. a country that no longer belongs in the developing world but, together with countries like China and India, forms a new country group located 'in between' the industrialized and developing world and which is therefore particularly suited to mediate between the North and South. Instead of creating a third category, the Brazilian government clearly ascribes itself to the group of developing countries and positions itself as a leading actor within this group. It repeatedly claims that it wants to serve as the voice of developing countries in global governance. According to Cooper & Flemes (2013:952) and Hurrell & Sengupta (2012:465), such a mismatch between self-portrayal and outer ascription not only applies to the Brazilian case, but to emerging powers in general.

The different emphasis placed on national sovereignty is in line with the government's experience of global governance interfering in its domestic policy priorities. The German administration has not experienced such inference with regard to domestic renewable energy policies. In Brazil, the construction of large hydropower plants and production of biofuels have been under the spotlight of international criticism. As such, national sovereignty is of central importance in the

Brazilian statements on global renewable energy governance, while it plays no major role in the Germans.

Involvement in transboundary policy-making on renewables merely contributes to the understanding of the prioritization of different global governors, as both governments only rank those global governors as important which they join as (active or even founding) members.

Division of competencies. The institutional context resulting from the governments' division of competencies affects the ideational differences of all analytical categories. Yet, the institutional context turns out to be less decisive for the understanding of individual ideational differences than the domestic policy context or the embeddedness in transboundary policy-making.

The division of competencies influences how the governments link transboundary policy-making on renewables to other policy issues. In Germany, the competencies for global renewable energy governance are split between various ministries. However, from 2002 to the end of 2013, the BMU clearly took the lead. This re-inforced the policy link between renewable energy promotion and environmental protection. The strong emphasis on environmental issues is particularly visible in the German deliberations on the global challenges and in the criteria of assessing different renewable energy options. In the Brazilian case, the Itamaraty is the ministry in charge and, as such, global renewable energy governance is closely embedded in foreign policy-making. The foreign policy background is particularly visible in the Brazilian deliberations on salient features, as North-South conflicts, power asymmetries and the emphasis of national sovereignty have been strong issues in Brazilian foreign policy. It furthermore reinforces the salience of trade issues and the emphasis on socio-economic development in its statements, as this is in line with the developmentalist tradition of the Itamaraty.

Both governments' concentration on different forms of discrimination is also shaped by the division of competencies. For the German government, the split competencies for renewable (BMU) and conventional energy (BMWi) went along with an institutional competition between the energy sources which might have intensified the perceived rivalry. In Brazil, it is common among diplomats to criticize the discriminatory trade policies of industrialized countries.

In the German case, the division of competencies furthermore affects the government perspective on biofuels. The ministries that are most active in global renewable energy governance, BMU and BMZ, are also the ministries that are most critical about the impacts of biofuels deployment. In addition, the little attention received by biofuels in the German deliberations might also be due to the fact that the BMU and BMZ departments in charge of global renewable energy governance were not responsible for biofuels. In line with the assessment by Scharpf (1997:39f.) that institutionalized responsibilities shape perceptions, they may have simply integrated those aspects into their statements that belong in the realm of their responsibilities.

At this point, a remark on ideational differences within each government should be made. While this study concentrates on ideational differences between the German and Brazilian governments, it does not conceal ideational differences between policy actors of the same government or across time. However, the deliberations on global renewable energy governance by different policy actors within each government rarely reveal expressed conflict. There is one notable exception on the German side. In contrast to the ministries that initiated IRENA's creation – BMU, BMZ and AA – the BMWi does not criticize the IEA as being biased towards conventional energy.[413] In interviews, BMWi representatives furthermore question IRENA's value added. Nevertheless, this assessment did not enter official statements because the BMWi was not in charge of IRENA during the time frame of this analysis. When the German government began its initiative to create IRENA, it did not enjoy the support of the BMWi which was responsible for IEA and did not want to create a competitor.

In the Brazilian case, ideational differences across time can be identified. Before commencing its active ethanol diplomacy, the Brazilian government advanced different ideas regarding the barriers and tasks for transboundary policy-making. Here, its assessments were very similar to those of its German counterpart. It pointed to inappropriate domestic policy frameworks and market distortions favoring fossil fuels and called for global targets on renewable energy deployments and improved policies and political frameworks. One reason behind the different assessment before the commencement of ethanol diplomacy might be that the Brazilian government did not have a marked agenda on global renewable energy governance at the time and thus primarily referred to issues which at that time were discussed at the global level (see chapter 3.1.3).

Self-interests. Taking the self-interests into account is crucial in order to understand the focus on different renewable energy sources and the ideational differences of the main barriers and tasks for transboundary cooperation.

Both administrations seek to position themselves as frontrunners in the worldwide promotion of renewables. The German government highlights its political achievements at domestic level, particularly in the electricity sector, and wishes to establish itself as a positive example of renewable energy policy-making. The Brazilian government emphasizes its pioneering activity in the promotion of biofuels. It regards the Brazilian experience of ethanol production as an important tool to leverage its influence in global governance. At the same time, it seeks to prevent

---

[413] This has changed after the end of this study's time frame. In December 2013, the competency for renewable energy was transferred from BMU to BMWi. This also implied a change in the institutional affiliation of the former BMU staff working on renewable energy.

external influence on internal matters. While the Brazilian government explicitly communicates its economic self-interests of enhancing ethanol exports, the German administration is more hesitant in linking its engagement for global renewable energy governance to economic advantage. This may also be due to the division of competencies, as promoting German exports or economic growth in Germany are not major institutional goals of BMU and BMZ. Overall, economic self-interests seem to be less decisive in understanding the German government's ideas regarding global renewable energy than suggested in the author's previous research (Acosta, Zilla 2011, Röhrkasten 2009:95, Röhrkasten, Zilla 2012).

The different involvement levels of private sector representatives in governmental action on global renewable energy governance affirm the different salience of economic self-interests. In Brazilian ethanol diplomacy, the private sector – particularly the Brazilian Sugarcane Industry Association, UNICA – has been a central player. German government initiatives have not met much interest within the country's renewable energy industry. This might be surprising at first glance. However, a closer look at the initiatives reveals that these are much more policy than business-oriented. Industry representatives did not expect there to be major economic benefits of engaging in the initiatives and the government uses other channels, such as the export initiative on renewables, to directly support the transboundary expansion of the domestic industry. Outside of the government, the only policy actor who has strongly shaped the German action on global renewable energy governance has been the late SPD parliamentarian Herrmann Scheer. He had been lobbying for the creation of an international organization for renewables since 1990, using his seat in the German Bundestag and the NGO Eurosolar – an NGO founded by himself that aims for the complete substitution of conventional energy – for these purposes.

Governmental self-interests strongly influence how barriers and tasks are constructed in global renewable energy governance. Here, the Brazilian deliberations mirror its economic interest in trade liberalization for biofuels, particularly for ethanol, while the German emphasis on domestic regulation and policy advice are in line with its political self-interest to present itself as a positive example of policy-making.

The Brazilian concentration on ethanol and biofuels is strongly influenced by its self-interests. While the government regards transboundary policy-making on biofuels as a suitable area to advance both its economic and political self-interests, it wants to avoid that transboundary policy-making focuses on hydropower as it fears that global governance could interfere with its domestic policy priorities. In the German case, self-interests are less decisive in order to understand its focus on the different renewable energy options. Here, they do not change but rather reinforce the focus that already prevails domestically.

The Brazilian emphasis on national sovereignty and non-interference touches central issues in academic discussions on global governance in general (see chapters 2.1.2 and 2.2.4) and global energy governance in particular (see chapter 3.3.1). As global governance research deviates from the Westphalian model of policy-making, underlining that spheres of political influence are not necessarily attached to territorial boundaries, an important question is to what degree states are willing to pool sovereignty (Reinicke 1998:71) and to transfer decision-making power to the global level. The present analysis reveals that through their involvement in global renewable energy governance, both the German and Brazilian governments aspire to expand their political influence – be it within or beyond their national borders (Held, McGrew 2003:191). What differs between the two is their exposure to outer interference and their perceived voice in global affairs. While the Brazilian government questions its influence in global policy-making due to the dominance of industrialized countries and fears that global governance could intrude into its domestic policy-making, this is not a concern for the German government.

Coherency between ideas. Due to the efforts of both governments to draft ideas in a coherent way, a number of ideational differences are caused by other ideational differences. The divergent perspectives on the salient features, for example, are clearly mirrored in each government's views on the barriers to a worldwide promotion of renewables. Here, the Brazilian administration raises confrontational issues and repeatedly criticizes policy actors in the industrialized world, while the German government follows a more cautious approach and tries to concentrate on consensual issues. This is particularly visible in the deliberations on 'knowledge barriers'. The German government merely points to the lack of knowledge and public awareness. Its Brazilian counterpart names policy actors that purposefully obstruct the promotion of renewables by circulating unjustified concerns and misinformation.

It is crucial to consider the focus of both governments on different renewable energy options in order to understand how they link the promotion of renewables to poverty reduction and economic development in developing countries. The production of biofuels is less technology-intensive and sophisticated than the production of solar panels, wind turbines etc. It requires input factors that are relatively abundant in many developing countries, such as fertile land and low-qualified labor. If poor people are involved in the production process, they can directly generate income from it. As opposed to this, it is much more difficult for poor people to generate a direct income from the production or provision of renewable electricity. Solar and wind energy, by contrast, facilitates the provision of electricity in remote areas without access to the grids.

The different assessments of biofuels are reflected in the perspectives regarding the purpose of internationally agreed sustainability standards and GBEP's thematic focus. The German government raises trade-offs and doubts with regard to biofuels and thus concentrates on risk prevention. As the Brazilian administration presents biofuels as a win-win option, it focuses on production expansion.

7.3 General Reflections and Suggestions for Further Research

This study demonstrates that ideas on global renewable energy governance are highly contested. It clearly confirms the assertion of Breitmeier, Young & Zürn (2006:191, 199-203) that transboundary cooperation is not only focused on solving transboundary problems, but also encourages new thinking about transboundary issues. All the analyzed global governors engage in the creation and diffusion of knowledge and social understandings. This also involves struggles over the prerogative of interpretation. Both governments suggest that 'false ideas' are being spread in global renewable energy governance. The Brazilian administration complains about unjustified concerns relating to biofuels and devotes much of its diplomatic efforts to tackling these concerns. The German government initiated the creation of REN21 and IRENA in order to challenge IEA's influence in global energy governance and to provide more 'renewables-friendly' analysis and policy advice.

The contested ideas concern aspects that are of crucial importance in global renewable energy governance. These not only affect how policy-makers understand their context of action and what policy directions they favor, but also what perspectives researchers take when analyzing global renewable energy governance: is global renewable energy governance primarily linked to global environment governance or to global governance on trade? Does it involve the electricity or transport sector? What factors are decisive when assessing the relationship between biofuels and hunger? Is global governance on renewables about improving domestic policy frameworks or tackling barriers at global level? What global governors deserve attention? What actors are endowed with the capacity to act and steer transboundary policy-making? There are no 'right' or 'wrong' answers. This study demonstrates that the answers chosen are strongly influenced by different policy contexts and may also reflect different policy goals. The questions furthermore illustrate that knowledge generation is never 'objective' or 'neutral'. Empirical descriptions and normative assessments are closely intertwined (Mehta 2011:33, Rein, Schoen 1977:243). Often, implicit assumptions are involved. Actors might take these for granted and not even notice if they are contested (Mehta 2011:34f.). However, the answers to these questions are highly relevant. They involve important trade-offs for policy-makers and researchers, as they decide on what topics to focus and where to direct limited capacities, such as money, attention and time.

If researchers simply assume common understandings and do not acknowledge that the definition of global governance issues is subject to political struggles, they fail to grasp a central dimension of transboundary policy-making. The unveiling of contested ideas is not only relevant from a theoretical point of view; it also has important political implications. It helps to enhance a deeper understanding of cooperation barriers and contributes to improving the functioning and outcomes of transboundary policy-making. Missing awareness of different problem definitions, for example, can easily lead to misunderstandings which hamper cooperation. Knowing the logics of one's counterparts, by contrasts, can be of crucial importance in order to find common ground and overcome cooperation deadlocks.

Further research should therefore aim at deepening the understanding of the scope, origins and consequences of contested ideas, not only in global renewable energy governance but also in other issue areas of transboundary policy-making. To achieve a more comprehensive assessment of the extent of contestation in global renewable energy governance and the reasons behind it, the ideas of further relevant policy actors should be analyzed. The theoretical-analytical framework of this study could be used to study the ideas of other national governments. Firstly, governments that have already taken an active stance in global renewable energy governance could be considered. Here, the UAE would be an especially interesting case. While it belongs to the group of OPEC countries which for a long time blocked global efforts to promote renewable energy, it actively supported IRENA's creation. Nowadays, it hosts IRENA's headquarters and is its most important financier. Another thought-provoking case could be the US government which nowadays promotes its 'fracking revolution' around the world. It exercises considerable influence in most of the global governors that were analyzed for this thesis. It, for example, founded CEM and is one of the most active players in GBEP. While it first opposed IRENA's creation as it did not want to weaken IEA's role, it now influences IRENA from within, trying to limit its budget and restrict its activities to developing countries only. Analyzing the ideas of the Chinese government could also provide interesting insights. China is the worldwide leader in renewable energy investments and installed renewable energy capacities (REN21 2013a:17). Yet, in this study's process tracing of global renewable energy governance, it did not appear as a very active or influential player. Like the Brazilian government, it communicates its discontent with the agenda-setting and issue-framing in transboundary policy-making on renewable energy (Röhrkasten 2012:342).

The theoretical-analytical framework of this study could furthermore be adapted to compare the ideas of non-state actors on global renewable energy governance. As this study suggests that NGO perspectives are highly context-dependent, a systematic comparison of NGOs could reveal particularly interesting insights. While the questions for the identification of ideational differences could remain unchanged, the quest for possible reasons behind ideational differences would require adaptation. With regard to the context factors, it would be interesting to study how the embeddedness of NGOs in transboundary policy-making, their geographic location, membership structure and origins of financing influence their ideas on global renewable energy governance.

For a deeper understanding of the politics behind the establishment of social understandings in global renewable energy governance, the political processes that accompany the work of global governors require further analysis. It would be informative to study in depth how the ideas of the analyzed global governors differ, which global governors are more successful in spreading their ideas and why. In addition, researchers should pay more attention to the politics behind the ideas spread by global governors. For example, it would be informative to investigate which policy actors exercise most influence on individual global governors and what issues are contested among the involved policy actors. Röhrkasten & Westphal (2013) focus on the politics behind IRENA's creation. Yet, further global governors require attention as well. As the IEA is confronted with the criticism that its modeling is based on untransparent assumptions and favors fossil fuels over renewable energy, it would be particularly interesting to investigate to what extent IEA findings are influenced by its member states and further policy actors. What policy actors are most influential and what cleavages arise between them? Another interesting case could be the contestation that evolved around SE4All, the major UN initiative on renewable energy so far. While it was launched by the UN Secretary-General, the UN member states refused to fully endorse it at the Rio plus 20 Conference in 2012.

Moreover, the analytical-theoretical framework that was developed for the purpose of this study could be easily adapted to study contested ideas in other areas of transboundary policy-making. Here, it could be particularly insightful to analyze the ongoing UN process on the development, adoption and implementation of Sustainable Development Goals as it touches upon a range of issue areas of transboundary policy-making and involves diverse stakeholders. Another aspect makes this process particularly interesting: while the preceding MDGs concentrated on action in developing countries, the Sustainable Development Goals are supposed to be universally applicable. As such, they shall also guide policy action in industrialized countries. This challenges the North-South direction of global action which proved to be so decisive in global renewable energy governance.

Last, but not least, this thesis wishes to seize on the goal formulated by Mayntz (2008:110) "to alert scholars to the contingent nature of their ways of perceiving, evaluating, and studying a given object of cognition." This is particularly relevant for global governance research, as analyses that concentrate on the 'common good' easily lose sight that the 'common good' may look very different, depending on the angle chosen. Researchers are likely to pick that which prevails in their own surroundings. If they do not reflect on the contingency of this angle, they run the risk of falsely generalizing the perspectives of their specific surroundings to the global level. This is particularly problematic when dealing with global governance, as it involves policy actors with highly diverse backgrounds and views. On top of that, researchers are also at risk of simply reproducing the ideas of the most powerful in global governance. They might particularly fall into this trap if their own surrounding is located on the 'stronger' or more influential side of global governance.

Bibliography

Acosta, S. & Zilla, C. 2011, *Markets and Minds. Trade and Value Conflicts over Biofuels*, SWP Research Paper 07/2011, Stiftung Wissenschaft und Politik, Berlin.
Adler, E. 2002, "Constructivism and International Relations" in *Handbook of International Relations*, ed. W. Carlsnaes, Sage Publishing, London, pp. 95-118.
Adler, E. 1997, "Seizing the Middle Ground: Constructivism in World Politics", *European Journal of International Relations*, vol. 3, no. 3, pp. 319-363.
Adler, E. 1992, "The emergence of cooperation: national epistemic communities and the international evolution of the idea of nuclear arms control", *International Organization*, vol. 46, no. 1, pp. 101-145.
Adler, E. & Haas, P.M. 1992, "Conclusion: epistemic communities, world order, and the creation of a reflective research program", *International Organization (Cambridge/Mass.)*, vol. 46, no. 1, pp. 367-390.
Alexandroff, A.S. & Cooper, A.F. 2010a, "Introduction" in *Rising States, Rising Institutions: Challenges for Global Governance*, eds. A.S. Alexandroff & A.F. Cooper, Brookings Institution Press, Baltimore, pp. 1-14.
Alexandroff, A.S. & Cooper, A.F. (eds) 2010b, *Rising States, Rising Institutions: Challenges for Global Governance*, Brookings Institution Press, Baltimore.
Altmaier, P. 2012, *Wir brauchen einen Klub der Energiewendestaaten*, Financial Times Deutschland, 27 August 2012.
Amado, A. 2010a, *O acerto da política do etanol*, O Globo, 25 October 2010.
Amado, A. 2010b, *O etanol e a diplomacia*, Valor Econômico, 15 March 2010.
Amorim, C. 2010, *Statement at the Opening of the General Debate of the 65th Session of the United Nations General Assembly*, New York, 23 September 2010.
Amorim, C. 2008, *Discurso no Segmento Intergovernamental de Alto Nível da Conferência Internacional de Biocombustíveis*, São Paulo, 20 November 2008.
Amorim, C. 2007a, *Discurso no Segmento de Alto Nível da 13a Conferência das Partes na Convenção-Quadro das Nações Unidas sobre Mudança do Clima e da 3a Conferência das Partes servindo como Reunião das Partes no Protocolo de Quioto*, Bali/Indonesia, 12 December 2007.
Amorim, C. 2007b, "Prefácio" in *Biocombustíveis no Brasil - Realidades e Perspectivas*, ed. MRE, Brasília, pp. 6-9.
Annan, K. 2002, "Toward a Sustainable Future", *Environment*, vol. 44, no. 7, pp. 10-15.
Armijo, L.E. 2007, "The BRICs countries (Brazil, Russia, India, and China) as analytical category", *Asian Perspective*, vol. 31, no. 4, pp. 7-42.
Avant, D.D., Finnemore, M. & Sell, S.K. 2010a, "Conclusion: authority, legitimacy, and accountability in global politics" in *Who Governs the Globe?*, eds. D.D. Avant, M. Finnemore & S.K. Sell, Cambridge University Press, Cambridge, pp. 356-370.
Avant, D.D., Finnemore, M. & Sell, S.K. (eds) 2010b, *Who Governs the Globe?*, Cambridge University Press, Cambridge.
Avant, D.D., Finnemore, M. & Sell, S.K. 2010c, "Who Governs the Globe?" in *Who Governs the Globe?*, eds. D.D. Avant, M. Finnemore & S.K. Sell, Cambridge University Press, Cambridge, pp. 1-31.
Barnett, M. & Duvall, R. (eds) 2005a, *Power in Global Governance*, Cambridge University Press, Cambridge.
Barnett, M. & Duvall, R. 2005b, "Power in Global Governance" in *Power in Global Governance*, eds. M. Barnett & R. Duvall, Cambridge University Press, Cambridge, pp. 1-32.
Barnett, M. & Finnemore, M. 2005, "The Power of Liberal International Organizations" in *Power in Global Governance*, eds. M. Barnett & R. Duvall, Cambridge University Press, Cambridge, pp. 161-184.

Barnett, M.N. & Finnemore, M. 1999, "The Politics, Power, and Pathologies of International Organizations", *International Organization*, vol. 53, no. 4, pp. 699-732.

Bastos Lima, M. 2012, "The Brazilian biofuel industry: achievements and geopolitical challenges" in *Secure Oil and Alternative Energy. The Geopolitics of Energy Paths of China and the European Union*, eds. M.P. Amineh & Y. Guang, Brill, Leiden,Boston, pp. 343-369.

Bastos Lima, M. & Gupta, J. 2013, "The Policy Context of Biofuels: A Case of Non-Governance at the Global Level?", *Global Environmental Politics*, vol. 13, no. 2, pp. 46-64.

Baumann, R., Rittberger, V. & Wagner, W. 2001, "Neorealist foreign policy theory" in *German foreign policy since unification: theories and case studies*, ed. V. Rittberger, Manchester University Press, Manchester, pp. 37-67.

Béland, D. & Cox, R.H. 2011, "Introduction: Ideas and Politics" in *Ideas and Politics in Social Science Research*, eds. D. Béland & R.H. Cox, Oxford University Press, Oxford, pp. 3-20.

Beneking, A. 2011, *Genese und Wandel der deutschen Biokraftstoffpolitik. Eine akteurszentrierte Policy-Analyse der Förderung biogener Kraftstoffe in Deutschland*, Institut für ökologische Wirtschaftsforschung, Berlin.

Bernstein, S. 2012, "Grand Compromises in Global Governance", *Government and Opposition*, vol. 47, no. 3, pp. 368-394.

Biermann, F. 2007, "Nord-Süd-Beziehungen in der Weltumweltpolitik: Globale Interdependenz und institutionelle Innovation" in *Politik und Umwelt*, eds. K. Jacob, F. Biermann, P. Busch & P.H. Feindt, VS Verlag, Wiesbaden, pp. 115-132.

Biermann, F. 1998, *Weltumweltpolitik zwischen Nord und Süd. Die neue Verhandlungsmacht der Entwicklungsländer*, Nomos, Baden-Baden.

Biswas, M.R. 1981, "United Nations Conference on New and Renewable Sources of Energy, Held in Nairobi, Kenya, During 10-21 August 1981", *Environmental Conservation*, vol. 8, no. 4, pp. 330-332.

Blyth, M. 2003, "Structures Do Not Come with an Instruction Sheet: Interests, Ideas, and Progress in Political Science", *Perspective on Politics*, vol. 1, no. 4, pp. 695-706.

BMU 2013a, *Representatives from ten pioneering countries establish Renewables Club*, BMU Press Release No. 075/13, 01 June 2013, Berlin.

BMU 2013b, *Expansion of renewable energies: international meeting of energy ministers and heads of government*, BMU Press Release No. 002/13, 14 January 2013, Berlin.

BMU 2013c, *Renewable Energy Sources in Figures. National and International Development*, BMU, Berlin.

BMU 2009a, *Environment State Secretary promotes IRENA*, BMU-Pressedienst No. 015/09, 19 January 2009, Berlin.

BMU 2009b, *A milestone for future-oriented energy supply. International Renewable Energy Agency being founded today in Bonn*, BMU-Pressedienst No. 023/09, 26 January 2009, Berlin.

BMU 2008, *Initiative der Bundesregierung für IRENA findet großen Zuspruch. Vorbereitungskonferenz für die Gründung einer Internationalen Agentur für Erneuerbare Energien*, BMU-Pressedienst Nr. 062/08, 11 April 2008, Berlin.

BMU 2007, *Nachhaltige Erzeugung ist Voraussetzung für den weiteren Ausbau der Bioenergie*, BMU Pressemitteilung Nr. 114/07, 25 April 2007, Berlin.

BMU & BMZ 2012a, *Altmaier and Niebel: Pivotal role for International Renewable Energy Agency (IRENA). IRENA Director-General Adnan Amin visits Germany*, BMU-Pressedienst No. 111/12, 27 August 2012, Berlin.

BMU & BMZ 2012b, *Sustainable Energy for Sustainable Development. The German Contributions*, Berlin/Bonn.

BMU & BMZ 2008, *One more step towards setting up an International Renewable Energies Agency. Workshop to prepare for the establishment of IRENA*, BMU-Pressedienst No.147/08, 01 July 2008, Berlin.

BMU, BMZ & AA 2008, *Breakthrough for the expansion of renewable energies. International Agency to be set up in January 2009*, BMU-Pressedienst No. 231/08, 25 October 2008, Berlin.

BMWi & Dena 2010, *renewables – Made in Germany. Die grüne Energieversorgung für heute und morgen*, Berlin.

BMZ 2013a, *Global Bioenergy Partnership. Hans-Jürgen Beerfeltz: "Bioenergien in Entwicklungsländern effizient nutzen"*, Aktuelle Meldung, 28 May 2013, Berlin.

BMZ 2013b, *Themenblatt Biokraftstoffe*, Bonn/Berlin.

BMZ 2011, *Biofuels. Opportunities and Risks for Developing Countries*, Bonn/Berlin.
BMZ 2008, *Development needs sustainable energy*, Bonn/Berlin.
BMZ 2007, *Sustainable Energy for Development*, Bonn/Berlin.
Boekle, H., Rittberger, V. & Wagner, W. 2001, "Constructivist foreign policy theory" in *German foreign policy since unification: theories and case studies*, ed. V. Rittberger, Manchester University Press, Manchester, pp. 105-137.
Bogner, A. & Menz, W. 2002, "Das theoriegenerierende Experteninterview. Erkenntnisinteresse, Wissensformen, Interaktion" in *Das Experteninterview. Theorie, Methode, Anwendung*, eds. A. Bogner, B. Littig & W. Menz, Leske und Budrich, Opladen, pp. 33-70.
Börzel, T.A. & Risse, T. 2005, "Public-Private Partnerships: Effective and Legitimate Tools of International Governance?" in *Complex Sovereignty: Reconstituting Political Authority in the Twenty-first Century*, eds. E. Grande & L.W. Pauly, Toronto University Press, Toronto, pp. 195-216.
Bradshaw, M. 2013, "Sustainability, Climate Change, and Transition in Global Energy" in *The Handbook of Global Energy Policy*, ed. A. Goldthau, Wiley-Blackwell, London, pp. 48-63.
Brand-Schock, R. 2010, *Grüner Strom und Biokraftstoffe in Deutschland und Frankreich. Ein Vergleich der Policy-Netzwerke*, Dissertation, Freie Universität Berlin.
Breitmeier, H., Young, O.R. & Zürn, M. 2007, "The International Regimes Database: Architecture, Key Findings, and Implications for the Study of Environmental Regimes" in *Politik und Umwelt*, eds. K. Jacob, F. Biermann, P. Busch & P.H. Feindt, VS Verlag, Wiesbaden, pp. 41-59.
Breitmeier, H., Young, O.R. & Zürn, M. 2006, *Analyzing International Environmental Regimes: from Case Study to Database*, MIT Press, Cambridge/Mass.
Brühl, T. & Rittberger, V. 2001, "From international to global governance: Actors, collective decision-making, and the United Nations in the world of the twenty-first century" in *Global governance and the United Nations system*, ed. V. Rittberger, United Nations University Press, Tokyo; New York; Paris, pp. 1-47.
Bündnis 90/Die Grünen 2002, *Grün wirkt! Unser Wahlprogramm 2002 – 2006*.
Bündnis 90/Die Grünen 1998, *Programm zur Bundestagswahl 98. Grün ist der Wechsel*.
Campbell, J.L. 2002, "Ideas, Politics, and Public Policy", *Annual Review of Sociology*, vol. 28, pp. 21-38.
Campbell, J.L. 1998, "Institutional Analysis and the Role of Ideas in Political Economy", *Theory and Society*, vol. 27, pp. 377-409.
Carbonnier, G. & Brugger, F. 2013, "The Development Nexus of Global Energy" in *The Handbook of Global Energy Policy*, ed. A. Goldthau, Wiley-Blackwell, London, pp. 64-78.
Castells, M. 2005, "Global Governance and Global Politics", *Political Science and Politics*, vol. 38, no. 1, pp. 9-16.
CDU, CDU & FDP 2009, *Growth. Education. Unity. The Coalition Agreement between the CDU, CSU and FDP*.
CDU, CDU & SPD 2005, *Gemeinsam für Deutschland – mit Mut und Menschlichkeit. Koalitionsvertrag*.
Chase, R., Hill, E. & Kennedy, P. 1996, "Pivotal States and U.S. Strategy", *Foreign Affairs*, January/February 1996.
Chasek, P. & Rajamani, L. 2003, "Steps Toward Enhanced Parity: Negotiating Capacity and Strategies of Developing Countries" in *Providing Global Public Goods: Managing Globalization*, eds. I. Kaul, P. Conceição, K. Le Goulven & R.U. Mendoza, Oxford University Press, New York, pp. 245-262.
Cherp, A. & Jewell, J. 2011, "The three perspectives on energy security: intellectual history, disciplinary roots and the potential for integration", *Current Opinion in Environmental Sustainability*, vol. 3, pp. 1-11.
Cherp, A., Jewell, J. & Goldthau, A. 2011, "Governing Global Energy: Systems, Transitions, Complexity", *Global Policy*, vol. 2, no. 1, pp. 75-88.
Christensen, S.F. 2013, "Brazil's Foreign Policy Priorities", *Third World Quarterly*, vol. 34, no. 2, pp. 271-286.
Coate, R.A. & Murphy, C.N. 1995, "Editors' Note", *Global Governance*, vol. 1, no. 1, pp. 1-2.
Coelho, S.T., & Goldemberg, J. 2013, "Global Energy Policy: A View from Brazil" in *The Handbook of Global Energy Policy*, ed. A. Goldthau, Wiley-Blackwell, London, pp. 457-474.

Colgan, J. 2009, "The International Energy Agency. Challenges for the 21st Century", *GPPi Policy Papers Series*, vol. 6.

Colgan, J., Keohane, R.O. & Van de Graaf, T. 2011, "Punctuated Equilibrium in the Energy Regime Complex", *Review of International Organizations*, published online 26 July 2011.

Colgan, J. & Van de Graaf, T. 2014, "Mechanisms of informal governance: Evidence from the IEA", *Journal of International Relations and Development*.

Commission on Global Governance 1995, *Our global neighbourhood*, Oxford Univ. Press, Oxford.

Conference on the Establishment of the International Renewable Energy Agency 2009, *Statute of IRENA*.

Conzelmann, T. & Faust, J. 2009, ""Nord" und "Süd" im globalen Regieren", *Politische Vierteljahresschrift*, vol. 50, no. 2, pp. 203-225.

Cooper, A.F., Antkiewicz, A. & Shaw, T.M. 2007, "Lessons from/for BRICSAM about South-North Relations at the Start of the 21st Century: Economic Size Trumps All Else?", *International Studies Review*, vol. 9, no. 3, pp. 673-689.

Cooper, A.F. & Flemes, D. 2013, "Foreign Policy Strategies of Emerging Powers in a Multipolar World: an introductory review", *Third World Quarterly*, vol. 34, no. 6, pp. 934-962.

Corrêa do Lago, A.A. 2009, *Stockholm, Rio, Johannesburg. Brazil and the Three United Nations Conferences on the Environment*, Fundação Alexandre de Gusmão, Brasília.

Costa, V. 2004, "Federalismo" in *Sistema Político Brasileiro: uma introdução*, eds. L. Avelar & A.O. Cintra, Fundação Konrad Adenauer; Fundação Editora da Unesp, Rio de Janeiro, São Paulo, pp. 173-184.

Coussy, J. 2005, "The Adventures of a Concept: Is Neo-Classical Theory Suitable for Defining Global Public Goods?", vol. 12, no. 1, pp. 177-194.

Dahl, R. 1957, "The concept of power", *Behavioral Science*, vol. 2, no. 3, pp. 201-215.

Dauvergne, P. & Farias, D.B. 2012, "The Rise of Brazil as a Global Development Power", *Third World Quarterly*, vol. 33, no. 5, pp. 903-917.

Dessler, D. 1999, "Constructivism within a positivist social science", *Review of International Studies*, vol. 25, no. 1, pp. 123-137.

Deutscher Bundestag 2003, *Antrag - Initiative zur Gründung einer Internationalen Agentur zur Förderung der Erneuerbaren Energien (International Renewable Energy Agency – IRENA)*, Drucksache 15/811, 15. Wahlperiode, 08 April 2003.

Dey, I. 1993, *Qualitative data analysis.A user-friendly guide for social scientists*, Routledge, London.

Dingwerth, K. & Pattberg, P. 2010, "How Global and Why Governance? Blind Spots and Ambivalences of the Global Governance Concept", *International Studies Review*, vol. 12, pp. 702-710.

Dingwerth, K. & Pattberg, P. 2009, "Actors, Arenas, and Issues in Global Governance" in *Palgrave Advances in Global Governance*, ed. J. Whitman, Palgrave Macmillan, Houndmills, pp. 41-65.

Dingwerth, K. & Pattberg, P. 2006a, "Global Governance as a Perspective on World Politics", *Global Governance*, vol. 12, no. 2, pp. 185-203.

Dingwerth, K. & Pattberg, P. 2006b, "Was ist Global Governance?", *Leviathan*, vol. 3/2006, pp. 377-399.

Dubash, N.K. 2011, "From Norm Taker to Norm Maker? Indian Energy Governance in Global Context", *Global Policy*, vol. 2, pp. 66-79.

Dubash, N.K. & Florini, A. 2011, "Mapping Global Energy Governance", *Global Policy*, vol. 2, pp. 6-18.

Ebinger, C. & Avasarala, G. 2013, "The "Gs" and the Future of Energy Governance in a Multipolar World" in *The Handbook of Global Energy Policy*, ed. A. Goldthau, Wiley-Blackwell, London, pp. 190-204.

Empresa de Pesquisa Energética 2012, *Balanço energético nacional 2012: Ano base 2011*, Empresa de Pesquisa Energética, Rio de Janeiro.

Erler, G. 2009, *The establishment of IRENA: "Kick-starting the third industrial revolution"*. Speech at the IRENA Founding Conference, 26 January 2009.

EUROSOLAR & World Council for Renewable Energy 2009a, "International Parliamentary Forum, Bonn, 2 June 2004" in *The long road to IRENA. From the Idea to the Foundation of the International Renewa-*

ble Energy Agency. Documentation 1990 - 2009., eds. EUROSOLAR & World Council for Renewable Energy, Ponte Press, Bochum, pp. 70-85.

EUROSOLAR & World Council for Renewable Energy 2009b, "The long road to IRENA - A chronology" in *The long road to IRENA. From the Idea to the Foundation of the International Renewable Energy Agency. Documentation 1990 - 2009.*, eds. EUROSOLAR & World Council for Renewable Energy, Ponte Press, Bochum, pp. 3-8.

Expert Group on Renewable Energy 2005, *Increasing Global Renewable Energy Market Share. Recent Trends and Perspectives.* Background Report Beijing International Renewable Energy Conference.

Feres, P.F.D. 2010, *Os biocombustíveis na matriz energética alemã: possibilidades de cooperação com o Brasil*, Fundação Alexandre de Gusmão, Brasília.

Figueroa, A. 2012, "International Energy Agency Implementing Agreements" in *Meeting global challenges through better governance: international co-operation in science, technology and innovation*, ed. OECD, pp. 131-148.

Finkelstein, L.S. 1995, "What is Global Governance?", *Global Governance*, vol. 1, no.3, pp. 369-372.

Finnemore, M. 1996, "Norms, culture, and world politics: insights from sociology's institutionalism", *International Organization*, vol. 50, pp. 325-347.

Finnemore, M. & Sikkink, K. 1998, "International norm dynamics and political change", *International Organization (Cambridge/Mass.)*, vol. 52, no. 4, pp. 887-917.

Fiori, M. 2008, *Brasil e potências emergentes reúnem-se para articular posições perante a cúpula do G8*, Agência Brasil, 07 July 2008.

Fischer, S. & Röhrkasten, S. 2013, *EU-Verkehrssektor: Ende der Biokraftstoffpolitik*, SWP-Aktuell 61/2013, Stiftung Wissenschaft und Poltik, Berlin.

Flemes, D. (ed) 2010, *Regional Leadership in the Global System. Ideas, Interests and Strategies of Regional Powers*, Ashgate Publishing, Fernham.

Florini, A. 2008, "Global Governance and Energy", *CAG Working Paper Series*, vol. 001.

Florini, A. & Sovacool, B.K. 2011, "Bridging the Gaps in Global Energy Governance", *Global Governance*, vol. 17, pp. 57-74.

Florini, A. & Sovacool, B.K. 2009, "Who governs energy? The challenges facing global energy governance", *Energy Policy*, vol. 37, no. 12, pp. 5239-5248.

Florini, A. & Dubash, N.K. 2011, "Introduction to the Special Issue: Governing Energy in a Fragmented World", *Global Policy*, vol. 2, pp. 1-5.

Flyvbjerg, B. 2006, "Five Misunderstandings About Case-Study Research", *Qualitative Inquiry*, vol. 12, no. 2, pp. 219-245.

Fonseca, M.B., Burrell, A., Gay, S.H., Henseler, M., Kavallari, A., M'Barek, R., Pérez Domínguez, I. & Tonini, A. 2010, *Impacts of the EU Biofuel Target on Agricultural Markets and Land Use. A Comparative Modelling Assessment*, Joint Research Centre of the European Commission, Sevilla.

Franck, T.M. 1990, *The Power of Legitimacy among Nations*, Oxford University Press, Oxford.

Frankfurt School - UNEP Centre & Bloomberg New Energy Finance 2013, *Global Trends in Renewable Energy Investment 2013*, Frankfurt am Main.

Freund, C. & Rittberger, V. 2001, "Utalitarian-liberal foreign policy theory" in *German foreign policy since unification: theories and case studies*, ed. V. Rittberger, Manchester University Press, Manchester, pp. 68-104.

G8 Global Renewable Energy Task Force 2001, *Final Report*, July 2001.

Gabriel, S. 2009, "IRENA: A pillar of clean energy supply. Speech held at the Founding Conference of IRENA, 26 January 2009" in *The long road to IRENA. From the Idea to the Foundation of the International Renewable Energy Agency. Documentation 1990 - 2009.*, eds. EUROSOLAR & World Council for Renewable Energy, Ponte Press, Bochum, pp. 103-104.

Galli, E. 2011, *Frame Analysis in Environmental Conflicts: The case of ethanol production in Brazil*, Dissertation, KHT - Royal Institute of Technology, School of Industrial Engineering and Management.

Garrison, J. 2007, "Constructing the "National Interest" in U.S.-China Policy Making: How Foreign Policy Decision Groups Define and Signal Policy Choices", *Foreign Policy Analysis*, vol. 3, no. 2, pp. 105-126.

Gehrke, B. & Schasse, U. 2013, *Position Deutschlands im Außenhandel mit Gütern zur Nutzung erneuerbarer Energien und zur Steigerung der Energieeffizienz*, Niedersächsisches Institut für Wirtschaftsforschung e.V., Hannover.

Giersdorf, J. 2011, *Politics and Economics of Ethanol and Biodiesel Production and Consumption in Brazil*, Dissertation, Freie Universität Berlin.

Gläser, J. & Laudel, G. 2010, *Experteninterviews und qualitative Inhaltsanalyse als Instrumente rekonstruierender Untersuchungen*, 4. Auflage edn, VS Verlag für Sozialwissenschaften, Wiesbaden.

Goldemberg, J. 2006, "The ethanol program in Brazil", *Environmental Research Letters*, vol. 1, pp. 1-5.

Goldemberg, J. 2002a, *The Brazilian Energy Initiative. Executive Summary*. World Summit on Sustainable Development, Johannesburg, South Africa, 26 August to 4 September 2002, 23 May 2002.

Goldemberg, J. 2002b, *Support Report for The Brazilian Energy Initiative*. World Summit on Sustainable Development, Johannesburg, South Africa, 26 August to 4 September 2002.

Goldstein, J. & Keohane, R.O. 1993, "Ideas and Foreign Policy: An Analytical Framework" in *Ideas and foreign policy. Beliefs, institutions, and political change*, eds. J. Goldstein & R.O. Keohane, Cornell Univ. Press, Ithaca/N.Y., pp. 3-30.

Goldthau, A. (ed) 2013a, *The Handbook of Global Energy Policy*, Wiley-Blackwell, London.

Goldthau, A. 2013b, "Introduction: Key Dimensions of Global Energy Policy" in *The Handbook of Global Energy Policy*, ed. A. Goldthau, Wiley-Blackwell, London, pp. 1-12.

Goldthau, A. 2012, "A Public Policy Perspective on Global Energy Security", *International Studies Perspectives*, vol. 13, pp. 65-84.

Goldthau, A. 2011, "Governing global energy: existing approaches and discourses", *Current Opinion in Environmental Sustainability*, vol. 3, pp. 213-217.

Goldthau, A. & Sovacool, B.K. 2012, "The uniqueness of the energy security, justice, and governance problem", *Energy Policy*, vol. 41, pp. 232-240.

Gosh, A. 2011, "Seeking Coherence in Complexity? The Governance of Energy by Trade and Investment Institutions", *Global Policy*, vol. 2, pp. 106-119.

Grant, C. 1995, "Equity in International Relations: a Third World Perspective", *International Affairs*, vol. 71, no. 3, pp. 567-587.

Groba, F. & Kemfert, C. 2011, "Erneuerbare Energien: Deutschland baut Technologie-Exporte aus", *DIW Wochenbericht*, vol. 45, pp. 23-29.

Gu, J., Humphrey, J. & Messner, D. 2008, "Global governance and developing countries: the implications of the rise of China", *World Development*, vol. 36, no. 2, pp. 274-292.

Gupta, J. 2012, "Global Energy Governance in the Twenty-First Century: Challenges and Opportunities" in *Secure Oil and Alternative Energy. The Geopolitics of Energy Paths of China and the European Union*, eds. M.P. Amineh & Y. Guang, Brill, Leiden,Boston, pp. 427-447.

Haas, P.M. 2002a, "UN Conferences and Constructivist Governance of the Environment", *Global Governance*, vol. 8, no. 1, pp. 73.

Haas, P.M. 2002b, "Constructing Environmental Conflicts from Resource Scarcity", *Global Environmental Politics*, vol. 2002, 2, 1, Feb, 1-11.

Haas, P.M. 1992, "Introduction: epistemic communities and international policy coordination", *International Organization (Cambridge/Mass.)*, vol. 46, no. 1, pp. 1-35.

Haas, P.M. & Haas, E.B. 2002, "Pragmatic constructivism and the study of international institutions", *Millennium (London)*, vol. 31, no. 3, pp. 573-601.

Hardy, C., Harley, B. & Phillips, N. 2004, "Discourse Analysis and Content Analysis: Two Solitudes?", *Qualitative Methods*, vol. 2, no. 1, pp. 19-22.

Harris, D., Moore, M. & Schmitz, H. 2009, *Country classifications for a changing world*, DIE Discussion Paper 09/2009, DIE, Bonn.

Hasenclever, A., Mayer, P. & Rittberger, V. 2000, "Integrating Theories of International Regimes", *Review of International Studies*, vol. 26, no. 1, pp. 3-33.

Hasenclever, A., Mayer, P. & Rittberger, V. 1996, "Interests, Power, Knowledge: The Study of International Regimes", *Mershon International Studies Review*, vol. 40, no. 2, pp. 177-228.

Hay, C. 2011, "Ideas and the Construction of Interests" in *Ideas and Politics in Social Science Research*, eds. D. Béland & R.H. Cox, Oxford University Press, , pp. 65-82.

Held, D., McGrew, A., Perraton, J. & Goldblatt, D. 1999, "Globalization", *Global Governance*, vol. 5, no. 4, pp. 483-496.

Held, D. & McGrew, A. 2003, "Political Globalization: Trends and Choices" in *Providing Global Public Goods: Managing Globalization*, eds. I. Kaul, P. Conceição, K. Le Goulven & R.U. Mendoza, Oxford University Press, New York, pp. 185-199.

Herrera, S. & Wilkinson, J. 2010, "Biofuels in Brazil: debates and impacts", *Journal of Peasant Studies*, vol. 37, no. 4, pp. 749-768.

Hewson, M. & Sinclair, T.J. 1999, "The Emergence of Global Governance Theory" in *Approaches to Global Governance Theory*, eds. M. Hewson & T.J. Sinclair, State University of New York Press, Albany, pp. 3-22.

Hirschl, B. 2009, "International renewable energy policy - between marginalization and initial approaches", *Energy Policy*, vol. 37, pp. 4407-4416.

Hirschl, B. 2008, *Erneuerbare Energien-Politik. Eine Multi-Level Policy-Analyse mit Fokus auf den deutschen Strommarkt*.

Houghton, D.P. 2007, "Reinvigorating the Study of Foreign Policy Decision Making: Toward a Constructivist Approach", *Foreign Policy Analysis*, vol. 3, pp. 24-45.

Humphrey, J. & Messner, D. 2008, "China and India in the Global Governance Arena" in *Governance and legitimacy in a globalized world*, eds. R. Schmitt-Beck, T. Debiel & K. Korte, edn, Nomos, Baden-Baden, pp. 187-206.

Hurrell, A. 2011, "The Theory and Practice of Global Governance: The Worst of All Possible Worlds?", *International Studies Review*, vol. 13, pp. 144-154.

Hurrell, A. 2010, "Brazil and the New Global Order", *Current History*, February 2010, pp. 60-66.

Hurrell, A. 2006, "Hegemony, liberalism and global order: what space for would-be great powers?", *International Affairs (Oxford)*, vol. 82, pp. 1-19.

Hurrell, A. 2005, "Power, institutions, and the production of inequality" in *Power in Global Governance*, eds. M. Barnett & R. Duvall, Cambridge University Press, Cambridge, pp. 33-58.

Hurrell, A. 2000, "Some Reflections on the Role of Intermediate Powers in International Institutions" in *Paths to Power: Foreign Policy Strategies of Intermediate States*, eds. A. Hurrell, A.F. Cooper, G. González González, R. Ubiraci Sennes & S. Sitaraman, Latin American Program Working Papers, The Woodrow Wilson International Center, pp. 1-11.

Hurrell, A. & Narlikar, A. 2006, "A New Politics of Confrontation? Brazil and India in Multilateral Trade Negotiations", *Global Society*, vol. 20, no. 4, pp. 415-433.

Hurrell, A. & Sengupta, S. 2012, "Emerging powers, Nort-South relations and global climate politics", *International Affairs*, vol. 88, no. 3, pp. 463-484.

Husar, J. & Maihold, G. 2010, "The EU and New Leading Powers: Analytical Approaches and Policy Options" in *Europe and New Leading Powers. Towards Partnership in Strategic Policy Areas*, eds. J. Husar, G. Maihold & S. Mair, Nomos, Baden-Baden, pp. 11-20.

Husar, J. & Maihold, G. 2009, "Einführung: neue Führungsmächte - Forschungsansätze und Handlungsfelder" in *Neue Führungsmächte: Partner deutscher Außenpolitik?*, eds. J. Husar, G. Maihold & S. Mair, Nomos, Baden-Baden.

Husar, J., Maihold, G. & Mair, S. (eds) 2010, *Europe and New Leading Powers. Towards Partnership in Strategic Policy Areas*, Nomos, Baden-Baden.

Husar, J., Maihold, G. & Mair, S. (eds) 2009, *Neue Führungsmächte: Partner deutscher Außenpolitik?*, Nomos, Baden-Baden.

IISD 2007, "Summary of the Fifteenth Session of the Commission on Sustainable Development: 30 April - 11 May 2007", *Earth Negotiation Bulletin*, vol. 5, no. 254.

IISD 2002, "Summary of the World Summit on Sustainable Development: 26 August - 4 September 2002", *Earth Negotiation Bulletin*, vol. 22, no. 51.

Initiative for an IRENA 2008, *The role of IRENA in the context of other international organisations and initiatives*, December 2008.

International Task Force on Global Public Goods 2006, *Meeting global Challenge: International Cooperation in the National Interest*, Stockholm.

IRENA 28.10.2011, *Proposed Medium-term Strategy of IRENA. 2012-2015. Report of the Director-General.* International Renewable Energy Agency, Second meeting of the Council, Abu Dhabi, 13 – 14 November 2011.

IRENA 2012 (16.12.2012), *Proposed Work Programme and Budget for 2013. Report of the Director-General.* International Renewable Energy Agency, Third session of the Assembly, Abu Dhabi, 13 – 14 January 2013.

Jacobsson, S. & Lauber, V. 2006, "The politics and policy of energy system transformation - explaining the German diffusion of renewable energy technology", *Energy Policy*, vol. 34, pp. 256-276.

Jacoby, K. 2009, "Energy Security: Conceptualization of the International Energy Agency (IEA)" in *Facing Global Environmental Change. Environmental, Human, Energy, Food, Health and Water Security Concepts*, eds. H.G. Brauch, N.C. Behera, P. Kameri-Mbote, et al, Springer, Berlin, pp. 345-354.

Johannesburg Renewable Energy Coalition 2005, *Information Note N°1: Members, Objectives and Roadmap.*

Johnson, É.C. 2010, *O etanol como alternativa energética e sua consolidação na política externa brasileira no governo Lula*, Thesis/Especialização em Relações Internacionais, Universidade de Brasília.

Kagan, R. 2004, *Of Paradise and Power: America and Europe in the New World Order*, Atlantic Books, London.

Karlsson-Vinkhuyzen, S.I. 2010, "The United Nations and global energy governance: past challenges, future choices", *Global Change, Peace & Security*, vol. 22, no. 2, pp. 175-195.

Karlsson-Vinkhuyzen, S.I., Jollands, N. & Staudt, L. 2012, "Global governance for sustainable energy: The contribution of a global public goods approach", *Ecological Economics*, vol. 83, pp. 11-18.

Karns, M.P. & Mingst, K.A. 2010, *International Organizations: The Politics and Processes of Global Governance*, 2. ed. edn, Rienner, Boulder/Colorado.

Katzenstein, P.J., Keohane, R.O. & Krasner, S.D. 1998, "International Organization and the study of world politics", *International Organization*, vol. 52, no. 4, pp. 645-685.

Kaul, I. 2008, "Providing (Contested) Global Public Goods" in *Authority in the Global Political Economy*, eds. V. Rittberger, M. Nettesheim & C. Huckel, Palgrave Macmillian, New York, pp. 89-115.

Kaul, I. & Conceição, P. (eds) 2006, *The new public finance: responding to global challenges*, Oxford Univ. Press, New York/N.Y.

Kaul, I., Conceição, P., Le Goulven, K. & Mendoza, R.U. 2003a, "How to Improve the Provision of Global Public Goods" in *Providing global public goods: managing globalization*, eds. I. Kaul, P. Conceição, K. Le Goulven & R.U. Mendoza, Oxford University Press, New York, pp. 21-57.

Kaul, I., Conceição, P., Le Goulven, K. & Mendoza, R.U. (eds) 2003b, *Providing Global Public Goods: Managing Globalization*, Oxford University Press, New York.

Kaul, I., Conceição, P., Le Goulven, K. & Mendoza, R.U. 2003c, "Why do global public goods matter today?" in *Providing global public goods: managing globalization*, eds. I. Kaul, P. Conceição, K. Le Goulven & R.U. Mendoza, Oxford University Press, New York, pp. 2-20.

Kaul, I., Grunberg, I. & Stern, M.A. 1999a, "Conclusion. Global Public Goods: Concepts, Policies and Strategies" in *Global Public Goods: International Cooperation in the 21st Century*, eds. I. Kaul, I. Grunberg & M.A. Stern, Oxford University Press, New York, pp. 450-507.

Kaul, I., Grunberg, I. & Stern, M.A. (eds) 1999b, *Global Public Goods: International Cooperation in the 21st Century*, Oxford University Press, New York.

Kaul, I. & Mendoza, R.U. 2003, "Advancing the concept of public goods" in *Providing global public goods: managing globalization*, eds. I. Kaul, P. Conceição, K. Le Goulven & R.U. Mendoza, Oxford University Press, New York, pp. 78-111.

Keller, S. 2010, "Sources of difference: In answer to the article about diverging paths of German and US policies for renewable energies", *Energy Policy*, vol. 38, pp. 4741-4742.

Keohane, R.O. 1989, *International Institutions and State Power. Essays in International Relations Theory*, Westview Press, Boulder.

Keohane, R.O. 1984, *After Hegemony: Cooperation and Discord in the World Political Economy*, Princton University Press, Princeton.
Keohane, R.O. & Nye, J.S. 1977, *Power and Interdependence: World Politics in Transition*, Little Brown, Boston.
Keohane, R.O. & Nye, J.S. 2000, "Globalization: What's new? What's not? (and so what?)", *Foreign Policy*, vol. 118, pp. 104-119.
Keohane, R.O. & Nye, J.S.J. 2002 (2000), "Governance in a globalizing world" in *Power and governance in a partially globalized world*, ed. R.O. Keohane, Routledge, London, pp. 193-215.
Ki-moon, B. 2011, *Sustainable Energy for All*. A Vision Statement by Ban Ki-moon, Secretary-General of the United Nations, November 2011.
Kindleberger, C.P. 1986, "International Public Goods without International Government", *The American Economic Review*, vol. 76, no. 1, pp. 1-13.
Kingdon, J.W. 2003, *Agendas, Alternatives, and Public Policies*, Longman, New York.
Kloss, E.C. 2012, *Transformação do etanol em commodity. Perspectivas para uma ação diplomática brasileira*. Fundação Alexandre de Gusmão, Brasília.
Knight, A.W. 2009, "Global Governance as a Summative Phenomenon" in *Palgrave Advances in Global Governance*, ed. J. Whitman, Palgrave Macmillan, Houndmills, pp. 160-188.
Koenig-Archibugi, M. 2006, "Introduction: Institutional Diversity in Global Governance" in *New modes of governance in the global system: exploring publicness, delegation and inclusiveness*, eds. M. Koenig-Archibugi & M. Zürn, Palgrave Macmillan, Houndmills, pp. 1-30.
Kohlhepp, G. 2010, "Análise da situação da produção de etanol e biodiesel no Brasil", *Estudos Avançados*, vol. 24, no. 68, pp. 223-253.
Kong, B. 2011, "Governing China's Energy in the Context of Global Governance", *Global Policy*, vol. 2, pp. 51-65.
Kracauer, S. 1952, "The Challenge of Qualitative Content Analysis", *Public Opinion Quarterly*, vol. 16, no. 4, pp. 631-642.
Krasner, S.D. (ed) 1983a, *International regimes*, Cornell University Press, Ithaca/N.Y.
Krasner, S.D. 1983b, "Structural causes and regime consequences: regimes as intervening variables" in *International regimes*, ed. S.D. Krasner, Cornell University Press, Ithaca/N.Y., pp. 1-22.
Kuik, O.J., Bastos Lima, M. & Gupta, J. 2011, "Energy Security in a Developing World", *Wiley Interdisciplinary Reviews: Climate Change*, vol. 2, no. 4, pp. 627-634.
Laird, F.N. & Stefes, C. 2009, "The diverging paths of German and United States policies on renewable energy: Sources of difference", *Energy Policy*, vol. 37, pp. 2619-2629.
Lesage, D. & Van de Graaf, T. 2013, "Thriving in Complexity? The OECD System's Role in Energy and Taxation", *Global Governance*, vol. 19, pp. 83-92.
Lesage, D., Van de Graaf, T. & Westphal, K. 2010, *Global Energy Governance in a Multipolar World*, Ashgate Publishing, Farnham.
Lesage, D., Van de Graaf, T. & Westphal, K. 2009, "The G8's Role in Global Energy Governance Since the 2005 Gleneagles Summit", *Global Governance*, vol. 15, pp. 259-277.
Leverett, F. 2010, "Consuming Energy: Rising Powers, the International Energy Agency, and the Global Energy Architecture" in *Rising States, Rising Institutions: Challenges for Global Governance*, eds. A.S. Alexandroff & A.F. Cooper, Brookings Institution Press, Baltimore, pp. 240-265.
Lula da Silva, L.I. 2009a, *Discurso na sessão plenária da Conferência das Partes da Convenção-Quadro das Nações Unidas sobre Mudança do Clima*, Copenhague, 17 December 2009.
Lula da Silva, L.I. 2009b, *Statement at the General Debate of the 64th Session of the United Nations General Assembly*. New York, 23 September 2009.
Lula da Silva, L.I. 2008a, *Discurso durante sessão plenária de encerramento da Conferência Internacional sobre Biocombustíveis*. São Paulo, 21 November 2008.
Lula da Silva, L.I. 2008b, *A Mudança do Clima e as Responsabilidades Globais*. Financial Times, 16 September 2008.
Lula da Silva, L.I. 2008c, *Speech at the UN Food and Agriculture Organization (FAO) conference on world food security*.

Lula da Silva, L.I. 2008d, *Speech at the special meeting of the UN Economic and Social Council (ECOSOC) on the world food crisis*. New York, 20 May 2008.
Lula da Silva, L.I. 2008e, *Statement at the General Debate of the 63rd Session of the UN General Assembly*. New York, 23 September 2008.
Lula da Silva, L.I. 2007a, *Speech by the President of Brazil, Luiz Inácio Lula da Silva, at the International Conference on Biofuels*. Brussels, 5 July 2007.
Lula da Silva, L.I. 2007b, *Statement at the General Debate of the 62nd Session of the United Nations General Assembly, "Opening the General Debate"*. New York, 25 September 2007.
Lula da Silva, L.I. 2007c, *The wonder of sugar-cane*. The Guardian, 1 June 2007.
Lula da Silva, L.I. 2006, *Lula presidente. Programa de governo 2007/2010*.
Lula da Silva, L.I. 2002, *Um Brasil para Todos. Crescimento, Emprego e Inclusão Social. Programa do Governo 2002*.
Mayntz, R. 2008, "Embedded theorizing: Perspectives on Globalization and Global Governance" in *Politikwissenschaftliche Perspektiven*, eds. S. Bröchler & H. Lauth, VS Verlag für Sozialwissenschaften, Wiesbaden, pp. 93-116.
Mayring, P. 2005, "Neuere Entwicklungen in der qualitativen Forschung und der Qualitativen Inhaltsanalyse" in *Die Praxis der Qualitativen Inhaltsanalyse*, eds. P. Mayring & M. Gläser-Zikuda, UTB, Stuttgart, pp. 7-19.
Mehta, J. 2011, "The Varied Roles of Ideas in Politics. From "Whether" to "How"" in *Ideas and Politics in Social Science Research*, eds. D. Béland & R.H. Cox, Oxford University Press, Oxford, pp. 23-46.
Merkel, A. 2012a, *Speech by Federal Chancellor Angela Merkel at the Petersberg Climate Dialogue III*. Berlin, 16 July 2012.
Merkel, A. 2012b, *Speech by Federal Chancellor Angela Merkel given at the international symposium "Towards Low-Carbon Prosperity: National Strategies and International Partnerships"*. German Advisory Council on Global Change, Berlin, 9 May 2012.
Mez, L. & Brunnengräber, A. 2011, "On the Way to the Future - Renewable Energies" in *After Cancún. Climate Governance or Climate Conflicts*, eds. E. Altvater & A. Brunnengräber, VS Verlag für Sozialwissenschaften, Springer Fachmedien, Wiesbaden, pp. 173-189.
MME 2012, *Electricity in the Brazilian Energy Plan PDE 2021*, Brasília.
MRE 2010a, *Balanço de Política Externa 2003/2010. Enegias Renováveis e Temas Correlatos: Levantamento dos Atos Internacionais do Brasil*, MRE, Brasília.
MRE 2010b, , *Balanço de Política Externa 2003/2010. Temas multilaterais - energias renováveis*, MRE, Brasília.
MRE 2007, *Repertório de política externa: posições do Brasil*, MRE, Brasília.
MRE 2003a, *Conferência Regional da América Latina e Caribe sobre Energias Renováveis*, Nota n°488, 28 October 2003.
MRE 2003b, *Plataforma de Brasília sobre Energias Renováveis*, Nota n°501, 31 October 2003.
Najam, A. 28.10.2010, *Renewing the International Renewable Energy Agency (IRENA)*, Triple Crisis, Global Perspectives on Finance, Development and Environment.
Najam, A. & Cleveland, C.J. 2005, "Energy and Sustainable Development at Global Environmental Summits: An Evolving Agenda" in *The World Summit on Sustainable Development. The Johannesburg Conference*, eds. L. Hens & B. Nath, Springer, Dordrecht, pp. 113-134.
Najam, A. & Robins, N. 2001, "Seizing the future: the South, sustainable development and international trade", *International Affairs*, vol. 77, no. 1, pp. 49-68.
Nakhooda, S. 2011, "Asia, the Multilateral Development Banks and Energy Governance", *Global Policy*, vol. 2, pp. 120-132.
Narlikar, A. 2010, *New Powers. How to Become One and How to Manage Them*, C.Hurst&Co Publishers, London.
Narlikar, A. 2008, "Bargaining for a raise: how new powers test their mettle in the international system", *Internationale Politik: Global Edition*, vol. 9, no. 3, pp. 96-101.
Narlikar, A. 2006, "Fairness in international trade negotiations", *The World Economy*, vol. 29, no. 8, pp. 1005-1029.

Nel, P. 2010, "Redistribution and Recognition: What Emerging Regional Powers Want", *Review of International Studies*, vol. 36, pp. 951-974.

Newell, P., Phillips, J. & Mulvaney, D. 2011, *Pursuing Clean Energy Equitably*. United Nations Development Program, Human Development Research Paper 2011/3.

Odingo, R.S. 1981, "Prospects for New Sources of Energy: A Report on the United Nations Conference on New and Renewable Sources of Energy, Nairobi, Kenya, 10-21 August 1981", *GeoJournal*, vol. Supplementary Issue 3, pp. 103-107.

OECD/IEA 2013, *World Energy Outlook 2013*, OECD/IEA, Paris.

Oliveira, A.d. 2012, "Transição para a Economia Verde: a Agenda Energética", *Breves Cindes*, vol. 67, Rio de Janeiro.

Olson, M. 1971, "Increasing the Incentives for International Cooperation", *International Organization*, vol. 25, pp. 866-874.

O'Neill, J. 2001, *Building Better Global Economic BRICs*, Goldman Sachs, Global Economics Paper No 66, 20 November 2001.

OPEC 2008, *World Oil Outlook 2008*, OPEC, Vienna.

Overbeek, H., Dingwerth, K., Pattberg, P. & Compagnon, D. 2010, "Forum: Global Governance: Decline or Maturation of an Academic Concept?", *International Studies Review*, vol. 12, pp. 696-719.

Pfahl, S., Oberthür, S., Tänzler, D., Kahlenborn, W. & Biermann, F. 2005, *Die internationalen institutionellen Rahmenbedingungen zur Förderung erneuerbarer Energien*, BMU, Berlin.

Pimentel, F. 2011, *O Fim da Era do Petróleo e a Mudança do Paradigma Energético Mundial: Perspectivas e Desafios para a Atuação Diplomática Brasileira*, Fundação Alexandre de Gusmão, Brasília.

Plagemann, J. & Röhrkasten, S. 2013, "Brasilien als globale Gestaltungsmacht in den Vereinten Nationen: zwischen Engagement und Skepsis" in *Brasilien: Auf dem Sprung zur Weltwirtschaftsmacht?*, ed. E.G. Fritz, Athena-Verlag, Oberhausen, pp. 91-106.

Pouliot, V. 2007, ""Sobjectivism": Toward a Constructivist Methodology", *International Studies Quarterly*, vol. 51, pp. 359-384.

Pousa, G.P.A.G., Santos, A.L.F. & Suarez, P.A.Z. 2007, "History and Policy of Biodiesel in Brazil", *Energy Policy*, vol. 35, pp. 5393-5398.

Price, R. & Reus-Smit, C. 1998, "Dangerous Liaisons? Critical International Theory and Constructivism", *European Journal of International Relations*, vol. 4, no. 3, pp. 259-294.

Putnam, R.D. 1988, "Diplomacy and domestic politics: the logic of two-level games", *International Organization*, vol. 42, no. 3, pp. 427-460.

Rein, M. & Schoen, D.A. 1977, "Problem Setting in Policy Research" in *Using social research in public policy making*, ed. C.H. Weiss, Lexington Books, pp. 235-251.

Reinicke, W.H. 1998, *Global public policy. Governing without government?*, Brookings Institution Press, Washington/D.C.

REN21 2013a, *Renewables 2013 Global Status Report*, REN21, Paris.

REN21 2013b, *Renewables Global Futures Report*, REN21, Paris.

REN21 2012, *Renewables 2012 Global Status Report*, REN21, Paris.

REN21 2005, *Renewables 2005 Global Status Report*, REN21, Paris.

Rittberger, V. 2004, *Approaches to the Study of Foreign Policy Derived from International Relations Theories*, Tübinger Arbeitspapiere zur Internationalen Politik und Friedensforschung 46.

Roberts, J.T. & Parks, B.C. 2007, *A Climate of Injustice. Global Inequality, North-South Politics, and Climate Policy*, Cambridge University Press, Cambridge/Mass.;London.

Röhrkasten, S. 2012, "Tagung: Internationale Kooperation im Bereich erneuerbare Energien. Ein Blick auf die Schwellenländer.", *Integration*, vol. 35, no. 4, pp. 339-346.

Röhrkasten, S. 2009, *Erneuerbare Energien in der globalen Strukturpolitik. Zum Potenzial einer strategischen Partnerschaft zwischen Deutschland und Brasilien.*, Eberhard-Karls-Universität Tübingen.

Röhrkasten, S. & Westphal, K. 2013, *IRENA and Germany's Foreign Renewable Energy Policy.*, Working Paper, Division Global Issues, Stiftung Wissenschaft und Poltik, Berlin.

Röhrkasten, S. & Westphal, K. 2012a, "Energy security and the transatlantic dimension: a view from Germany", *Journal of Transatlantic Studies*, vol. 10, no. 4, pp. 328-342.

Röhrkasten, S. & Westphal, K. 2012b, *IRENA: Stay the Course!*, SWP Comments 37, Stiftung Wissenschaft und Poltik, Berlin.

Röhrkasten, S. & Zilla, C. 2012, "O comércio de biocombustível e conversas entre Brasil e Eu" in *Economia, parlamentos, desenvolvimento e migrações: as novas dinâmicas bilaterais entre Brasil e Europa*, ed. O. Jacob, Konrad Adenauer Stiftung, Rio de Janeiro, pp. 85-104.

Rosenau, J.N. 2009 (1995), "Governance in the twenty-first century" in *Palgrave Advances in Global Governance*, ed. J. Whitman, Palgrave Macmillan, Houndmills, pp. 7-40.

Rosenau, J.N. 2009, "Introduction. Global Governance or Global Governances?" in *Palgrave Advances in Global Governance*, ed. J. Whitman, Palgrave Macmillan, Houndmills, pp. 1-6.

Rosenau, J.N. 2003, "Governance in a New Global Order" in *Governing globalization. Power, Authority and Global Governance*, eds. D. Held & A. McGrew, Polity Press, Cambridge, pp. 70-86.

Rosenau, J.N. 1999, "Toward an Ontology for Global Governance" in *Approaches to Global Governance Theory*, eds. M. Hewson & T.J. Sinclair, State University of New York Press, Albany, pp. 287-301.

Rosenau, J.N. 1992, "Governance, order, and change in world politics" in *Governance without government: order and change in world politics*, eds. J.N. Rosenau & E. Czempiel, Cambridge University Press, Cambridge, pp. 1-29.

Rosenau, J.N. & Czempiel, E. (eds) 1992, *Governance without government: Order and change in world politics*, Cambridge University Press, Cambridge.

Rousseff, D. 2011a, *Discurso durante cerimônia de criação da Comissão e do Comitê Nacional de Organização da Conferência das Nações Unidas sobre o Desenvolvimento Sustentável*, Brasília, 07 June 2011.

Rousseff, D. 2011b, *Discurso durante cerimônia de encerramento do Fórum Empresarial Brasil-União Europeia*, Brussels, 04 October 2011.

Rowlands, I.H. 2005, "Renewable Energy and International Politics" in *Handbook of Global Environmental Politics*, ed. P. Dauvergne, Edward Elgar, Cheltenham, Northampton, pp. 78-94.

Ruggie, J.G. 1998, "What Makes the World Hang Together? Neo-Utilitarianism and the Social Constructivist Challenge", *International Organization*, vol. 52, no. 4, pp. 855-885.

Ruggie, J.G. 1995, "Peace in our Time? Causality, Social Facts and Narrative Knowing", *Proceedings of the Annual Meeting (American Society of International Law)*, vol. 89, pp. 93-100.

Russett, B.M. & Sullivan, J.D. 1971, "Collective Goods and International Organization", *International Organization*, vol. 25, pp. 845-865.

Sanchez Badin, M.R. & Godoy, D.H. 2013, *International trade regulatory challenges for Brazil and some lessons from the promotion of ethanol*. From the SelectedWorks of Michelle R Sanchez-Badin, February 2013.

Scarlat, N. & Dallemand, J. 2011, "Recent developments of biofuels/bioenergy sustainability certification: A global overview", *Energy Policy*, vol. 39, pp. 1630-1646.

Scharpf, F.W. 1997, *Games real actors play: Actor-centered institutionalism in policy research*, Westview Press, Boulder/Colo.

Scheer, H. 2009 [2004], "Report from the International Parliamentary Forum to Delegates of the International Governmental Conference "Renewables 2004", June 2004" in *The long road to IRENA. From the Idea to the Foundation of the International Renewable Energy Agency. Documentation 1990 - 2009.*, eds. EUROSOLAR & World Council for Renewable Energy, Ponte Press, Bochum, pp. 84-85.

Scheer, H. 2009 [2001], "Towards an International Treaty to Promote the Use of Renewable Energy" in *The long road to IRENA. From the Idea to the Foundation of the International Renewable Energy Agency. Documentation 1990 - 2009.*, eds. EUROSOLAR & World Council for Renewable Energy, Ponte Press, Bochum, pp. 43-48.

Scheer, H. 2009 [2000], "Memorandum for the Establishment of an International Renewable Energy Agency (IRENA)" in , eds. EUROSOLAR & World Council for Renewable Energy, Ponte Press, Bochum, pp. 24-33.

Scheer, H. 2009 [1990], "Memorandum for the Establishment of an International Solar Energy Agency (ISEA) within the United Nations" in *The long road to IRENA. From the Idea to the Foundation of the International *Renewable Energy Agency. Documentation 1990 - 2009.*, eds. EUROSOLAR & World Council for Renewable Energy, Ponte Press, Bochum, pp. 9-19.

Scheibe, E.F. 2008, *Biocombustíveis e Política Externa Brasileira: Coerência Histórica entre Política Energética e Política Externa e o Papel dos Grupos de Interesse na Questao dos Biocombustíveis*, Thesis/ Bacharel, Universidade Federal do Rio Grande do Sul.

Schirm, S.A. 2010, "Leaders in need of followers: Emerging powers in global governance", *European Journal of International Relations*, vol. 16, no. 12, pp. 197-221.

Schmidt, M.G. 2004, *Wörterbuch zur Politik*, Alfred Kröner Verlag Stuttgart, Stuttgart.

Schmidt, V.A. 2011, "Reconciling Ideas and Institutions through Discursive Insitutionalism" in *Ideas and Politics in Social Science Research*, eds. D. Béland & R.H. Cox, Oxford University Press, Oxford, pp. 47-64.

Schmidt, V.A. 2008, "Discursive Institutionalism: The Explanatory Power of Ideas and Discourse", *Annual Review of Political Science*, vol. 11, pp. 303-326.

Schreurs, M. 2013, "Orchestrating a Low-Carbon Energy Revolution Without Nuclear: Germany's Response to the Fukushima Nuclear Crisis", *Theoretical Inquiries in Law*, vol. 14, no. 1, pp. 83-108.

Schröder, G. 2004, *Speech at the International Conference for Renewable Energies in Bonn on June 3*, Renewables 2004 Conference, Bonn.

Schröder, G. 2002a, *Regierungserklärung des Bundeskanzlers am 29. Oktober 2002 vor dem Deutschen Bundestag in Berlin*, bpa-bulletin/ Bulletin der Bundesregierung, 29 October 2002.

Schröder, G. 2002b, *Speech given at a conference "Strategy for Germany – Mission for Johannesburg" held by the Council on Sustainable Development*, Berlin, 13 May 2002.

Schröder, G. 2002c, *Statement at the World Summit on Sustainable Development*, Johannesburg, 2 September 2002.

Schubert, S. & Gupta, J. 2013, "Comparing Global Coordination Mechanisms on Energy, Environment, and Water", *Ecology and Society*, vol. 18, no. 2.

Schuppert, G.F. 2008, "Governance - auf der Suche nach Konturen eines "anerkannt uneindeutigen Begriffs"" in *Governance in einer sich wandelnden Welt*, eds. G.F. Schuppert & M. Zürn, VS Verl. für Sozialwissenschaften, Wiesbaden, pp. 13-40.

Schutte, G.R. & Barros, P.S. 2010, "A Geopolítica do Etanol", *Boletim de Economia e Política Internacional*, vol. 1, pp. 33-44.

Searle, J.R. 1995, *The Construction of Social Reality*, The Free Press, New York.

Selbmann, K. & Kaup, F. 2010, *Biofuels in Germany - Historical Trends and Impact Factors Set Course for Prospective Design of Policies*, For Presentation at the SGIR 7th Pan-European International Relations Conference, Stockholm, 9-11 September 2010.

Sikkink, K. & Finnemore, M. 2001, "Taking Stock: The Constructivist Research Program in International Relations and Comparative Politics", *Annual Review of Political Science*, vol. 4, pp. 391-416.

Simmons, P.J. & Jonge Oudraat, C.d. 2001, "Managing Global Issues: An Introduction" in *Managing global issues: lessons learned*, eds. P.J. Simmons & C.d. Jonge Oudraat, Carnegie Endowment for International Peace., Washington/D.C., pp. 3-22.

Simões, A.J.F. 2007a, "Biocombustíveis: a experiência brasileira e o desafio da consolidação do mercado internacional" in *Biocombustíveis no Brasil - Realidades e Perspectivas*, ed. MRE,Brasília, pp. 11-33.

Simões, A.J.F. 2007b, *Biofuels will help fight hunger*, New York Times, 6 August 2007.

Soares de Lima, M.R. & Hirst, M. 2006, "Brazil as in intermediate state and regional power: action, choice and responsibilities", *International Affairs*, vol. 82, no. 1, pp. 21-40.

Späth, K. 2005, "Inside Global Governance: New Borders of a Concept" in *Critizising Global Governance*, eds. M. Lederer & P.S. Müller, Palgrave McMillian, New York, pp. 21-44.

SPD 2002, *Erneuerung und Zusammenhalt - wir in Deutschland. Regierungsprogramm 2002 - 2006.*

SPD 1998, *"Arbeit, Innovation und Gerechtigkeit". SPD-Programm für die Bundestagswahl 1998.*

SPD & Bündnis 90/Die Grünen 2002, *Koalitionsvertrag 2002 – 2006: Erneuerung – Gerechtigkeit – Nachhaltigkeit.*

SPD & Bündnis 90/Die Grünen 1998, *Aufbruch und Erneuerung – Deutschlands Weg ins 21. Jahrhundert. Koalitionsvereinbarung zwischen der Sozialdemokratischen Partei Deutschlands und BÜNDNIS 90/DIE GRÜNEN*.

Stamm, A. 2004, *Schwellen- und Ankerländer als Akteure einer globalen Partnerschaft. Überlegungen zu einer Positionsbestimmung aus deutscher entwicklungspolitischer Sicht*, DIE Discussion Paper 01/2004, Deutsches Institut für Entwicklungspolitik, Bonn.

Steiner, A., Wälde, T., Bradbrook, A. & Schutyser, F. 2006, "International Institutional Arrangements in Support of Renewable Energy" in *Renewable Energy. A Global Review of Technologies, Policies and Markets*, ed. D. Assmann, Earthscan Publications, London, pp. 152-165.

Stoker, G. 1998, "Governance as Theory: Five Propositions", *International Social Science Journal*, vol. 50, no. 1, pp. 17-28.

Suding, P.H. & Lempp, P. 2007, "The Multifaceted Institutional Landscape and Processes of International Renewable Energy Policy", *International Association for Energy Economics Newsletter*, vol. Second Quarter, pp. 4-9.

Sustainable Energy for All 2013, *Global Tracking Framework*.

Szulecki, K., Pattberg, P. & Biermann, F. 2010, "The Good, the Bad, and the Even Worse: Explaining Variation in the Performance of Energy Partnerships", *Global Governance Working Paper*, vol. 39, January 2010.

Szulecki, K. & Westphal, K. 2014, "The cardinal sins of European energy policy: Non-Governance in an uncertain global landscape", *Global Policy*, vol 5, October 2014, pp. 38-51.

The Government of the Federal Republic of Germany 2013, *Club der Energiewendestaaten*. Antwort der Bundesregierung auf die Kleine Anfrage der Abgeordneten, Dr. Hermann E. Ott, Bärbel Höhn, Hans-Josef Fell, weiterer Abgeordneter und der Fraktion BÜNDNIS 90/Die Grünen. Drucksache 17/14038.

The Government of the Federal Republic of Germany 2008, *The Case for an International Renewable Energy Agency (IRENA)*, Preparatory Conference for the Foundation of IRENA, Berlin, April 10 - 11, 2008.

The Government of the Federal Republic of Germany & Initiative for an IRENA 2008, *Founding an International Renewable Energy Agency (IRENA). Promoting renewable energy worldwide*, BMU, Berlin, October 2008.

The Independent Commission on International Development Issues 1980, *North-South: A programme for survival*, Pan Books, London.

Tickner, A.B. 2013, "Core, periphery and (neo)imperialist International Relations", *European Journal of International Relations*, vol. 19, no. 3, pp. 627-646.

Trittin, J. 2004a, *"Renewables2004 - Make it real"*. *The Age of Renewables has now begun*, Welcome Address at the International Conference for Renewable Energies, Bonn, 01 June 2004.

Trittin, J. 2004b, *Closing Address at Plenary Session D*, The International Conference for Renewable Energies, Bonn, 04 June 2004.

Trittin, J. 2004c, *We are making good progress. The Age of Renewables has now begun*, Speech at the Plenary Session A, International Conference for Renewable Energies, Bonn, 03 June 2004.

United Nations 2002, *Report of the World Summit on Sustainable Development*, United Nations, New York.

United Nations 1981, *UN Yearbook 1981*, United Nations, New York.

United Nations Commission on Sustainable Development 2001a, *Report on the Ninth session, 5 May 2000 and 16-27 April 2001*.

United Nations Commission on Sustainable Development 2001b, *Report of the Ad Hoc Open-ended Intergovernmental Group of Experts on Energy and Sustainable Development*.

United Nations Conference on Environment & Development 1992, *AGENDA 21*.

Van de Graaf, T. 2013a, "Fragmentation in Global Energy Governance: Explaining the Creation of IRENA", *Global Environmental Politics*, vol. 13, no. 3, pp. 14-33.

Van de Graaf, T. 2013b, *The Politics and Institutions of Global Energy Governance*, Palgrave Macmillan, New York.

Van de Graaf, T. 2012a, "How IRENA is reshaping the global energy architecture", *European Energy Review*, [Online].
Van de Graaf, T. 2012b, "Obsolte or resurgent? The International Energy Agency in a changing global landscape", *Energy Policy*, vol. 48, pp. 233-241.
Van de Graaf, T. & Westphal, K. 2011, "The G8 and G20 as Global Steering Committees for Energy: Opportunities and Constraints", *Global Policy*, vol. 2, pp. 19-30.
Vargas, E. 2004, *Remarks on Renewable Energy at the 30 Session of the Economic Commission for Latin America and the Caribbean*, San Juan, Puerto Rico, 30 June 2004.
Vieira, M. 2013, "Brazilian Foreign Policy in the Context of Global Climate Norms", *Foreign Policy Analysis*, vol. 9, pp. 369-386.
Viola, E. 2013, "Brazilian Climate Policy since 2005: Continuity, Change and Prospective", *CEPS Working Document*, no. 373/February 2013.
Wang, H. & Rosenau, J.N. 2009, "China and Global Governance", *Asian Perspective*, vol. 33, no. 3, pp. 5-39.
Weber, M. 1964 (1947), *The Theory of Social and Economic Organization*, The Free Press, New York.
Weldes, J. 1996, "Constructing National Interests", *European Journal of International Relations*, vol. 2, no. 3, pp. 275-318.
Wendt, A. 1992, "Anarchy Is What States Make of it: the Social Construction of Power Politics", *International Organization*, vol. 46, no. 2, pp. 391-425.
Westerwelle, G. 2012, *Rede aus Anlass der 3. IRENA Ratssitzung in Abu Dhabi, 06.06.2012*.
Westphal, K. & Röhrkasten, S. 2013, "Energieversorgung: Vom Umgang mit internationalen und vernetzten Versorgungsrisiken" in *Der "Nexus" Wasser - Energie - Nahrung. Wie mit vernetzten Versorgungsrisiken umgehen?*, ed. M. Beisheim, Stiftung Wissenschaft und Politik, Berlin, pp. 35-43.
Wieczorek-Zeul, H. 2009 [2001], "Economic Co-Operation for Renewable Energy" in *The long road to IRENA. From the Idea to the Foundation of the International Renewable Energy Agency. Documentation 1990 - 2009.*, eds. EUROSOLAR & World Council for Renewable Energy, Ponte Press, Bochum, pp. 34-42.
Wieczorek-Zeul, H. 2009, "Closing speech held at the Founding Conference of IRENA, 26 January 2009" in *The long road to IRENA. From the Idea to the Foundation of the International Renewable Energy Agency. Documentation 1990 - 2009.*, eds. EUROSOLAR & World Council for Renewable Energy, Ponte Press, Bochum, pp. 108-109.
Wieczorek-Zeul, H. 2004a, *Closing Address at Plenary Session D*, International Conference for Renewable Energies, Bonn, 04 June 2004.
Wieczorek-Zeul, H. 2004b, *Opening Speech at Plenary Session A*, International Conference for Renewable Energies, Bonn, 03 June 2004.
Wieczorek-Zeul, H. 2004c, *Welcome Address at the Plenary Session I*, International Conference for Renewable Energies, Bonn, 01 June 2004.
Williams, M. 1993, "Re-articulating the Third World coalition", *Third World Quarterly*, vol. 14, no. 1, pp. 7-29.
Yergin, D. 2006, "Ensuring Energy Security", *Foreign Affairs*, vol. 85, pp. 69-82.
Young, O.R. 1997, "Rights, Rules, and Resources in World Affairs" in *Global governance: Drawing insights from the environmental experience*, ed. O.R. Young, MIT Press, Cambridge/Mass.;London, pp. 1-23.
Zelli, F., Pattberg, P., Stephan, H. & van Asselt, H. 2013, "Global Climate Governance and Energy Choices" in *The Handbook of Global Energy Policy*, ed. A. Goldthau, Wiley-Blackwell, London, pp. 340-357.
Ziegler, J. 2007, *Report of the Special Rapporteur on the right to food*, United Nations General Assembly, 22 August 2007.
Zilla, C. 2011a, *"Demokratie" im Diskurs politischer Parteien. Argentinien und Chile im Vergleich*, Nomos, Baden-Baden.
Zilla, C. 2011b, *Brasilianische Außenpolitik. National Tradition, Lulas Erbe und Dilmas Optionen*, SWP-Studie 29/2011, Stiftung Wissenschaft und Politik, Berlin.

Zürn, M. & Koenig-Archibugi, M. 2006, "Conclusion II: The Modes and Dynamics of Global Governance" in *New modes of governance in the global system: exploring publicness, delegation and inclusiveness*, eds. M. Koenig-Archibugi & M. Zürn, Palgrave Macmillan, Houndmills, pp. 236-251.

If you have any concerns about our products,
you can contact us on
ProductSafety@springernature.com

In case Publisher is established outside the EU,
the EU authorized representative is:
**Springer Nature Customer Service Center GmbH
Europaplatz 3, 69115 Heidelberg, Germany**

Printed by Libri Plureos GmbH
in Hamburg, Germany